Carnivoran Ecology

Carnivoran Ecology

The Evolution and Function of Communities

Steven W. Buskirk

Professor Emeritus, University of Wyoming

OXFORD

UNIVERSITY PRESS

OXFORD
UNIVERSITY PRESS

Great Clarendon Street, Oxford, OX2 6DP,
United Kingdom

Oxford University Press is a department of the University of Oxford.
It furthers the University's objective of excellence in research, scholarship,
and education by publishing worldwide. Oxford is a registered trade mark of
Oxford University Press in the UK and in certain other countries

Published in the United States of America by Oxford University Press
198 Madison Avenue, New York, NY 10016, United States of America

British Library Cataloguing in Publication Data

Data available

Library of Congress Control Number: 2022944176

ISBN 978–0–19–286324–9
ISBN 978–0–19–286325–6 (pbk.)

DOI: 10.1093/oso/9780192863249.001.0001

Printed and bound by
CPI Group (UK) Ltd, Croydon, CR0 4YY

Links to third party websites are provided by Oxford in good faith and
for information only. Oxford disclaims any responsibility for the materials
contained in any third party website referenced in this work.

Preface

This book arose from my several decades of interest in and research on carnivorans. Like many young people with naturalist tendencies, I was particularly intrigued by my early encounters with wild carnivorans—their rarity, elusiveness, and implicit threat to my well-being. I vividly recall seeing my first bear track on a solo backpacking trip in the Sierra Nevada of California, followed by a sleepless night spent contemplating my fate. I recall the horror of my older sister when I asked her to stop our parents' car so that I could retrieve a road-killed domestic cat and add its skull to my collection. In the summers of my college years, working as a tour guide in Mount McKinley (now Denali) National Park, Alaska, I observed regular interactions involving wolves, caribou, brown bears, moose, red foxes, and other species. Some of these observations were dramatic and photogenic, but curious as well. Why did 150-kg brown bears invest so much effort to capture 200-g ground squirrels? Why did wolves give birth in the same dens for decades on end, even though they were well known and prone to human disturbance? Why were coyotes deathly afraid of being anywhere near wolves, while red foxes merely stayed out of their grasp?

My PhD research and subsequent faculty appointment in Zoology and Physiology at the University of Wyoming gave me opportunities to pursue these kinds of questions. While most carnivoran research of that time gravitated to intraspecific and predator–prey interactions, my interests ranged more widely. My research addressed various mechanisms by which carnivorans might be limited: thermal energetics, allometry, tooth morphology, fasting endurance, genetic variability, and interspecific competition. I learned that predator–prey interactions were a small part of our growing understanding of what limits the distributions and abundances of carnivorans.

During those early years, I depended on the standard academic references of the day: journal articles, monographs, book chapters, theses, dissertations, and R. F. Ewer's (1973) *The carnivores*. Ewer's monograph was the definitive source for carnivoran biology—especially paleontology and behavior. Writing the current account nearly fifty years on, I benefitted from a wealth of high-quality material facilitated by the revolution in publishing—a proliferation of journal titles, many of them open access, expanded opportunities for scientists in developing countries to publish their work, and many more women working and publishing as scientists. The period since 2000 has offered unprecedented opportunities to write a book such as this, and the pandemic of 2020–22 gave my isolation and focus new purpose.

What qualifies me to write a book about all carnivorans? I am not the most prolific author of carnivoran papers, nor the one with the broadest geographic experience. Others have spent more time in the field, or are more quantitative than I. However, my interests extend in multiple directions, all of which relate to how carnivorans succeed or fail in the wild. I am as intrigued by answers to big ecological questions as by solutions to specific conservation problems. I am as satisfied by understanding of some ecological puzzle as by watching a large carnivore stalk an ungulate. I appreciate new discoveries in natural history as much as I do studies of functional genomics. I also value those who study carnivorans and share their findings with scientists and the public. This group overlaps strongly with those committed to assuring the presence and importance of carnivorans in future communities. This book is really about, for, and the result of the work of carnivoran biologists, conservationists, and naturalists, both professional and amateur. It is about their passion. Last of all, hopefully it is a motivation for others to follow in their tracks.

Steven W. Buskirk
Professor Emeritus, University of Wyoming
December 10, 2022
Silver City, New Mexico, US

References cited

Ewer, R.F. (1973). *The carnivores*. Ithaca: Cornell University Press.

Acknowledgments

Many people supported and aided me in writing this book. My wife Beth encouraged me at every stage and tolerated many interruptions to our routine so that I could spend time writing. Close colleagues Dennis Knight and Carlos Martinez del Rio were reliable discussants and sources of encouragement early in the conception and early writing phases. Hannah Sease produced all graphic arts work, keeping pace with my requests throughout her graduate studies in graphic arts.

The William Robertson Coe Library at the University of Wyoming provided outstanding support, filling scores of requests for materials, and the J. Cloyd Miller Library at Western New Mexico University provided additional assistance. My academic home, the Department of Zoology and Physiology at the University of Wyoming, provided office space and other support for most of duration of the project.

I benefitted from reviews by over thirty scientists, eight of them commissioned by Oxford University Press, and others solicited by me, who generously reviewed chapters or shorter sections: Benjamin Allen, Rudy Boonstra, Jeff Bowman, Joseph Bump, Emiliano Donadio, Jacob Goheen, Henry Harlow, Dennis Knight, Serge Lariviere, Paul Leberg, Jason Lillegraven, Carlos Martinez del Rio, Sterling Miller, Robert Naiman, Richard Ostfeld, Jonathan Pauli, James D. Rose, Oswald Schmitz, John Schoen, Qian-Quan Sun, Blaire Van Valkenburgh, Lars Werdelin, and Andrzej Zalewski. Collectively, the reviewers corrected errors of fact, identified important omissions, provided more pertinent or more recent references, and improved the organization and presentation. Without their help the project would not have been possible.

The illustrations strengthen the narrative throughout, and some of the finest artwork and photography were contributed gratis or licensed at reduced rates. I particularly thank Justin Binfet, Darin Croft, Walton Ford, Stan Gehrt, Don Gutoski, Esperanza Iranzo, Jeffrey Kerby, Janet Kessler, Débora Kloster, Susan McConnell, Chris Mills, Larissa Nituch, Velizar Simeonovski, Alejandro Travaini, Juan Zanón, the Philmont Scout Ranch, and Kasmin Gallery. Although I have tried to select photos taken under natural conditions, I cannot assure that none was staged or in some way contrived.

The staff of Oxford Press were most supportive and encouraging. Ian Sterling was immediately receptive when I approached him with my proposal, and consistently improved my presentation as well as my understandings of how the book could be made most useful. Charlie Bath provided excellent suggestions on draft chapters, and Katie Lakina shepherded the project through the editorial and production processes.

Contents

CHAPTER 1

Introduction to *Carnivoran Ecology*

Order Carnivora represents one of the most species-rich, phenotypically diverse, widely distributed, and ecologically influential mammalian lineages. Its extant members live on all continents and in all oceans and range from vole-sized (c. 30 g) to larger than a rhinoceros (c. 4,000 kg). They eat diverse foods including leaves, fruits, insects, honey, marine invertebrates, and mammals larger than themselves. Some species live below ground for weeks at a time, a few live mostly in the forest canopy, and others live at sea for months on end. Many are wilderness dwellers, wary of humans and their activities, while some non-domestic species thrive in major cities, largely dependent on humans for food, shelter, or protection from larger, wilder carnivorans. No other mammalian order approaches Carnivora in the breadth of adaptive suites shown and ecological niches occupied.

Humans have always been keenly interested in carnivorans. Our hominin ancestors were preyed on by carnivorans and competed with them for food. Both species hunted the same prey and drove each other away from prey carcasses (Figure 1.1) (Espigares *et al.*, 2013). Paleolithic humans converted a potential competitor, the wolf, to a partner. The resultant dog was the first domesticated animal, and became essential to human lives, providing a food source, vigilance against intruders, transport of possessions, and assistance in hunting and herding (Figure 1.2). Dogs became so important to early humans that they are credited with shaping human evolution as much as humans shaped theirs (Pierotti and Fogg, 2017).

The life-or-death nature of predation elicits strong human emotions—either the predator eats and lives, or the prey escapes and survives. These emotions are stronger when we imagine ourselves or animals we own as prey. As a result, we have long imbued carnivorans with spiritual powers to match their impressive physical abilities, and our ancestors represented carnivorans in some of the earliest figurative art (Figure 1.3) (Hart and Sussman, 2008; Azéma, 2015). Today, we continue to use carnivorans to symbolize wildness, ferocity, and independence in visual and literary arts. Every pocket and fold of most human cultures—languages, parables, spiritual beliefs, and symbols—is rich with carnivoran references.

1.1 "Carnivoran" vs. "carnivorous"

The terms "mammalian predator," "mammalian carnivore," and "carnivoran" are not precisely synonymous. Predators are animals that kill and consume multicellular animals (Taylor, 1984), whether one at a time, or filtered from water by the thousands. Eagles, dragonflies, and blue whales are predators. By contrast, a carnivore consumes the flesh of animals, whether it kills or scavenges it; vultures, snakes, and Venus fly traps are carnivores *and* carnivorous. Carnivorans, the subject of this book, are exclusively mammals in Order Carnivora—a branch of the tree of life. Species in the order may pursue and kill vertebrates, eat termites excavated from soil, or subsist entirely on plant parts (Figure 1.4). "Hypercarnivore" sometimes indicates a species with a diet exceeding some threshold level of vertebrate prey, typically killed rather than scavenged. Most members of the cat family, the Felidae, are considered hypercarnivores. Other terms, such as "mesopredator," "apex predator," and "keystone predator," have imprecise meanings that I parse as they arise in the book.

Carnivoran Ecology. Steven W. Buskirk, Oxford University Press. © Steven W. Buskirk (2023). DOI: 10.1093/oso/9780192863249.003.0001

Figure 1.1 The first contact between humans and the saber-toothed *Smilodon* in America, interpreted by Velizar Simeonovski. Carnivorans have been competitors with and potential predators of hominins from before we became humans through to today.
Painting: © Velizar Simeonovski.

Figure 1.2 Lions (right) depicted on the walls of Salle du Fond in Chauvet-Pont d'Arc Cave, France, dating to 32,000–30,000 years ago. The lions appear to be watching a rhinoceros on a distant panel. Carnivorans were subjects of some of the earliest figurative art. Vertical scratches in the lower right and on the rhinoceros painting were made by cave bears trapped in deep chambers.
Photo: J. Clottes in Azéma (2015), CC 4.0.

As is often the case in biology, the defining traits of the Carnivora—the traits held by all members of the lineage, both living and extinct—are difficult to state without qualification. Traditionally, biologists have cited the presence of cheek teeth that have shearing functions, comprising the upper fourth premolar (P^4) and lower first molar (M_1). All living carnivorans have ancestors with this trait. However, living genets, bears, and seals have secondarily lost the shearing function of those teeth, and at least one species has lost those teeth completely. Some prehistoric linages of other carnivorous mammals had shearing cheek teeth, but they occupied other positions in the tooth row, for example M_1 and M_2. They were not homologous to modern carnassials, and the lineages that exhibited them have no living descendants. The fused scaphoid and lunate carpal bones (the scapholunate) are sometimes regarded as

defining the group, because all modern species have that trait, but some early carnivorans did not.

Predatory placental mammals—those that primarily kill other animals for food—occur in several orders, including shrews and moles (Order Soricomorpha), bats (Chiroptera), whales (Cetacea), pangolins (Pholidota), and hairy anteaters (Pilosa). If we broaden our frame of reference to include marsupial predators, several additional orders must be included as mammalian carnivores. Even some obligate herbivores, among them ungulates and rodents, prey on or scavenge vertebrates opportunistically (Boonstra *et al.*, 1990; Dudley *et al.*, 2016). Finally, many extinct non-carnivoran lineages were at least partially carnivorous. Clearly, Order Carnivora is not unique among mammals in killing and eating vertebrates. This book is about a mammalian lineage, not a foraging style or trophic niche.

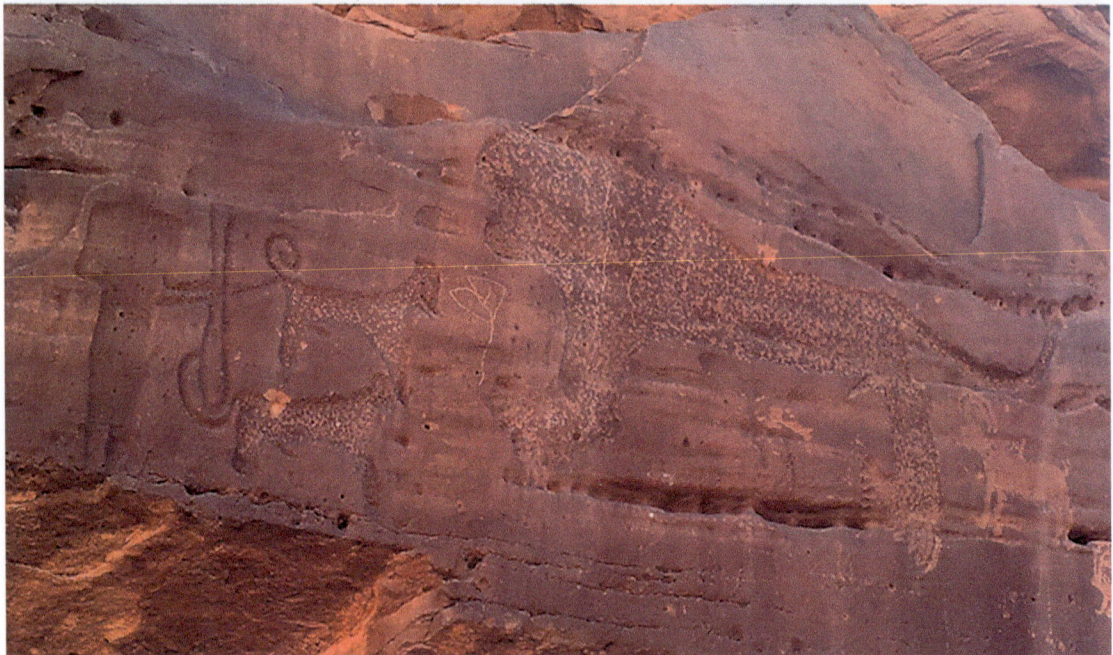

Figure 1.3 Rock art in northwestern Saudi Arabia showing dogs resembling modern Canaan dogs assisting with lion hunting. Various glyphs from this site include the earliest depictions of dogs on leashes, 12,000–10,000 years old. The upper image is shaded to show relief that is less visible in the lower, unaltered image.
Photo: Guagnin et al. (2018, Figure 10) by permission.

Herbivorous
mammals

Carnivora

Carnivorous
mammals

Figure 1.4 Venn diagram of the relationships between carnivory, herbivory, and Order Carnivora. Herbivorous mammals overlap with the Carnivora, and the Carnivora overlap with carnivorous mammals. Elephants are herbivorous non-carnivoran mammals. The giant panda is an herbivorous carnivoran. The brown bear is an omnivorous carnivoran, neither fully herbivorous nor carnivorous. The tiger is a carnivorous carnivoran, and the killer whale is a carnivorous non-carnivoran mammal. In this book I use "carnivorous" to denote diet, and "carnivoran" to denote phylogeny—inclusion in Order Carnivora.

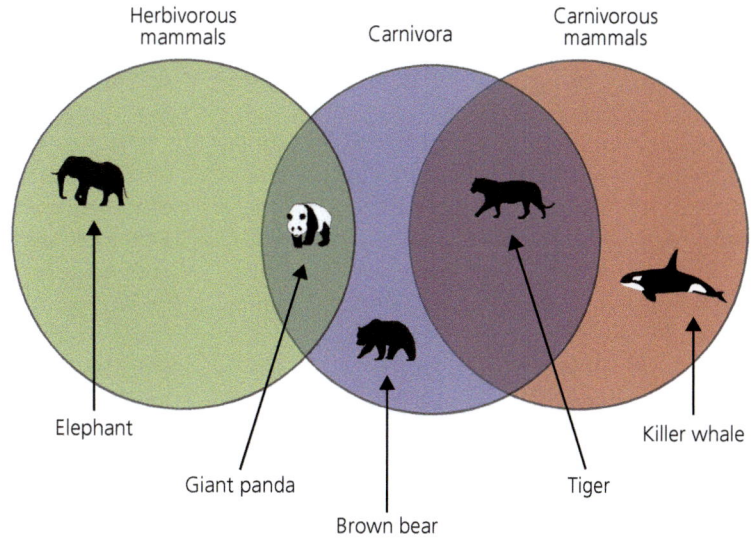

Elephant

Killer whale

Giant panda

Tiger

Brown bear

1.2 The carnivorans—who and where?

The approximately 287 extant species of carnivorans are organized into fifteen currently recognized families (Appendix I) and occur on all continents, if we include seals that haul out on Antarctic beaches. These numbers change as we learn more about how mammal lineages are related. For example, the neotropical olinguito recently has been identified as distinct from other olingos, and the African golden wolf was judged a separate species from the golden jackal, (Gaubert *et al.*, 2012; Helgen *et al.*, 2013). On the other hand, the long-recognized red wolf of eastern North America is now regarded as an ancient hybrid of the wolf and the coyote, with uncertain endangered species status (vonHoldt *et al.*, 2016, but see Hohenlohe *et al.*, 2017). At the level of taxonomic families, the skunks and stink badgers were grouped with weasels and otters in the Mustelidae before Family Mephitidae was recognized as warranting recognition (Dragoo and Honeycutt, 1999). With each such discovery, our knowledge of carnivoran classification becomes more reflective of evolutionary history, as revealed through genetic and morphological studies.

Carnivoran species are distributed unevenly across lineages and continents. Family Mustelidae holds sixty species, whereas the Nandiniidae, Ailuridae, and Odobenidae hold a single species each. The Felidae, Canidae, and Mustelidae are nearly cosmopolitan, found on all continents except Antarctica and Australia before humans transported them. On the other hand, the eight species of Eupleridae occur only on Madagascar Island, their common ancestor having rafted there from the African mainland around 20 million years ago (Ma). The thirty-three extant species of seals, sea lions, and walrus make up the pinniped (ear foot) group—a lineage comprising two or three families that arose from a single aquatic ancestor. "Fissiped" (split foot), on the other hand, denotes the remaining, mostly terrestrial, non-pinniped carnivorans.

1.3 The growth of knowledge

Before 1900, understanding of carnivoran ecology (as opposed to natural history) was limited, often based on lore and conjecture. Much of our knowledge of their genetics, behavior, and reproduction at that time resulted from observing domestic cats, ferrets, and dogs, and farmed minks and foxes (Figure 1.5). Very little scientific knowledge about wild carnivorans existed, and most interest centered on the value of their furs or other body parts and the threat they posed to agriculture. They were difficult to study directly because of their low densities, elusive behaviors, and constant persecution near humans. Well into the twentieth century, the leading ecological questions about carnivorans dealt with how many ungulates, waterfowl, and other valued vertebrates they killed and how to mitigate those losses (Leopold, 1933). In the late twentieth century, however, perspectives shifted, tools improved, and ecological research on this group surged. The number of scientific journal articles indexed by Web of Science with "Carnivora" as a topic increased by a factor of eleven from 1992 to 2016, compared with a factor of four for "mammal" and six for "Mammalia." New understandings of carnivoran biology, especially ecology, began to unfold in the 1960s, when carnivorans gained significance in conservation issues, either as threatened taxa or as agents of endangerment of other taxa or communities. For example, the severe contraction in distribution and abundance of the brown bear in the contiguous United States from 1850 to 1950 resulted in greatly expanded research in bear biology, broadly cast. Of 159 scientific journal articles with the topics "grizzly bear + Yellowstone" indexed by Web of Science, all but two were published subsequent to the initial 1975 listing of the Yellowstone population under the Endangered Species Act. Similarly, no articles indexed by Web of Science dealt with the population genetics of the cheetah before the discovery by O'Brien and colleagues (1983) that cheetahs exhibited low genetic variability. This finding had such strong conservation implications that 153 subsequent articles reported on genetic traits of cheetahs related to evolutionary or ecological processes. Genetic depletion in other carnivoran species became a research focus as well.

The American mink is an example of a carnivoran that has stimulated research as a result of the great ecological harm it causes. Introduced to Europe and South America during the twentieth century for fur production, it today poses threats on both continents, spurring intensive study (Bonesi and Palazon, 2007; Crego *et al.*, 2016). Over the past twenty-five years, published studies on the ecology of invasive American minks outside of North America have greatly outnumbered those conducted within the species' native range.

This expansion of knowledge reflects that humans need to know much more about carnivoran ecology than they did seventy years ago. For example, public health planners now must consider whether the most rapidly emerging infectious diseases of humans are influenced by the diversity and abundance of wild carnivorans (Section 11.3.1) (Levi *et al.*, 2012, Hofmeester *et al.*, 2017). The current incidence and severity of such debilitating zoonoses as avian and swine influenza, Lyme disease, and tick-borne encephalitis may be mediated by the presence and abundance of predators—many of them carnivorans—that kill intermediate or alternate hosts (Thulin *et al.*, 2015).

Figure 1.5 Farmed black-phase red foxes at Fromm Fox Farm, Wisconsin, US, in the early twentieth century. Knowledge of carnivoran reproductive and nutritional physiology during that era came largely from observing domesticated and farmed animals.
Photo: Fromm Historical Society.

Even the viral pandemic COVID-19 has been linked to transmission of the SARS-CoV-2 virus among various wild and domestic mammals—particularly carnivorans—and humans. The carnivorans include domestic dogs and cats, captive tigers, and American minks (Vinodh Kumar *et al.*, 2020; Hammer *et al.*, 2021).

At the same time, carnivorans are credited with performing valuable ecological services not understood decades ago. In Chapter 6 I show that bears are recognized for transporting marine-derived nutrients in salmon carcasses from spawning streams to neighboring forests along the North Pacific Rim. Leopards and pumas protect some plants from overuse by herbivores. Fruit-eating carnivorans transport seeds away from parent plants, in some cases more effectively than birds or herbivorous mammals. Leopards in India even receive credit for reducing human fatalities inflicted by feral dogs, although leopards themselves kill a small number of humans. Each of these functions and services has been recognized or better understood recently, contributing to a much richer and more nuanced picture of how carnivorans affect human lives and well-being.

1.4 Purpose and organization of the book

This book is intended as a text for a college course in community ecology or predation ecology, and as a reference for students (academic or otherwise) of ecology who have some background in biological concepts and vocabulary. It emphasizes documented, mechanistic explanations for the ecology of carnivorans and species they interact with. Importantly, I do not review all biological knowledge that applies to the Carnivora—only those aspects that set carnivorans apart; many aspects of carnivoran biology resemble those of other mammalian orders. For example, the vibrissae of carnivorans are well developed and important for tactile sensation, but the primary research model for this organ has been the laboratory rat, not a carnivoran. An analogous situation exists for gut fermentation—it is rare in the Carnivora and is better understood from studies of other mammalian orders. To help with vocabulary issues, I provide a glossary of technical terms.

Because of my focus on community ecology, the reader will find only passing mention of some topics central to carnivoran biology in the past. These include intraspecific behavioral interactions: sociality, mating systems, parental care, and territoriality. These topics were at the forefront of carnivoran ecology during the 1970s—the heyday of interest in kin selection—but they are peripheral to community interactions. Ecological systems are complex, with indistinct boundaries between components. Predation relates to competition, and competition affects population growth. Population density affects dispersal, which, in turn, drives colonization and biogeography. This complexity makes ecology difficult to compartmentalize, and my chapter organization requires the reader to navigate specific topics via the index, in addition to the table of contents. For example, the reader interested in population biology will find carnivoran-induced changes in prey populations covered in Chapter 8, but the demography of carnivorans themselves is treated in Chapter 10. Dental adaptations are mostly covered in Chapter 2, but some other aspects of digestive morphology fit better in Chapter 4, the chapter on physiology. The responses of prey species to carnivorans can be behavioral, physiological, demographic, or have uncertain mechanisms, and these processes are covered in various sections, best located via the index. Chemical defenses against carnivorans by prey and other carnivorans, and by plants against herbivores, as well as detoxification of venoms by carnivorans, are each treated separately, in sections best located in the index. Habitat ecology is a traditional and highly diffuse topic that permeates wildlife biology. Scarcely a section of this book does not have habitat aspects. However, no aspect of carnivoran habitat ecology seems unique to the order, so I fold discussions of habitat in with others.

This account is based almost entirely on the peer-reviewed scientific literature. I use both primary and secondary sources, relying on community-level analyses, meta-analyses, or reviews where possible. I have tried to be comparative throughout, contrasting carnivoran families with each other and carnivorans with other vertebrate carnivores, including marsupials, reptiles, and birds. I have tended to prefer mechanistically based studies to those based only on correlative results or modeling and have favored widely accessible publications to more obscure ones.

While I have tried to enhance the geographic and taxonomic diversity of the case studies that I cited, I recognize that the literature is biased toward studies from developed countries and on high-profile or endangered carnivorans. Therefore, my presentation no doubt includes cultural biases that affect the generality of my conclusions. In some cases where examples support a generalization and multiple published examples illustrate the point, I have tended to cite a study from a region or ecosystem that is less well represented in the literature, rather than an equally illustrative study from North America or western Europe.

Some unifying themes connect the various facets of carnivoran ecology, and I return to them frequently. These factors explain much of the great diversity of form and function across the Carnivora, as well as many of the differences between carnivorans and other mammalian orders. The most important are body size, metabolic rate, and trophic level. Allometry is the study of body size and its consequences, and the reader will note the recurring importance of allometry in many processes at physiological and community levels (Calder, 1984). Metabolic rate is a function of body size, body temperature, and mitochondrial density, and has strong explanatory power in carnivoran ecology. Trophic level is correlated with these factors; carnivorans that eat large mammalian prey exemplify the constraints imposed and benefits conferred by high metabolic rate and large bodies—high foraging costs and maintenance costs—but food availability that is more seasonally consistent than for herbivores or predators of ectotherms. By watching for the recurring mention of body size, metabolic rate, and trophic level, the reader can appreciate how much carnivoran diversity arises from only a few principles.

1.5 Context in carnivoran ecology

While I search for pattern in carnivoran ecology, I make scarcely a generalization about this group without a qualification or caveat. Contingencies have caused the group to radiate into an astonishing range of phenotypes. Trophic niche determines dentition, digestive process, and gut passage. Size of prey species affects hunting behavior, frequency of predation events, and competitive interactions with scavengers. Context lies at the intellectual core of carnivoran ecology, and I have embraced it fully.

Conservation also is an applied arena in which context is all-important. Carnivorans are widely regarded as one of the most threatened mammalian lineages, with severe challenges to species, subspecies, and populations across Earth. On the other hand, carnivorans also cause or exacerbate conservation problems for other species that are rare or threatened. Further complicating the picture, humans have benefited some carnivoran species (or stopped persecuting them), and others are reoccupying their former geographic ranges with or without human assistance. Carnivore conservation does not merely represent a sad list of decline, dysfunction, and disappearance, but examples—admittedly anomalous—of restoration and independent recovery. Ecological and socio-economic context determine how carnivorans are faring in the modern world. All told, the biology of the Carnivora is an extraordinarily rich subdiscipline, full of pattern, nuance, contingency, and relevance to human culture and livelihoods.

1.6 Nomenclature

For the current scientific nomenclature of carnivorans, I have modified Wilson and Reeder (2005) to reflect recent taxonomic revisions (Appendix I). Common names are more problematic, because all are local or regional, and using one requires selecting from among those used by various indigenous groups or colonizing nations, or the native tongue of the original naming authority. Because I write in English, I default to my language, but I have tried to use the common name applied most geographically broadly where the species occurs. For example, *Puma concolor* occurs from Patagonia, South America to Yukon Territory, North America, with many locally used names over its range. However, the common name applied over most of the geographic range is "puma," which I use here.

References

Azéma, M. (2015) "Animation and graphic narration in the Aurignacian," *Palethnology*, 7, pp. 256–79.

Bonesi, L. and Palazon, S. (2007) "The American mink in Europe: status, impacts, and control," *Biological Conservation*, 134, pp. 470–83.

Boonstra, R., Krebs, C.J. and Kanter, M. (1990) "Arctic ground squirrel predation on collared lemmings," *Canadian Journal of Zoology*, 68, pp. 757–60.

Calder, W.A. III. (1984). *Size, function and life history*. Cambridge: Harvard University Press.

Crego, R.D., Jiménez, J.E. and Rozzi, R. (2016) "A synergistic trio of invasive mammals? Facilitative interactions among beavers, muskrats, and mink at the southern end of the Americas," *Biological Invasions*, 18, pp. 1923–38.

Dragoo, J.W. and Honeycutt, R.L. (1999) "Systematics of mustelid-like carnivores," *Journal of Mammalogy*, 78, pp. 426–43.

Dudley, J.P. *et al.* (2016) "Carnivory in the common hippopotamus *Hippopotamus amphibius*: implications for the ecology and epidemiology of anthrax in African landscapes," *Mammal Review*, 46, pp. 191–203.

Espigares, M.P. *et al.* (2013) "*Homo* vs. *Pachycrocuta*: earliest evidence of competition for an elephant carcass between scavengers at Fuente Nueva-3 (Orce, Spain)," *Quaternary International*, 295, pp. 113–25.

Gaubert, P. *et al.* (2012) "Reviving the African wolf *Canis lupus lupaster* in North and West Africa: a mitochondrial lineage ranging more than 6,000 km wide," *PLoS ONE*, 7, p. e42740.

Guagnin, M., Perri, A.R. and Petraglia, M.D. (2018) "Pre-Neolithic evidence for dog-assisted hunting strategies in Arabia," *Journal of Anthropological Archaeology*, 49, pp. 225–36.

Hammer, A.S. *et al.* (2021) "SARS-CoV-2 transmission between mink (*Neovison vison*) and humans, Denmark," *Emerging Infectious Diseases*, 27, pp. 547–51.

Hart, D. and Sussman, R.W. (2008) *Man the hunted: primates, predators, and human evolution*. Expanded edn. Boulder: Westview Press.

Helgen, K.M. *et al.* (2013) "Taxonomic revision of the olingos (*Bassaricyon*), with description of a new species, the olinguito," *ZooKeys*, 324, pp. 1–83.

Hofmeester, T.R., Jansen, P.A., Wijnen, H.J., Coipan, E.C., Fonville, M., Prins, H.H.T., Sprong, H., and van Wieren, S.E. 2017. Cascading effects of predator activity on tick-borne disease risk. *Proceedings of the Royal Society B* 284:20170453.

Hohenlohe, P.A. *et al.* (2017) "Comment on 'Whole genome sequence analysis shows two endemic species of North American wolf are admixtures of the coyote and gray wolf'," *Science Advances*, 3, p. e1602250.

Leopold, A. (1933) *Game management*. New York: Charles Scribner's Sons.

Levi, T. *et al.* (2012) "Deer, predators, and the emergence of Lyme disease," *Proceedings of the National Academy of Sciences of the United States of America*, 109, pp. 10942–7.

O'Brien, S.J. *et al.* (1983) "The cheetah is depauperate in genetic variation," *Science*, 221, pp. 459–62.

Pierotti, R. and Fogg, B.R. (2017) *The first domestication: how wolves and humans coevolved*. New Haven: Yale University Press.

Taylor, R.J. (1984) *Predation*. New York: Chapman and Hall.

Thulin, C.-G., Malmsten, J. and Ericsson, G. (2015) "Opportunities and challenges with growing wildlife populations and zoonotic diseases in Sweden," *European Journal of Wildlife Research*, 61, pp. 649–56.

Vinodh Kumar, O.R. *et al.* (2020) "SARS-CoV-2 (COVID-19): zoonotic origin and susceptibility of domestic and wild animals," *Journal of Pure and Applied Microbiology*, 14, pp. 741–7.

VonHoldt, B.M. *et al.* (2016) "Whole-genome sequence analysis shows that two endemic species of North American wolf are admixtures of the coyote and gray wolf," *Science Advances*, 2, p. e1501714.

CHAPTER 2

Functional morphology

Functional morphology considers how organ- and tissue-level structure and coloration are the basis of function. At smaller physical scales, "histology," "cell structure," and "molecular biology" are more common terms. Here I consider structure with direct ecological relevance, important for locomotion, prey handling, crypsis, communication, and reproduction. "Ecomorphology" is a synonym, and living and fossil carnivorans are well represented as subjects. The carnivoran-specific literature on this subject focuses mostly on three morphological regions: the skull, the forelimbs, and the hindlimbs. The skull is significant because of the great range of functions it performs and the spatial trade-offs between them. The forelimb is important because its structure differs across locomotor and foraging styles: swimming, digging through soil, sprinting, and climbing all leave strong adaptive signatures. To a lesser extent, the hindlimb also reflects adaptations to locomotor modes.

2.1 The skull

The skull is the bony structure with the most complex design trade-offs and constraints of any vertebrate body part. In carnivorans its components have key functions in display-defense, prey capture, mastication, vision, hearing, balance, olfaction, cranial enervation, and endocrine function, as well as housing the brain. The largest body of work dealing with skull morphology—dentition aside—concerns how trophic specialization affects non-dental skull features. Moore (2009, p. 197) explored how the ancestral carnivoran jaw joint diverged in two trajectories for dietary specializations. The jaw of most carnivorans features a deep mandibular fossa on the temporal (= "squamosal") bone that limits the lateral motion of the mandible. The shearing function of the carnassials occurs on one side at a time, and the mandible shifts toward the shearing side for any single bite. More herbivorous carnivorans—the giant panda and spectacled bears—do not have shallower mandibular fossae as might be expected for the greater range of motion required for grinding leaves and stems. Instead, their fossae are even deeper than those of more predaceous carnivorans and the mandible shifts laterally during a single bite to achieve the grinding effect. This chewing motion differs from that found in all other herbivorous mammals.

The adaptive trajectory toward killing prey larger than the predator requires further modification to the skull. This niche is occupied by large felids, canids, some ursids, and some hyaenids. Canine teeth may seem remarkably uniform in shape across extant carnivorans, but in fact show strong modification for various roles. The single most biomechanically challenging function of the skull and teeth is applying jaw force to the tips of the canine teeth while handling struggling large prey (Penrose et al., 2020). However, that task differs across carnivoran families. Slender, sharp canines are found in taxa (e.g. felids) that kill by quick stabs to the cervical region (dorsal spine or neck vessels). More robust canines occur in predators that take longer to kill prey that struggle, or that consume bony material. Canids tend to have more curved canine teeth, plausibly to hold struggling prey for longer times (Pollock et al., 2021), and not only must the canine teeth be stout, but also the jaw adductor muscles and associated bony structures must be strengthened—otherwise, the braincase is vulnerable. These adductors, originating from the sagittal crest, the nuchal crest, and the zygomatic arch,

Carnivoran Ecology. Steven W. Buskirk, Oxford University Press. © Steven W. Buskirk (2023). DOI: 10.1093/oso/9780192863249.003.0002

are especially powerful in predators of large prey. The sagittal crest runs antero-posteriorly along the dorsal midline and is the site of origin for the temporalis muscle, the largest jaw adductor. It inserts on the coronoid process of the mandible and closes the jaw. Tall and massive sagittal crests, resembling the sails of sailfish—are found in several carnivoran lineages: pinnipeds, hyaenids, ursids, and some mustelids. The masseter muscle, another powerful jaw adductor, originates from the zygomatic arch and inserts onto the lateral mandible. The nuchal crest, which follows the dorso-posterior margin of the occipital bone, is the site of insertion of neck extensor muscles that originate from the cervical vertebrae, powering head movements used in subduing large prey. Among carnivoran predators of large prey, all of these structures are large relative to the floor of the braincase (Penrose *et al.*, 2016).

The bony labyrinth is a complex structure that reflects feeding and non-feeding forces. Encased in the temporal bone, it comprises organs of balance (the semicircular canals with ampullae) and hearing (the cochlea). While not feeding structures per se, the canals are key to maintaining equilibrium and spatial orientation by measuring angular acceleration in three planes. It is therefore an important sense for mammals, especially carnivorans, with active locomotor and foraging styles. That is, they must stabilize the head in order to allow the brain to process visual inputs while pursuing and subduing prey. Schwab and colleagues (2019) showed that all dimensions of the semicircular canals were larger in ambushing carnivorans than in omnivores or pounce hunters. A strong phylogenetic signal was apparent as well. Other skull features are closely tied to sensory modalities (Section 5.1.2).

2.1.1 Dentition

Much study of modern and prehistoric diets of carnivorans and other carnivorous mammals has concerned prey handling (gripping and killing) and masticating. Basal carnivorans had the primitive eutherian dental formula—I^3/I_3, C^1/C_1, P^4/P_4, M^3/M_3—and teeth in some of these positions evolved more quickly than in others. A landmark in carnivoran dentition is the carnassial (shearing)

pair: P^4/M_1. This feature allows severing of muscle and connective tissue, so that parts of large carcasses can be removed and either swallowed or consumed away from the prey carcass, thereby reducing conflict. The earliest carnivorans had less prominent carnassial P^4/M_1 structures than do modern predaceous forms. All dental evolution involved changes in size and shape, as well as some tooth loss; no carnivoran lineage shows an increase in tooth number. Incisors have undergone relatively little change in morphology or number, and the canines are retained even in species that do not use them to handle prey. This reflects the dual functions of canines: prey handling and display-defense. Aardwolves, bat-eared foxes, and pandas—none of them primarily predators of vertebrates—have canines similar in shape and only slightly smaller to those of vertebrate-killing relatives. In at least five predaceous mammalian lineages (two metatherian, three carnivoran), canine teeth became so elongated that they extended outside the oral cavity. Most of these taxa showed the saber-toothed condition, using canines for prey killing, whereas the oversized canines of walruses function primarily in display and defense.

Consistent with adaptive changes to the carnassials themselves, other tooth positions reflect the importance of killing large prey vs. other trophic strategies (Figure 2.1). With specialization on predation and a meat diet came the reduction in size and number of the post-carnassial teeth (M^{1-3}, M_{2-3}). In felids, the best extant examples, premolars P^{1-2} and P_{1-2} are lost as well, because a diet of animal soft tissue requires little mastication before swallowing—duodenal lipases and peptidases are sufficient to initiate digestion. The second trajectory—a trend toward omnivory, frugivory, and folivory—retained the post-carnassials and modified them for grinding, giving them multi-cusped, rounded (bunodont) shapes. Bunodont teeth crush invertebrate exoskeletons and shells (e.g. in the walrus and otter), and coarsely grind plant material (e.g. in bears). In the highly carnivorous spotted hyena, the pre-carnassial premolars are enlarged for bone crushing—important for a scavenger of large ungulate carcasses. Piscivorous pinnipeds tend to retain large, sharp canines, but have lost the position-specific functions of ancestral cheek teeth, including the carnassial pair (Box 2.1). Instead, they tend

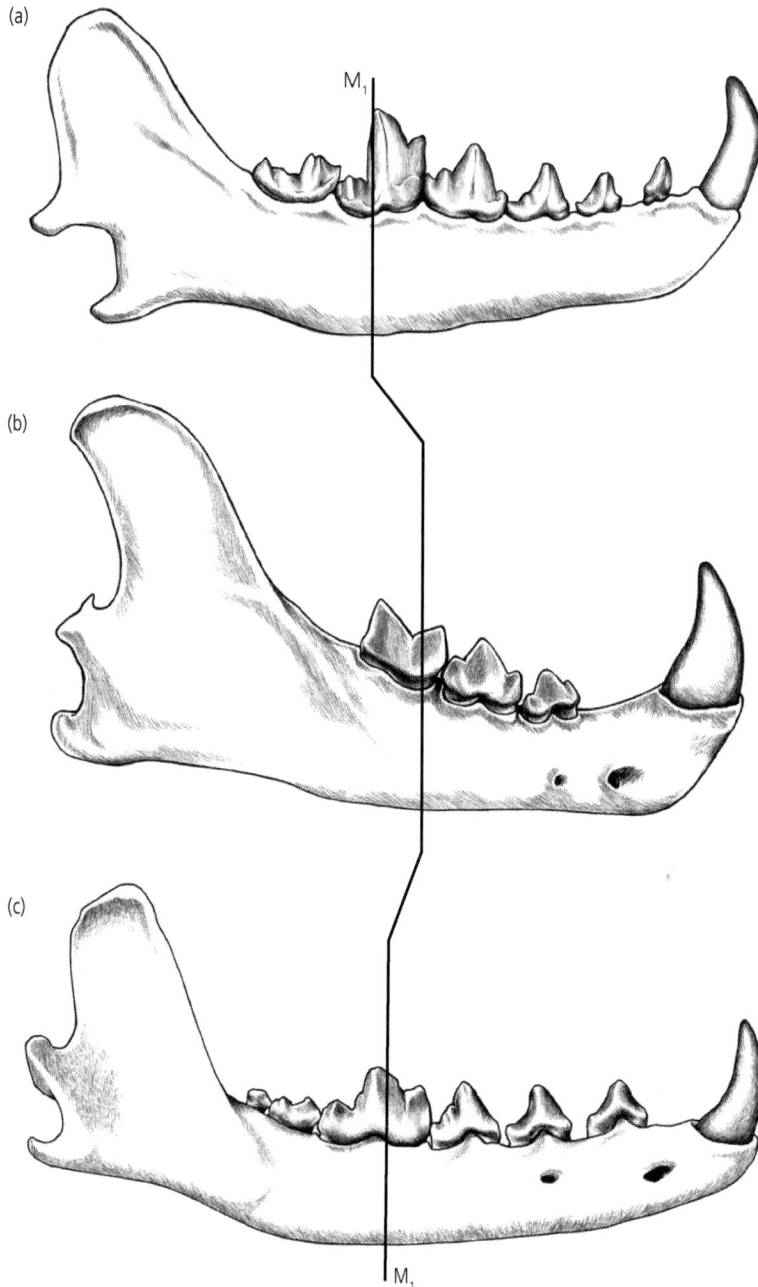

Figure 2.1 Lower dentitions of the earliest and modern carnivorans, showing enlargement, reduction, and loss of cheek teeth with diverging diets. A: *Protictis* (Viverravidae), an insectivore-carnivore of the Paleocene (M_3 lost). B: Modern lion, a strict carnivore (M_{2-3} lost). C: Red fox, an omnivore (M_2 retained, bunodont, M_3 vestigial). D: Giant panda, an obligate folivore (M_{2-3} bunodont). E: Aardwolf, an obligate termitivore (M_{2-3} variably vestigial or lost). F: California sea lion, an obligate piscivore (M_{2-3} lost, post-canines reduced and homodont).

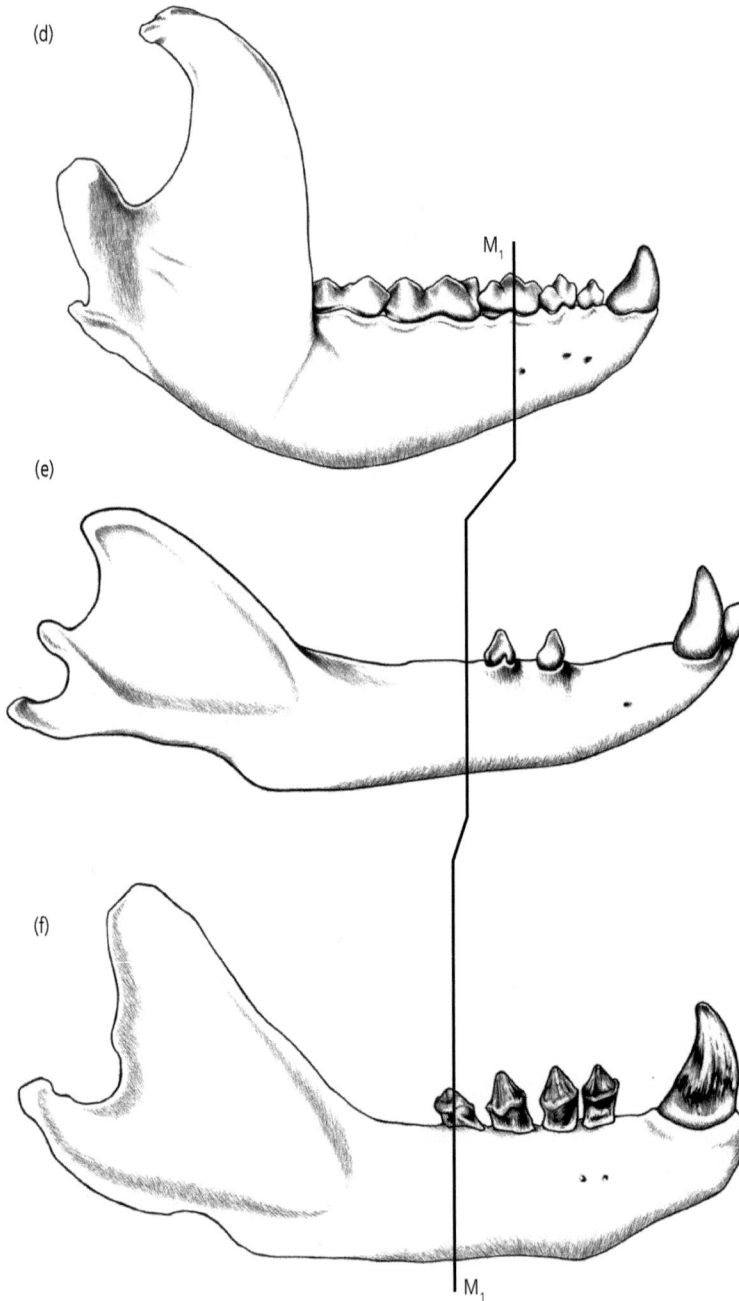

(d)

(e)

(f)

M_1

Figure 2.1 *continued*

toward generalized, nearly homodont cheek teeth (Figure 2.1F), consistent with swallowing fish whole or severing them into large pieces (Kienle and Berta, 2016). Importantly, carnivoran dental specializations have tended to be irreversible over evolutionary time; they disappear via lineage extinction,

Box 2.1 Filter-feeding carnivorans

Several pinnipeds have evolved highly specialized cheek teeth that facilitate the filtering of aquatic invertebrates. The water is gulped into the oral cavity and pharynx, then ejected past the projections on the cheek teeth (Figure 2.2). The trait has arisen several times in the phocids and in both freshwater and marine forms. It reaches its most extreme expression in the crabeater seal of the Antarctic Ocean, which eats almost entirely finger-sized krill. The Baikal seal of Russia shows a more moderate expression of the same trait—it feeds heavily on freshwater amphipods (Watanabe *et al.*, 2020).

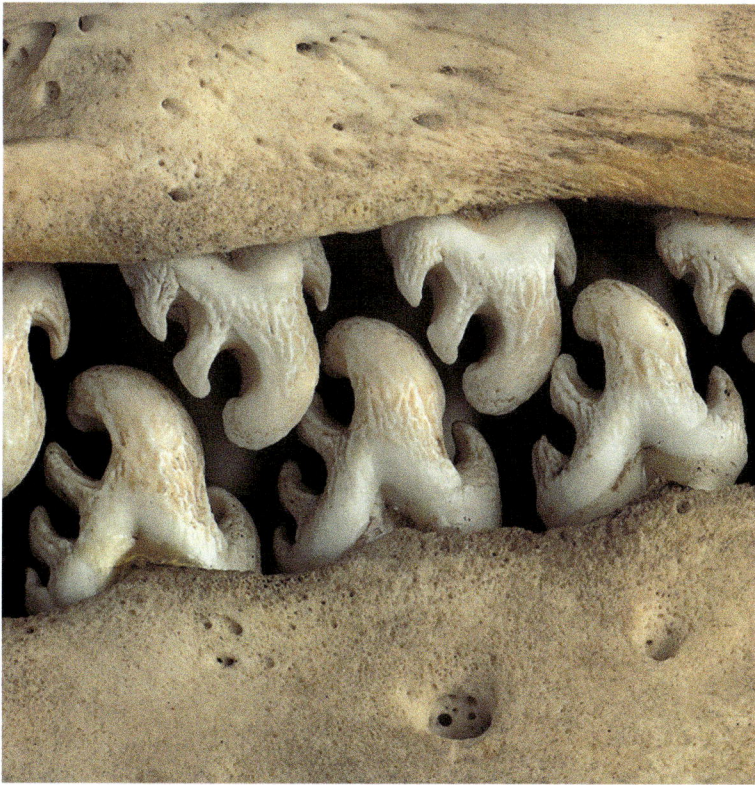

Figure 2.2 The cheek teeth of the filter-feeding crabeater seal, showing dental projections that filter krill from sea water. Photo: © Te Papa Tongarewa Museum of New Zealand.

rather than by restoring lost tooth functions. This is similar to mammalian dental evolution generally, but contrasts with that of squamate reptiles, in which multiple tooth cusps arose to grind vegetation. In some lineages, these cusps disappeared, and cheek teeth reverted to unicuspid form as diets returned to carnivory (Lafuma *et al.*, 2021).

2.2 Post-cranial skeleton

No other mammalian order matches the range of morphological adaptations for locomotion and foraging observed in the Carnivora. Several locomotor modalities are recognizable, and some species exhibit more than one. Basal carnivorans were locomotor generalists, walking, running, and climbing, but specializations arose early in several lineages.

2.2.1 Fossorial movement

The Mustelidae and Mephitidae include several obligatory below-ground foragers, that is, they must move soil to forage. These include American badgers, European badgers, honey badgers, hog badgers, ferret badgers, ferrets, polecats, and stink badgers. Mustelids smaller than ferrets also forage underground, but do not move soil. Fossorial forms have forelimbs modified for digging, including longer claws and elongated olecranon processes of the ulna, which strengthened the digging stroke (Rose et al. 2014). They also exhibit enlarged scapular surfaces and shortened limbs to allow moving in constricted spaces (Figure 2.3) (Van Valkenburgh, 1987). The distal limbs tend toward the plantigrade condition, in which the podial-metapodial joint touches the ground, as seen in the hind foot of a bear. As well, the plane of the forefoot bends medially (supinates) in the American badger to allow excavated soil to pass beneath the abdomen and between the hind legs. The clavicle is vestigial or completely lost, allowing greater range of forelimb motion. Domestic ferrets walk in tunnels by lowering and flattening the lumbar spine, which serves to decrease the effective lengths of the limbs. These adaptations allow ferrets to walk at the same speed along subterranean burrows as above ground (Horner and Biknevicius, 2010). A slender snout in some species facilitates seizing prey in tight spaces, and the pinnae are reduced. In addition to burrow-dwellers, several small and mid-sized mustelids forage or occupy dens in talus fields, hollow tree boles or other confined spaces (Harris and Steudel, 1997). Many small canids, herpestids, and viverrids occupy burrows for protection from enemies or weather, but not necessarily for foraging.

2.2.2 Running and walking

Most terrestrial carnivorans trot, bound, or run. I distinguish between two patterns of cursation (gaits faster than a walk): coursing and sprinting. All carnivorans that run fast tend toward elongated metapodials and proximal phalanges, which increase stride length. The long bones of their limbs are as long as predicted from body mass (Figure 2.3) (Harris and Steudel, 1997), with prey captured in the jaws or forelimbs. The clavicle is lost, allowing scapular rotation along the dorso-ventrally elongated ribcage, and the digitigrade stance further lengthens the stride. Coursers such as the canids, wolverine, and spotted hyena trot, gallop, or bound long distances at sustained intermediate speeds, whereas sprinters, typically ambush predators, run fast in short bursts, typically while pursuing prey. Some coursers have limited hip and elbow rotation, which increases running efficiency at the expense of maneuvering and manipulating prey. Distinct suites of adaptations are associated with coursing vs. ambushing. Coursing species cover their home ranges at a walk or trot, detecting prey via sight, hearing, or scent. Once a prey animal is encountered, the courser may seize and kill it with a bite (e.g. small canids), disable it by exhaustion and blood loss, injure its cervical spine, or asphyxiate it by crushing the trachea. Their running muscles have high concentrations of myoglobin-rich, slow-oxidative fibers, with dense mitochondria, which are fatigue-resistant at sustained intermediate speeds (see Goldspink, 1977 for canid–felid comparison, Hill *et al.*, 2012). These muscles tend to have high densities of capillaries essential for sustained oxygen delivery (Sjøgaard, 1982). Coursers cannot easily grasp prey with the forelimbs, having little elbow rotation and claws that cannot protract.

Sprinters, by contrast, traverse their home ranges at a walk, lie in wait for longer periods, tend to detect prey by vision rather than olfaction, and capture them after brief pursuits, in other words, ambushing. Most felids are such predators, and are physiologically similar to coursers, but run faster for briefer periods and tire more quickly. Their running muscles are mostly fast glycolytic and have higher rates of contraction, less myoglobin, and fewer mitochondria than do coursers. They metabolize anaerobically during peak exertion, accumulating lactic acid in muscle cells. Cats and other ambush

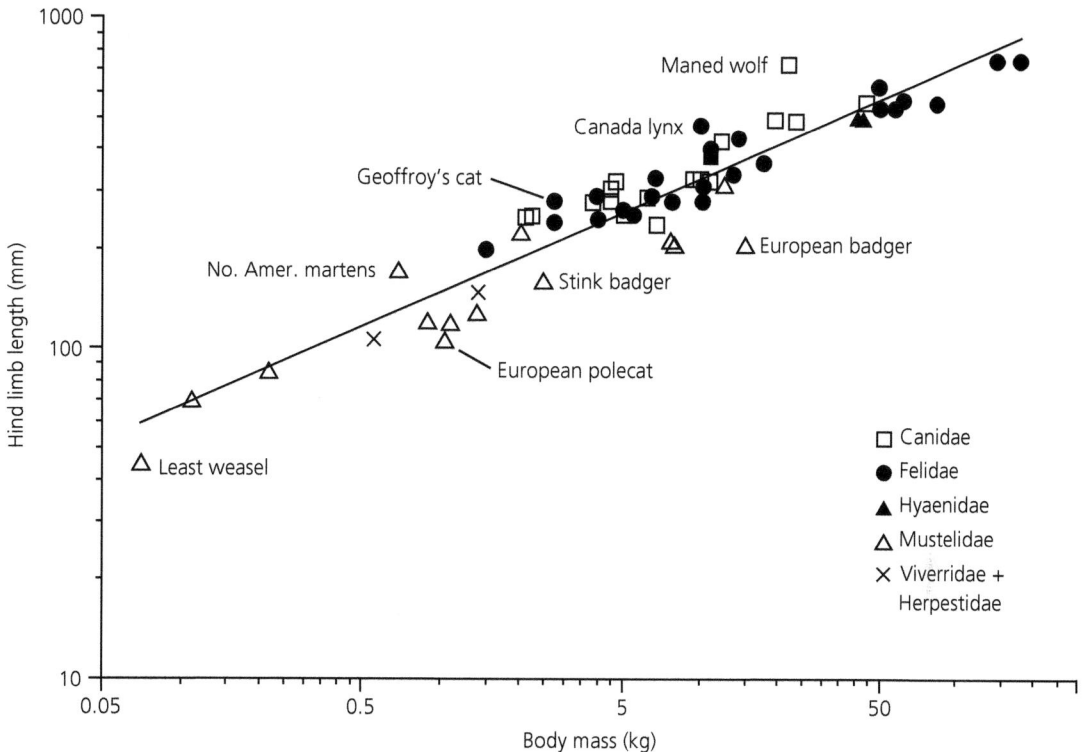

Figure 2.3 Hind limb length in relation to body mass for sixty-three terrestrial carnivoran species. Mustelids that excavate soil have proportionally short legs. Canada lynx and boreal forest martens, which walk on snow in winter, have long legs.
Modified from Harris and Steudel (1997).

carnivorans rotate the radius well, consistent with forelimb grasping, and the claws—sharper than for coursers or diggers—are protractible. Elbow rotation and sharp claws also facilitate tree climbing (Figure 2.4) (Stein *et al.*, 2015). Cheetahs are intermediate between coursers and ambushers in these traits; they have highly flexible spines that increase stride length. The scapholunate bone, the largest carpel bone in the carnivoran wrist, represents the fused scaphoid and lunate (separate bones in predecessors to the Miacoidea (Rose, 2006, p. 129)). This fusion was earlier regarded as adaptive specifically for running, which is intuitive inasmuch as reduction in the size and number of distal limb elements is a common cursorial adaptation. Analyses of shape across carnivoran foraging types show that the scapholunate has been modified for pure cursation vs. grappling with prey in the respective hunting types (Dunn *et al.*, 2019).

2.2.3 Climbing

Most carnivoran families include species that range from scansorial—semi-arboreal—to highly arboreal. These species also climb cliffs and buildings, and most are small-bodied (Van Valkenburgh, 1987). The most arboreal include procyonids (e.g. kinkajou and olingos), a few viverrids and feliforms (e.g. palm civets and linsangs), and euplerids of Madagascar. Tree climbing is carried out in various ways, involving differing adaptive suites. Two species (kinkajou and binturong) benefit from partially prehensile tails that feature some muscular control of the distal caudal vertebrae (Figure 2.5). Protractible claws allow grasping of the tree bole, and rotation of the hip allows some species to descend the bole head-first. Small (and juvenile) bears are scansorial, having recurved claws on the forelimbs and a sesamoid carpal bone encased in tendons attached to the radius (Ewer, 1998; Fisher, 2011).

Figure 2.4 A caracal rotates an elbow to lick the plantar surface of a forefoot. Elbow rotation is important for grasping prey and for climbing trees. The claws of the forelimb are normally retracted.
Photo: © four oaks/Canstock.

The latter trait is well developed in both species of panda of and the kinkajou and is so well developed in the giant panda as to suggest adaptation for grasping bamboo during feeding or climbing (Gould, 1978). Climbing bears also have robust subscapularis muscles, which insert along the posterior margin of the scapula and stabilize the shoulder when bears pull themselves up by their forelimbs (Davis, 1949). Among canids, only the scansorial gray foxes have somewhat sharp and recurved forelimb claws, climbing mostly by hopping from limb to limb, rather than by hoisting themselves. Smaller felids ascend trees by either branch-hopping or by grasping the bole. In addition to various musteloids and procyonids, several Oligocene felids were in this group; the latter likely used trees for escape, rather than for foraging (Van Valkenburgh, 1987).

2.2.4 Swimming and deep diving

Mammals returned to the marine environment seven times over their evolution, and of the five surviving marine lineages, three (pinnipeds, polar bear, and sea otter) are carnivoran (Uhen, 2007). Among pinniped families, we see the strongest adaptations to swimming and deep diving in the Phocidae. Their hind limbs have lost all locomotor function on land but are effective propulsive organs in water. By contrast, the Otariidae and Odobenidae support their posterior bodies on hind limbs to walk awkwardly on land, and dive effectively to shallower depths than do phocids. The less derived semi-aquatic forms include (in approximately decreasing order of aquatic adaptation) the sea otter, other otters, minks, the otter civet, the aquatic genet, raccoons, and the fishing cat. Fishing cats, jaguars, and raccoons forage by wading and

Figure 2.5 The binturong (shown), a viverrid, and the kinkajou, a procyonid, are among the most arboreal of carnivorans and have prehensile tails, like some arboreal primates.
Photo: © Pakhnyushchyy/iStock.

feeling for prey with their forelimbs, and swim on the surface with limbs better suited for walking. Jaguars can subsist on an almost exclusively aquatic diet, while being equally well adapted to living in arid habitats with scarce free water; their aquatic adaptations are entirely behavioral (Eriksson *et al.*, 2022). At the other carnivoran extreme, elephant seals spend 8–10 months/year at sea, dive to depths of over 1500 meters for durations of over 100 minutes, and come ashore only to give birth, mate, and molt (DeLong and Stewart, 1991). The highly marine habits of large seals create major life history trade-offs. Their large bodies and fat reserves reduce mass-specific thermal losses in cold water and prolong fasting endurance. While on land, however, they are nearly immobile, and vulnerable to

predation, injuries from short falls, and crushing by conspecifics on haul-outs or rookeries (Udevitz *et al.*, 2013).

2.3 Other adaptations to aquatic living

All adapted carnivoran swimmers display some degree of streamlining, insulation, webbed or fin-like distal limbs, reduced pinnae, and cardiopulmonary adaptations to prolonged apnea. Enlarged propulsive structures (webbing of digits, modification of limbs to fin shapes) are blended with coordinated trunk motion, so that the entire axial skeleton and associated muscles contribute to swimming motion (Williams, 1999).

Among air-breathing vertebrates, carnivorans live at a uniquely wide range of ambient pressures. They occur at elevations above sea level as high as 6000 meters in the case of the snow leopard and 5100 meters for the Andean cat (Alianza Gato Andino, unpublished data). The partial pressure of oxygen at 6000 meters is around half its value at sea level. By contrast, the southern elephant seal dives to 2400 meters below the ocean surface, where ambient pressures create gas pressures in respiratory passages many times higher than at sea level. The snow leopard lacks any known biochemical adaptations to high-altitude hypoxia (Janecka et al. 2015); however, diving as deep as elephant seals do requires extreme morphological and biochemical adaptations. They include rib articulations that allow thoracic collapse, frontal sinuses being reduced or lost, and lung alveoli that compress elastically to reduce gas exchange between blood and pulmonary passages at depth (Ponganis, 2015; Curtis *et al.*, 2015). Ambient pressure increases linearly by one atmosphere per ten meters of water depth, so that a diving mammal at 2400 meters is exposed to around 240 atmospheres of pressure. The structures of a non-adapted mammal's pulmonary tract, from the pharynx to the alveoli, cannot withstand the pressures of such dives without trauma. In deep-diving pinnipeds these structures collapse and rebound without injury.

Duration of dives is generally correlated with depth of dives because of transit times to depth, but the adaptations to deep vs. long-duration dives are distinct. Extended apnea is facilitated by high blood volume, high hematocrit, high O_2 binding of hemoglobin, and concentrated myoglobin in somatic muscles. Most oxygen is stored in muscle and blood, rather than in air in respiratory passages. Further, the spleen stores oxygenated red cells, which are released into the general circulation under hypoxia. In addition, long-duration divers experience bradycardia, increase their anaerobic metabolism, and temporarily reduce blood flow to organs not essential during dives. Blood supply to and aerobic metabolism in the head, adrenal glands, and placenta are maintained (Bininda-Emonds *et al.*, 2001; Ponganis, 2015).

2.4 Gut morphology

In comparison with other mammals, and setting aside deep divers, the carnivoran digestive tract is remarkable for its simple form. It has no volume-enhancing chambers, sacculation, or structures to house fermenting microbes (Figure 2.6). This is so even for species that evolved obligate herbivory millions of years ago; phylogeny shows a stronger signal than diet in carnivoran gut morphology. Several factors complicate comparisons, however. Most variation in gut length lies in the small intestine; colon length varies little with diet. Mustelids have total intestinal (small + large) lengths about 1.4 times that for terrestrial carnivorans generally, for unclear reasons. The carnivoran caecum is rudimentary (e.g. in Canidae and Felidae) or absent (e.g. in Ursidae, Mustelidae, and Procyonidae), unlike that of hindgut-digesting herbivores (e.g. rabbits, horses, and elephants). Carnivorans that eat more plant leaves and stems tend toward longer caecae, but no substantial microbial fermentation has been reported. The curiously long caecum of the domestic dog—but not of wild canids—is consistent with the increase in grains in the human diet during the agricultural revolution, which coincided with dog domestication (McGrosky *et al.*, 2016). This suggests potential post-gastric fermentation, but how the products could be absorbed downstream of the caecum is not known.

Marine carnivorans represent a special case of digestive morphology, having intestinal lengths 1.4–2.8 times that predicted from body size (Williams *et al.*, 2001). Sea otters have small intestines—more like pinnipeds than terrestrial mustelids—that are long for their body sizes. The adaptive value of small intestines of such length has long puzzled physiologists (Maxwell, 1967). The current hypothesis suggests this result is due to the redirection of blood away from the gut under anoxic diving conditions. This suspends digestive processes for extended periods and lengthened small intestines compensate for lost time when divers return to the surface and circulation to the gut is restored (Duque-Correa *et al.*, 2021).

2.5 The integument

The adaptive value of fur color and markings has been considered for various species and multiple hypotheses exist, of which crypsis and aposematism are the most common. Various carnivorans, like other mammals, feature pelage that matches their background—dark in forest environments (Gloger's rule), light tan on arid, sandy backgrounds, and white in snowy habitats (Caro and Mallarino, 2020). Some pinniped species whelp white neonates where they are born on snowy backgrounds, but dark ones in caves or on predator-free islands. Four carnivorans, three of them *Mustela* spp., undergo seasonal color changes to white in winter. Two of those species do so only in the parts of their ranges with winter-long snow cover. Crypsis is also suggested as the adaptive purpose for bold black and white patches of the giant panda (Nokelainen et al. 2021). Aposematic coloration has received special consideration in Carnivora because defensive weapons are so widespread (Section 7.7.5; Newman *et al.*, 2005; Howell *et al.*, 2021).

Many mammals reduce thermal losses via adaptive fur traits, and fur affects buoyancy in aquatic species. Among small and mid-sized semiaquatic carnivorans, the fur is dense, but subcutaneous fat is slight or intermediate. Dense fur traps air that is insulative and retains its volume during the shallow dives made by minks, civets, and otters. The sea otter, the smallest marine carnivoran, has the densest underfur reported for any mammal (Figure 2.7), with over 1100 hairs/mm^2 (Fish *et al.*, 2002). Sea otter fur is the most buoyant of any fur reported, which is important for reducing the contact area between the integument and water while resting on

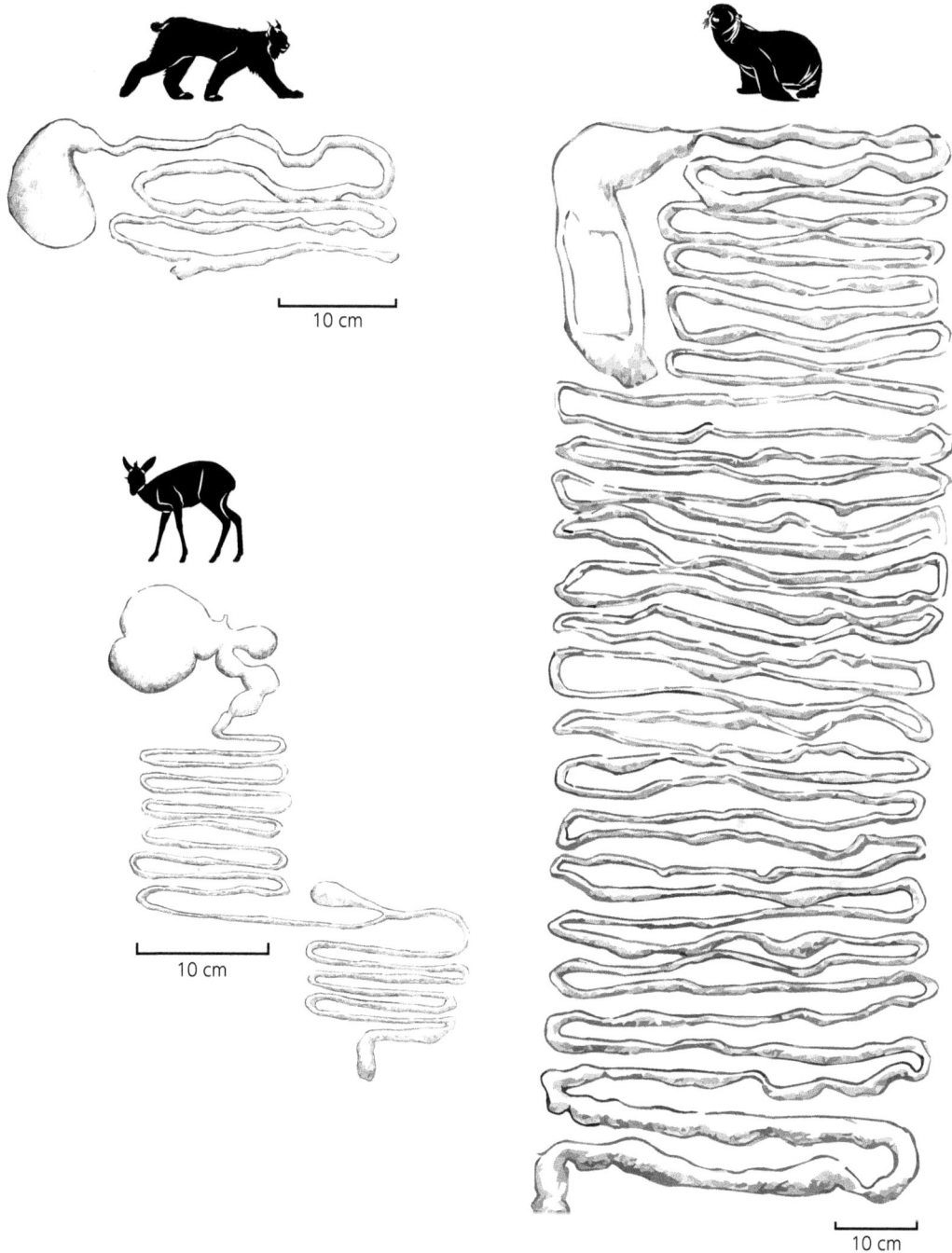

Figure 2.6 Digestive tract length varies highly across mammals, changing partly with diet. The Canada lynx has a shorter digestive tract than a ruminant herbivore of comparable body size (e.g. dik-dik). Pinnipeds (e.g. California sea lion) have especially long digestive tracts, even accounting for body size. Scale bar is for digestive tracts.
Modified from Stevens and Hume (1995) and McGrosky *et al.* (2016).

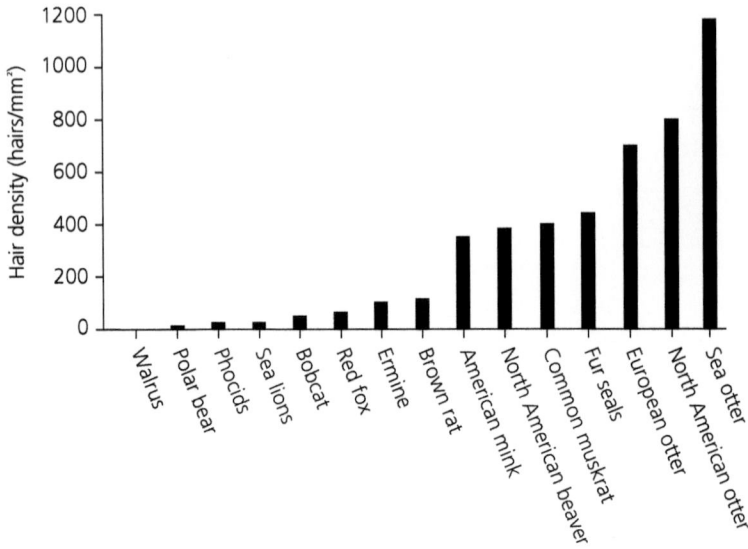

Figure 2.7 Hair density of mammalian species varies with aquatic living and with body size. Air trapped between hairs contributes both thermal insulation and buoyancy. The largest-bodied pinnipeds have low hair densities, whereas small-bodied otters have high ones. This partly reflects the dependence of buoyancy on water depth; deep divers lose insulative effect of fur under high ambient pressures. Values for pinnipeds are means of species means (phocids, $n = 13$; sea lions, $n = 5$; fur seals, $n = 6$). Data are from Liwanag *et al.* (2012, Table S4) and Fish *et al.* (2002, Figure 4).

the surface. This is especially so for females supporting neonates on their torsos (Figure 2.8). Fur is a less-effective insulator for deep divers, because of gas compression at depth—the thickness of an insulating air layer at 500 m depth is 2% what it is at the surface. The deepest pinniped divers have short, stiff hairs, rather than air-trapping fur. In addition, the largest marine carnivorans—elephant seals, sea lions, and walruses—minimize mass-specific thermal losses via their large body sizes. Blubber is a fair thermal insulator and has the added benefit of energy storage and retaining its insulation and buoyancy at depth (Liwanag *et al.*, 2012). Subcutaneous fat is especially important for deep-diving pinnipeds, which spend a lot of diving time drifting passively through the water column. Body condition (and therefore the blubber depot) affects drift rates strongly. As elephant seals increase blubber depots, they become more buoyant but more hydrodynamically resistant. The relationship is so strong that body condition can be estimated from drift rates, as recorded from pressure transducers (Biuw *et al.*, 2003).

In the 1980s polar bears were proposed to represent a special mechanism of solar warming via translucent guard hairs that transmit visible and ultraviolet light down the length of the hair shaft, where it is absorbed by the darkly pigmented skin (Grojean et al. 1980). This "optical fiber" model of solar heating held sway, was thought to be unique

to polar bears, and was considered a possible model for winter clothing design. However, Koon (1998) pointed out that the scattering of light in the hair shaft was so great that little light travelled down a single hair shaft to the skin. More recently, Khattab and Tributsch (2015) explained that the mechanism observed in polar bears involves light transmission via guard hair shafts collectively, not individually, so that there is some radiation that makes its way to the skin. However, the warming effect on an endotherm as large and well insulated as the polar bear is modest. The fiber optic model has lost explanatory power for how polar bears keep warm, and no longer inspires designers of cold-weather apparel.

2.6 The major ecomorphotypes

Functional morphologists have tended to clump fossil and living mammals into groups of discrete types ("ecomorphs," vs. "ecotypes"), an approach proposed by White and Keller (1984) and adapted to the Carnivora by Martin (1989). These types are based mostly on locomotor and dental traits and reflect major adaptive spaces. Examples include species requiring burrowing vs. tree-climbing, exhibiting dietary generalization vs. specialization, or ambushing vs. other means of prey capture. Some of these suites of traits have appeared only once in the fossil record, whereas others have

Figure 2.8 Buoyancy is a critical feature of the fur of sea otters, allowing them to float on the surface with minimal contact with cold water. Here a female supports a neonate on its torso, above water level. Resting adults and neonates metabolize at elevated rates via mitochondrial proton leak (Section 4.3.1).
Photo: © Jeanninebryan/CanStock.

arisen repeatedly. The strong marine adaptations of the pinnipeds, specifically those for deep diving and exposure to the open sea, arose only once in the Carnivora (Arnason *et al.*, 2006). By contrast, the "sabertoothed" condition of enlarged upper canines and associated skull modifications appeared in multiple feliform lineages, none of which persisted to the present.

2.6.1 Scansorial ecomorph

This small- to mid-sized ecomorphotype, the predominant one in the earliest carnivorans, is associated with forested habitats, traveling on the ground or on tree branches, and eating small vertebrates, insects, and soft mast. The body plan includes a somewhat elongated skull rostrum, long tail, elbows with intermediate rotation for climbing and clamoring, and in highly frugivorous forms, crushing cheek teeth. Modern representatives occur among the Viverridae, Mustelidae, Procyonidae, and Nandiniidae (see Ercoli and Youlatos, 2016 for discussion of the tayra).

2.6.2 Dog-like ecomorph

Dog-like carnivorans have ranged in size from small to over 75 kg. They have had limbs of intermediate length and long tails, and were coursers—sustaining trotting for long durations, but limited

in peak running speeds, in climbing ability, in maneuverability around obstacles, and in ability to rotate the elbow (Anderrson and Werdelin, 2003). Early representatives included creodonts and some amphicyonids (bear-dogs) of the Miocene (Viranta, 1996). The most obvious extant representatives are canids and hyaenids, although the latter have intermediate tails and limited cursorial abilities. The rise of this type coincided with the appearance, at temperate latitudes, of savannah beyond the margins of moist riparian zones in the mid-Miocene. This major shift in temperate habitats initially produced, in temperate North America, woodland steppe, followed by grass-dominated prairie by the early Pliocene. Steppe-like habitats provided the adaptive space for abundant large grazing herbivores, which promoted this carnivoran ecomorphotype (Webb, 1977; Janis *et al.*, 2002).

2.6.3 Cat-like ecomorph

These animals occur in two discrete habitat types with associated adaptations. Forest-dwelling cats have tended toward small bodies and forelimbs with short, stout bones for climbing. By contrast, open-country cats tend toward larger bodies, longer, thinner bones and greater use of crouching and ambush (Schellhorn and Sanmugaraja, 2015). Cat-like carnivorans have short rostra with strong binocular vision, variable tails, and recurved, protractible claws. The elbow joint rotates well for grasping prey and tree boles (Anderrson and Werdelin, 2003). One extant exception is the cheetah, which has reduced forelimb rotation and semi-retractile claws, and uses longer chases to capture prey than do pure ambushers. Cat-like animals have tended to be the largest-bodied carnivorans in many fossil assemblages; small-bodied extant species remain common (Van Valkenburgh, 2007). Prey can be larger or smaller than the predator, killed by either a bite to the dorsal cervical spine or (in the case of saber-toothed forms) laceration of the neck vessels. The saber-tooted dentition evolved twice outside of the Carnivora—in *Machaeroides* (Creodonta) and *Thylacosmilus* (Sparassodonta)—and multiple times within the Feliformia. Two major saber-toothed subtypes are recognized. The scimitar-toothed feliforms had upper canines intermediate in length and elongated dorso-ventrally, while flattened laterally. These teeth featured distinct crenulations along their edges. Their legs were long and optic regions of the brain enlarged, adaptive for bursts of speed and visually localizing prey. By contrast, "dirk-toothed" animals had especially long upper canines, in *Smilodon* serrated along both margins. Their body plans featured shorter legs and enlarged olfactory bulbs in the brain in comparison with scimitar-toothed forms. Analyses of bone density and stress modeling of the rostra of both subtypes and modern lions have revealed differences in prey-killing styles (Figueirido *et al.*, 2018). The dirk-toothed *Smilodon* seems to have used its shorter, more powerful forelimbs to grasp prey while stabbing the vascularized neck with its dirk-like canines. Modern lions, with their much shorter and more conical canines, are better suited to sustained multi-direction stresses experienced when they grasp and hold struggling prey. Scimitar-toothed forms were intermediate between *Smilodon* and modern lions in their adaptations for stabbing vs. withstanding lateral shaking. Some fossil feliforms combined elements of the two subtypes, featuring scimitar-like teeth, but stout, short limbs (Martin *et al.*, 2000).

2.6.4 Scavenger ecomorph

This ecomorph was characterized by a robust skeleton adapted to trotting and running, but not at high speed. Its most distinctive features were cranial and dental: powerful jaw musculature and large, pyramidal premolars for cracking bones. Among extinct species, we see examples in the Borophaginae (Canidae), some creodonts, and some early Laurasiatherian forms (Arctocyonia) that may have had ungulate or carnivoran ancestors. *Australohyaena*, a South American marsupial sparassodont, converged on the same ecomorph (Forasiepi *et al.*, 2015). Among extant species, it is exemplified in three species of hyaenids, the wolverine, and the ursids.

2.6.5 Semi-fossorial ecomorph

These mostly small and mid-sized species either excavate burrows to forage or enlarge those made by other species. I do not include the many species of several families that use and modify underground dens or burrows to thermoregulate, avoid threats, or give birth. Among extinct forms, this type was exemplified in several genera of Miocene mustelids not ancestral to the modern analogues (Hochstein, 2007). In extant forms, it is more prevalent among the Caniformia than in the Feliformia and shows specific adaptations for either digging or entering tight spaces: slender bodies, short legs, or elongated claws. Frontal sinuses tend to be lost to conserve skull volume (Curtis *et al.*, 2015). Associated with these characteristics are altricial young, geographic ranges that extend to high latitudes, and invertebrate diets. Species that excavate their own burrows tend toward group denning, as opposed to group hunting (Noonan *et al.*, 2015). Modern examples include aardwolves, badgers, meerkats, polecats, some *Mustela* spp., and some mephitids.

2.6.6 Semi-aquatic ecomorph

These are the otters of subfamily Lutrinae, which exhibit many standard aquatic adaptations: shortened limbs and pinnae, elongated bodies, streamlining, webbed digits, and dentition adapted for seizing and severing parts of fish, as well as crushing the exoskeletons of invertebrates. Most swim powerfully and nimbly to depths of scores of meters, but return to land to rest, and all but the sea otter give birth in sheltered dens. European and

American minks are less derived for aquatic living, eating slower-moving fish, some small mammals, and intertidal or shallow-water invertebrates.

2.6.7 Marine ecomorph

The seals, sea lions, and walruses exhibit extreme marine adaptations among the Carnivora: large, streamlined bodies, limbs modified as flippers for swimming at the expense of walking, and dental modifications for seizing fish before swallowing them whole, or for crushing invertebrates. Adaptations to deep and long-duration diving are extensive: skeletal morphology, streamlining, oxygen storage systems, circulatory redistribution, and endocrine adaptations to deep or long-duration dives. The sea otter combines elements of the semi-aquatic and marine ecomorphs, diving only to shallow depths, but being more aquatic-adapted than other otters in other respects.

2.6.8 Intermediate and unique ecomorphs

Some extant species do not fit neatly into only one of these named categories, and we can assume that some extinct ones did not either. Among living forms, the boreal forest martens (five species of *Martes*) forage near the ground or on the snow surface, investigating burrows (or subnivean access points) by scent. However, they also can forage, rest, and den in the forest canopy, and rely seasonally on foods of aquatic or marine origin. Mustelids are typically regarded as short-legged, but martens are among the longest-legged carnivorans for their body size (Figure 2.3). Long limbs permit them to traverse home ranges that are large for their body sizes, even by carnivoran standards (Buskirk and McDonald, 1989), but flexible spines allow them to traverse narrow passages and negotiate tree branches. Raccoons (Procyonidae) alternate between shallow-water and riparian environments for foraging omnivorously, but climb trees for soft mast, predator escape, and denning. Giant pandas (Ursidae) and red pandas (Ailuridae) convergently evolved specialized morphologies for bamboo herbivory and tree climbing. Finally, the aardwolf

represents a unique carnivoran phenotype: a burrowing termite eater with locomotor adaptations similar to canids but derived from a lineage that radiated into ungulate predators and bone-cracking scavengers.

Key points

- Functional morphology examines relationships between structure and various kinds of performance at the organ or tissue scale. The most active areas of carnivoran research have concerned the skull, fore- and hind limbs, and gastrointestinal tract.
- Dental adaptations to hunting and vertebrate carnivory include refinement of the shearing function of P^4-M_1, hyper-elongation of the canines in some extinct lineages, and reduction or loss of premolars and post-carnassials. The incisors have changed little through carnivoran evolutionary history.
- Omnivory, folivory, and a diet of mollusks tend to select for rounded cheek teeth and loss of the shearing function and retention of premolars.
- Locomotor derivations from the early carnivoran condition include suites of adaptations for digging and fossorial foraging, coursing, ambushing, climbing, swimming, and diving to great depths.
- The guts of carnivorans are short and simple, and tend toward evolutionary conservatism, with only slight adaptation to folivory in extant folivorous forms. Pinnipeds have long guts for their body sizes.
- Cardiopulmonary and circulatory adaptations in deep-diving pinnipeds are dramatic and relate to two discrete issues: hypoxia resulting from extended duration dives, and pressure-related effects of dives to great depths.
- Ecomorphotypes (ecomorphs) are groups of species that tend toward similar morphologies and niches, based mostly on locomotor morphology and dentition. For carnivorans, these types comprise tree-climbing forms, dog-like coursers, ambushers of several types, scavengers, semifossorial forms, marine dwellers, and intermediates.

References

Anderrson, K. and Werdelin, L. (2003) "The evolution of cursorial carnivores in the Tertiary: implications of elbow-joint morphology," *Proceedings of the Royal Society of London B (Suppl.)*, **270**, pp. S163–5.

Arnason, U. *et al.* (2006) "Pinniped phylogeny and a new hypothesis for their origin and dispersal," *Molecular Phylogenetics and Evolution*, **41**, pp. 345–54.

Bininda-Emonds, O.R., Gittleman, J.L. and Kelly, C.K. (2001) "Flippers versus feet: comparative trends in aquatic and non-aquatic carnivores," *Journal of Animal Ecology*, **70**, pp. 386–400.

Biuw, M. *et al.* (2003) "Blubber and buoyancy: monitoring the body condition of free-ranging seals using simple dive characteristics," *Journal of Experimental Biology*, **206**, pp. 3405–23.

Buskirk, S.W. and McDonald, L.L. (1989) "Analysis of variability in home-range size of the American marten," *Journal of Wildlife Management*, **53**, pp. 997–1004.

Caro, T. and Mallarino, R. (2020). "Coloration in mammals," *Trends in Ecology and Evolution*, **35**, pp. 357–66.

Curtis, A.A. *et al.* (2015) "Repeated loss of frontal sinuses in arctoid carnivorans," *Journal of Morphology*, **276**, pp. 22–32.

Davis, D.D. (1949) "The shoulder architecture of bears and other carnivores," *Fieldiana Zoology*, **31**, pp. 285–305.

DeLong, R.L. and Stewart, B.S. (1991) "Diving patterns of northern elephant seal bulls," *Marine Mammal Science*, **7**, pp. 369–84.

Dunn, R.H. *et al.* (2019) "Locomotor correlates of the scapholunar of living and extinct carnivorans," *Journal of Morphology*, **280**, pp. 1197–1206.

Duque-Correa, M.J. *et al.* (2021) "Mammalian intestinal allometry, phylogeny, trophic level and climate," *Proceedings of the Royal Society B: Biological Sciences*, **288**, pp. 20202888.

Ercoli, M.D. and Youlatos, D. (2016) "Integrating locomotion, postures and morphology: the case of the tayra, *Eira barbara* (Carnivora, Mustelidae)," *Mammalian Biology*, **81**, pp. 464–76.

Eriksson, C.E. *et al.* (2022) "Extensive aquatic subsidies lead to territorial breakdown and high density of an apex predator," *Ecology*, **103**, pp. e03543.

Ewer, R.F. (1998) *The carnivores*. Ithaca: Cornell University Press.

Figueirido, B. *et al.* (2018) "Distinct predatory behaviors in scimitar- and dirk-toothed sabertooth cats," *Current Biology*, **28**, pp. 3260–6.

Fish, F.E. *et al.* (2002) "Fur does not fly, it floats: buoyancy of pelage in semi-aquatic mammals," *Aquatic Mammals* **28**, pp. 103–12.

Fisher, R.E. (2011) "Red panda anatomy," in Glatston, A.R. (ed.) *Red panda: biology and conservation of the first panda*. London: Elsevier, pp. 89–100.

Forasiepi, A.M., Babot, M.J. and Zimicz, N. (2015) "*Australohyaena antiqua* (Mammalia, Metatheria, Sparassodonta), a large predator from the Late Oligocene of Patagonia," *Journal of Systematic Palaeontology*, **13**, pp. 503–25.

Goldspink, G. (1977) "Design of muscles in relation to locomotion," in Alexander, R.M. and Goldspink, G. (eds.) *Mechanics and energetics of animal locomotion*. London: Chapman and Hall, pp. 1–22.

Gould, S.J. (1978) "Panda's peculiar thumb," *Natural History*, **87**, pp. 20–30.

Grojean, R.E., Sousa, J.A. and Henry, M.C. (1980) "Utilization of solar radiation by polar animals: an optical model for pelts," *Applied Optics*, **19**, pp. 339–46.

Harris, M.A. and Steudel, K. (1997) "Ecological correlates of hind-limb length in the Carnivora," *Journal of Zoology*, **241**, pp. 381–408.

Hill, R.W., Wyse, G.A. and Anderson, M. (2012) *Animal physiology*. 3rd edn. Sunderland: Sinauer Associates.

Hochstein, J.L. (2007) "A new species of *Zodiolestes* (Mammalia, Mustelidae) from the early Miocene of Florida," *Journal of Vertebrate Paleontology*, **27**, pp. 532–4.

Horner, A.M. and Biknevicius, A.R. (2010) "A comparison of epigean and subterranean locomotion in the domestic ferret (*Mustela putorius furo*: Mustelidae: Carnivora)," *Zoology*, **113**, pp. 189–97.

Howell, N. *et al.* (2021) "Aposematism in mammals," *Evolution*, **75**, pp. 2480–93.

Janecka, J.E. *et al.* (2015) "Genetically based low oxygen affinities of felid hemoglobins: lack of biochemical adaptation to high-altitude hypoxia in the snow leopard," *Journal of Experimental Biology*, **218**, pp. 2402–9.

Janis, C.M., Damuth, J. and Theodor, J.M. (2002) "The origins and evolution of the North American grassland biome: the story from the hoofed mammals," *Palaeogeography, Palaeoclimatology, Palaeoecology*, **177**, pp. 183–98.

Khattab, M.Q. and Tributsch, H. (2015) "Fibre-optical light scattering technology in polar bear hair: a re-evaluation and new results," *Journal of Advanced Biotechnology and Bioengineering*, **3**, pp. 38–51.

Kienle, S.S. and Berta, A. (2016) "The better to eat you with: the comparative feeding morphology of phocid seals (Pinnipedia, Phocidae)," *Journal of Anatomy*, **228**, 396–413.

Koon, D.W. (1998) "Is polar bear hair fiber optic?", *Applied Optics*, **37**, pp. 3198–200.

Lafuma, F. *et al.* (2021) "Multiple evolutionary origins and losses of tooth complexity in squamates," *Nature Communications*, **12**, p. 6001.

Liwanag, H.E.M. *et al.* (2012) "Morphological and thermal properties of mammalian insulation: the evolution of fur for aquatic living," *Biological Journal of the Linnean Society*, **106**, pp. 926–39.

Martin, L.D. (1989) "Fossil history of the terrestrial Carnivora," in Gittleman, J.L. (ed.) *Carnivore behavior, ecology and evolution*. Ithaca: Cornell University Press, pp. 536–68.

Martin, L.D. *et al.* (2000) "Three ways to be a saber-toothed cat," *Naturwissenschaften*, **87**, pp. 41–4.

Maxwell, G. (1967) *Seals of the world*. Boston: Houghton Mifflin.

McGrosky, A. *et al.* (2016) "Gross intestinal morphology and allometry in Carnivora," *European Journal of Wildlife Research*, **62**, pp. 395–405.

Moore, W.J. (2009). *The mammalian skull*. Cambridge: Cambridge University Press.

Newman, C., Buesching, C.D. and Wolff, J.O. (2005) "The function of facial masks in 'midguild' carnivores", *Oikos*, **108**, pp. 623–33.

Nokelainen, O. *et al.* (2021) "The giant panda is cryptic," *Scientific Reports*, **11**, p. 21287.

Noonan, M.J. *et al.* (2015) "Evolution and function of fossoriality in the Carnivora: implications for group-living," *Frontiers in Ecology and Evolution*, **3**, p. 116.

Penrose, F. *et al.* (2020) "Functional morphology of the jaw adductor muscles in the Canidae," *Anatomical Record*, **303**, pp. 2878–2903.

Penrose, F., Kemp, G.J and Jeffery, N. (2016) "Scaling and accommodation of jaw adductor muscles in Canidae," *Anatomical Record*, **299**, pp. 951–66.

Pollock, T.I., Hocking, D.P. and Evans, A.R. (2021) "The killer's toolkit: remarkable adaptations in the canine teeth of mammalian carnivores," *Zoological Journal of the Linnean Society*, zlab064, https://doi.org/10.1093/zoolinnean/zlab064

Ponganis, P.J. (2015) *Diving physiology of marine mammals and seabirds*. Cambridge: Cambridge University Press.

Rose, J. *et al.* (2014) "Functional osteology of the forelimb digging apparatus of badgers," *Journal of Mammalogy*, **95**, pp. 543–58.

Rose, K.D. (2006) *The beginning of the age of mammals*. Baltimore: Johns Hopkins University Press.

Schellhorn, R. and Sanmugaraja, M. (2015) "Habitat adaptations in the felid forearm," *Paläontologische Zeitschrift*, **89**, pp. 261–9.

Schwab, J.A. *et al.* (2019) "Carnivoran hunting style and phylogeny reflected in bony labyrinth morphometry," *Scientific Reports*, **9**, p. 70.

Sjøgaard, G. (1982) "Capillary supply and cross-sectional area of slow and fast twitch muscle fibres in man," *Histochemistry*, **76**, pp. 547–55.

Stein, A.B., Bourquin, S.L. and McNutt, J.W. (2015) "Avoiding intraguild competition: leopard feeding ecology and prey caching in northern Botswana," *African Journal of Wildlife Research*, **45**, pp. 247–57.

Stevens, C.E., and Hume, I.D. (1995). *Comparative physiology of the vertebrate digestive system*. Second edition. Cambridge: Cambridge University Press.

Udevitz. M.S. *et al.* (2013) "Potential population-level effects of increased haulout-related mortality of Pacific walrus calves," *Polar Biology*, **36**, pp. 291–8.

Uhen, M.D. (2007) "Evolution of marine mammals: back to the sea after 300 million years," *Anatomical Record*, **290**, pp. 514–22.

Van Valkenburgh, B. (1987) "Skeletal indicators of locomotor behavior in living and extinct carnivores," *Journal of Vertebrate Paleontology*, **7**, pp. 162–82.

Van Valkenburgh, B. (2007) "Déjà vu: the evolution of feeding morphologies in the Carnivora," *Integrative and Comparative Biology*, **47**, pp. 147–63.

Viranta, S. (1996) "European Miocene Amphicyonidae—taxonomy, systematics and ecology," *Acta Zoologica Fennica*, **204**, pp. 1–61.

Watanabe, Y.Y., Baranov, E.A. and Miyazaki, N. (2020) "Ultrahigh foraging rates of Baikal seals make tiny endemic amphipods profitable in Lake Baikal," *Proceedings of the National Academy of Sciences of the United States of America*, **117**, pp. 31242–8.

Webb, S.D. (1977) "A history of savanna vertebrates in the New World. Part I: North America," *Annual Review of Ecology and Systematics*, **8**, pp. 355–80.

White, J.A. and Keller, B.L. (1984) "Evolutionary stability and ecological relationships of morphology in North American Lagomorpha," *Carnegie Museum of Natural History Special Publication*, **9**, pp. 58–66.

Williams, T.M. (1999) "The evolution of cost efficient swimming in marine mammals: limits to energetic optimization," *Philosophical Transactions of the Royal Society of London B: Biological Sciences*, **354**, pp. 193–201.

Williams, T.M. *et al.* (2001) "A killer appetite: metabolic consequences of carnivory in marine mammals," *Comparative Biochemistry and Physiology Part A*, **129**, pp. 785–96.

CHAPTER 3

Evolution and historical biogeography

The major processes of evolution above the species level—macroevolution—are basic to understanding the structure and function of modern carnivoran ecology. Some modern carnivorans occupy similar ecological niches to those of the earliest members of the order. Others are highly derived, having morphological traits or ecological functions that arose—in some cases multiple times across lineages—over the course of carnivoran evolution. Competition structured past carnivoran communities as it does modern ones, although the inferential processes used by ecologists vs. paleoecologists differ. Here, I consider phylogenetic differentiation at the levels of subclass and family and the importance of geography in producing the diversity and distributions of carnivorans we see in fossil and modern forms.

3.1 Evidence for mammalian phylogeny

A major consideration in the reconstruction of carnivoran evolutionary history is the inferential utility of fossil vs. genetic evidence. Fossils provide evidence of morphology, geographic location, and geological context. The morphology of preserved body parts—particularly teeth, skulls, and long bones—provides information on how the animal lived, particularly its diet and style of locomotion. The geological contexts of fossil deposits provide estimates of minimum ages of various lineages and their relatedness. However, fossil representation of past vertebrate diversity declines with increasing age, and phylogenetic relatedness can be confused with trait convergence. As a result, divergence times and inferred relatedness of early carnivoran lineages are less reliable than for those that are more recent. Genetic methods, particularly since 1990, have used DNA from living and recently extinct forms to infer phylogeny to estimate divergence times. Regions of the genome that change rapidly, including those that do not code proteins and are under weak (or no) selective pressure, change rapidly enough that populations—or even individuals—of the same species can be discerned. Other regions, which code for the basic architecture of form and function, evolve slowly. If a focal branch of the tree of life has modern descendants that provide undegraded DNA, divergence times of branches can be estimated. The combination of fossil and molecular evidence is a powerful tool for telling the evolutionary history of the Carnivora, but the two lines of evidence can paint contradictory pictures (Gura, 2000). Published phylogenies based on molecular evidence are now calibrated to fossil-based divergence dates, so that dated trees reflect both kinds of evidence (e.g. Dornburg *et al.*, 2012; Nyakatura and Bininda-Emonds, 2012). This is the current gold standard for vertebrate phylogenies.

3.2 Early mammals

The first mammals, as partially defined by a jaw joint in which the dentary bone of the lower jaw articulates with the squamosal (temporal) bone of the cranium, have been identified from the Late Triassic Period, around 205 Ma (Figure 3.1). They descended from and strongly resembled an amniote lineage that began to display mammalian traits—heterodont dentition, two occipital condyles, and distal limbs oriented antero-posteriorly—over the previous 100 million years. These traits arose multiple times in the nearest non-mammalian lineage, Cynodontia, only one branch of which gave rise to mammals. The earliest mammals persisted until eutherian mammals, recognized in fossils by

Carnivoran Ecology. Steven W. Buskirk, Oxford University Press. © Steven W. Buskirk (2023). DOI: 10.1093/oso/9780192863249.003.0003

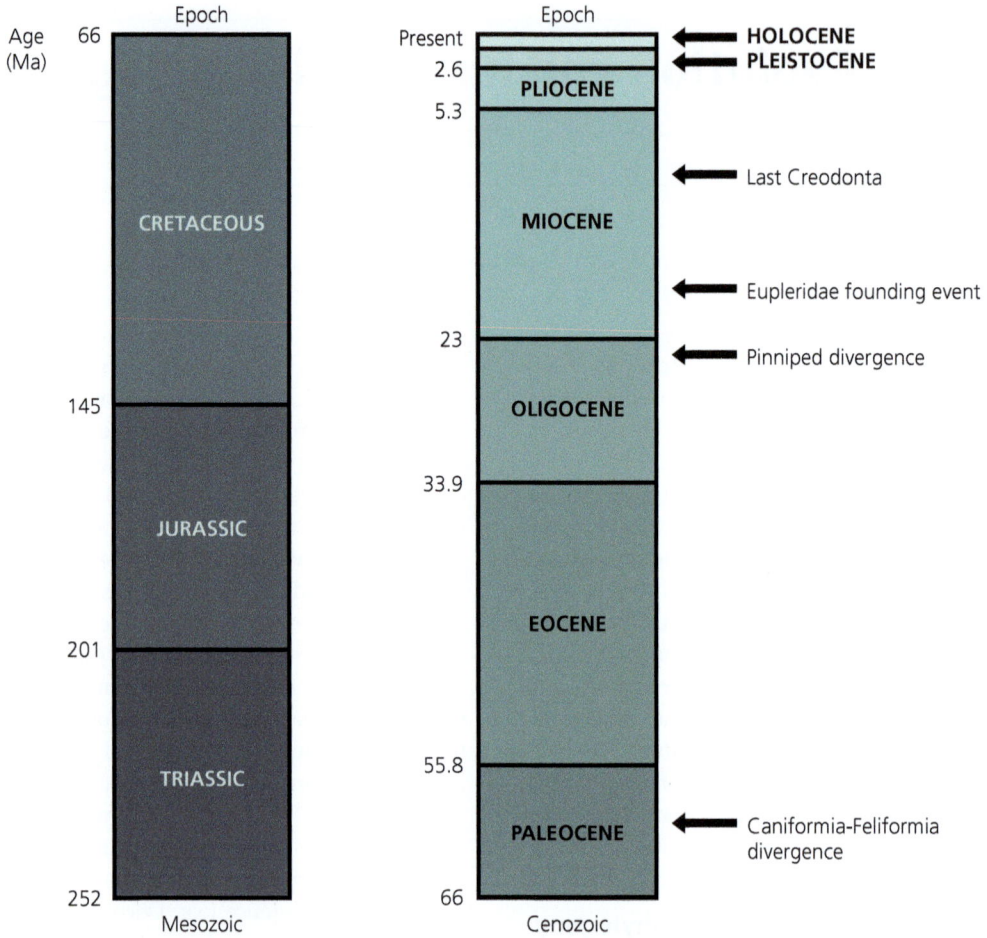

Figure 3.1 Geologic time scales for the Mesozoic and Cenozoic eras, with key events in carnivoran evolution.
Ma = million years before present. The boundary between Mesozoic and Cenozoic eras, formerly denoted the K–T (Cretaceous–Tertiary) boundary, is now called the K–Pg (Cretaceous-Paleogene) boundary.

their forelimb bones and, except in the earliest forms, absence of epipubic bones, diverged from the metatherian lineage that led to marsupials about 160 Ma (Luo *et al.*, 2011). The earliest eutherian for which we have well-preserved skull and postcranial bones had five upper and four lower incisors on each side, single upper and lower canines, and five premolars and three molars in each jaw (I^5/I_4, C^1/C_1, P^5/P_5, M^3/M_3). Its three-cusped cheek teeth were consistent with a diet dominated by insects; it had the limb joints of a scansorial tree-climber and was mouse-sized. Eutherians remained rare and small-bodied for the remainder of the Mesozoic Era. Most were omnivores no larger than rats, although descriptions exist of badger-sized predators that ate small dinosaurs (Hu *et al.*, 2005).

Land mammals held ecologically minor roles compared to land reptiles, the latter being larger-bodied, more diverse, and more abundant. The predaceous dinosaurs likely caused mammals to remain small, nocturnal, and rare.

These ecological circumstances changed abruptly at the Cretaceous–Paleogene (K–Pg, formerly K–T) boundary, about 66.5 Ma, when an extraterrestrial bolide struck the coastline of modern Yucatan, Mexico. The resulting pulse of heat, lasting mere hours, killed nearly all forms of terrestrial vertebrate life that could not shelter in water, in underground burrows, or in caves, over most of the planet (Robertson *et al.*, 2004). How many mammalian lineages survived the event is unclear, but molecular tools have identified four major continental centers for the early post-K–Pg mammalian radiation. In Africa, Superorder Afrotheria gave rise to proboscideans, hyraxes, manatees, and golden moles. South America saw Superorder Xenarthra diversify into armadillos, sloths, and carnivorous sparassodonts. Eurasia was the center of evolution of lagomorphs, rodents, and primates—Superorder Euarchontoglires. Most central to our subject, in North America the eutherian survivors of the K–Pg event represented Superorder Laurasiatheria, the lineage that gave rise to shrews, bats, ungulates, pangolins, and the Carnivora (Figure 3.2).

3.3 Early carnivorans

3.3.1 Continental biogeography

Continental connections are key to understanding carnivoran phylogeny at the family level, because carnivoran colonization of any non-coastal island by a means other than walking, swimming, or human intervention has not been inferred—with two exceptions. One is rafting by a mongoose-like species that founded the Eupleridae lineage on Madagascar Island about 20 Ma (Yoder *et al.*, 2003). The second involves rafting between North America and Eurasia on ice flows by ice-adapted species, which has occurred often and continues today. Antarctica was an important filter route for the colonization of Australia by cold-tolerant metatherians from Patagonia, but Antarctica shows no evidence of a carnivoran-analogue radiation of its own and lost all land mammals as it drifted southward and cooled. South America, after splitting from Africa about 110 Ma and drifting westward, saw the evolution of two metatherian lineages with carnivoran-like traits. Borhyaenidae of Order Sparassodonta (Figure 3.3A) arose in the Paleocene and developed hyena-like, fox-like, and saber-toothed members during the approximate 35-million-year isolation of South America. Evolving with these families were a single genus of armadillos (*Macroeuphractus*, Chlamyphoridae) that displayed clearly carnivorous dental adaptations (Figure 3.4), and a large carnivorous opossum (*Thylophorops*, Didelphidae) (Vizcaino and Iuliis, 2003; Goin *et al.*, 2009). None of these predatory lineages attained the diversities or abundances of carnivorans to the north; South America seems to have lacked a guild of effective cursorial predators for over 50 million years (Croft, 2006; Faurby and Svenning, 2016).

The sparassodonts died out around when carnivorans arrived from North America during the Great American Biotic Interchange (GABI)—about 9 Ma for Procyonidae and 3 Ma for Felidae (Prevosti and Soibelzon, 2012; Forasiepi *et al.*, 2014). A number of other carnivoran-like xenarthran and marsupial species appeared briefly during this time, making the causes of the sparassodont extinction unclear. This synchrony of carnivoran colonization and sparassodont disappearance has elicited explanations invoking competitive displacement; however, Prevosti and Soibelzon (2012) have questioned the order of events, and the suspicion is that climatic and geophysical factors played roles (e.g. Tarquini *et al.*, 2022). The effect of the biotic exchanges on the North American carnivoran fauna was nil, while that on South American carnivores was dramatic. Sparassodonts made no fossil-evidenced inroads into Central or North America, but the carnivoran families that colonized southward radiated rapidly, driving many old South American mammalian lineages to extinction (Figure 3.3A). Surviving South American lineages tended to be arboreal species that could escape ground predators, and herbivorous species protected by their large bodies—some weighed over 1000 kg (Faurby and Svenning, 2016). By the early Pleistocene, six carnivoran families had colonized South America and radiated.

Competition is inferred somewhat differently between ecologists and paleontologists. For the former, demonstrating competition usually involves some combination of evidence of behavioral avoidance, interspecies aggression, lack of local sympatry, and overlapping resource needs. For paleontologists, by contrast, the sequence and overlap in the geological distributions of similar taxa is the basis for inferring competition. If two lineages become co-occurring geographically and temporally and one becomes extinct, competition commonly is inferred. However, if one lineage disappears synchronously or well before the appearance of a similar one, "replacement" commonly describes the process.

Carnivorous metatherians arose in Australia as well (Figure 3.3B). Various lineages of Patagonian origin colonized Australia via Antarctica during the late Cretaceous and earliest Paleocene, with Australia separating from Antarctica

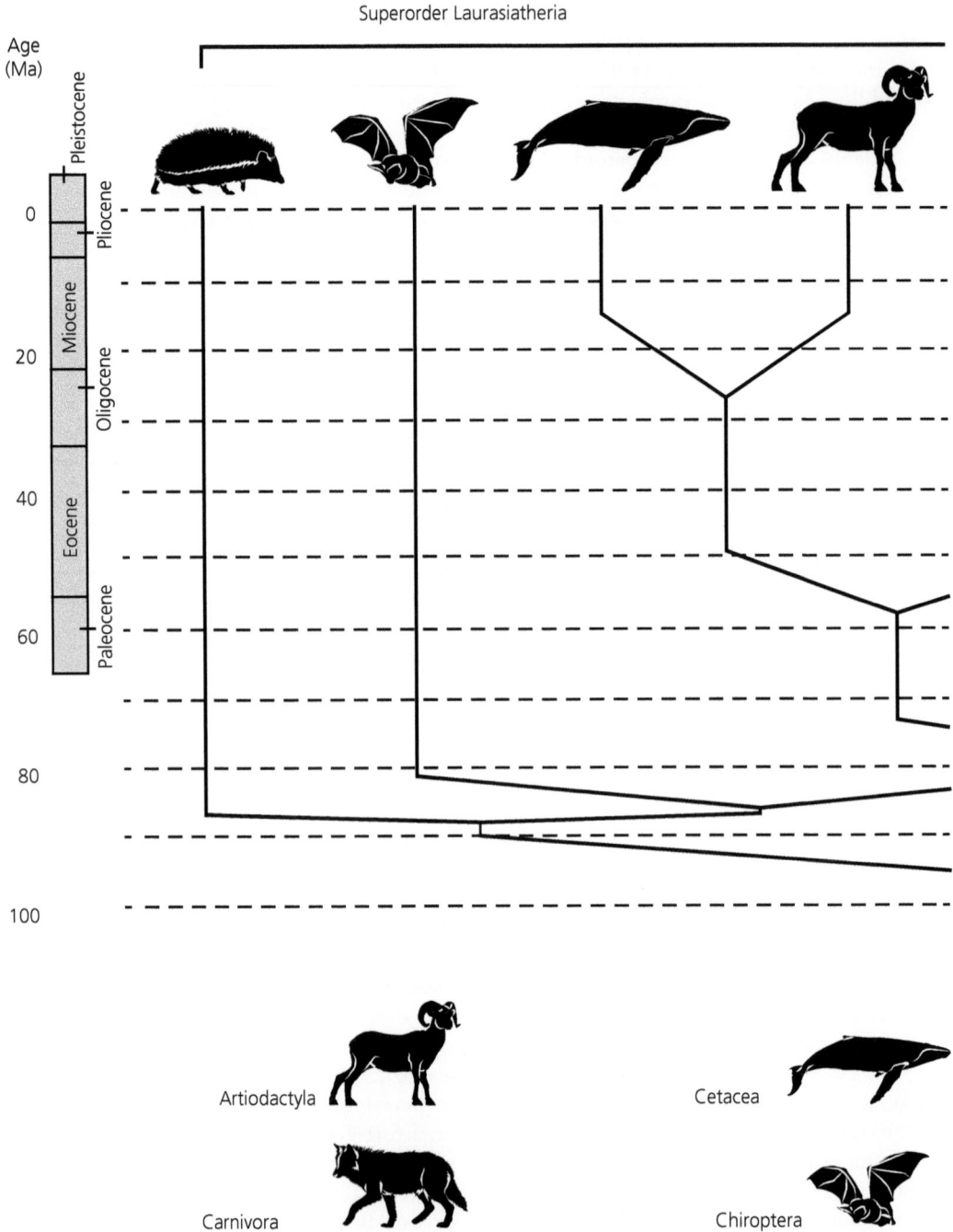

Figure 3.2 Superorder Laurasiatheria diversified on what is now North America, having diverged about 95 Ma from Superorder Euarchontoglires in Eurasia (represented by Lagomorpha, far right). Divergence times are from DNA-based estimates. The clade comprising Carnivora–Pholidota is designated Ferae (Luo *et al.*, 2012), and diverged about 70.5 Ma. The divergence time of Eulipotyphla (far left) from other Laurasiatherians is not aligned with other divergences but precedes that of Chiroptera. The divergence time of Carnivora from Pholidota is from Springer et al. (2011).

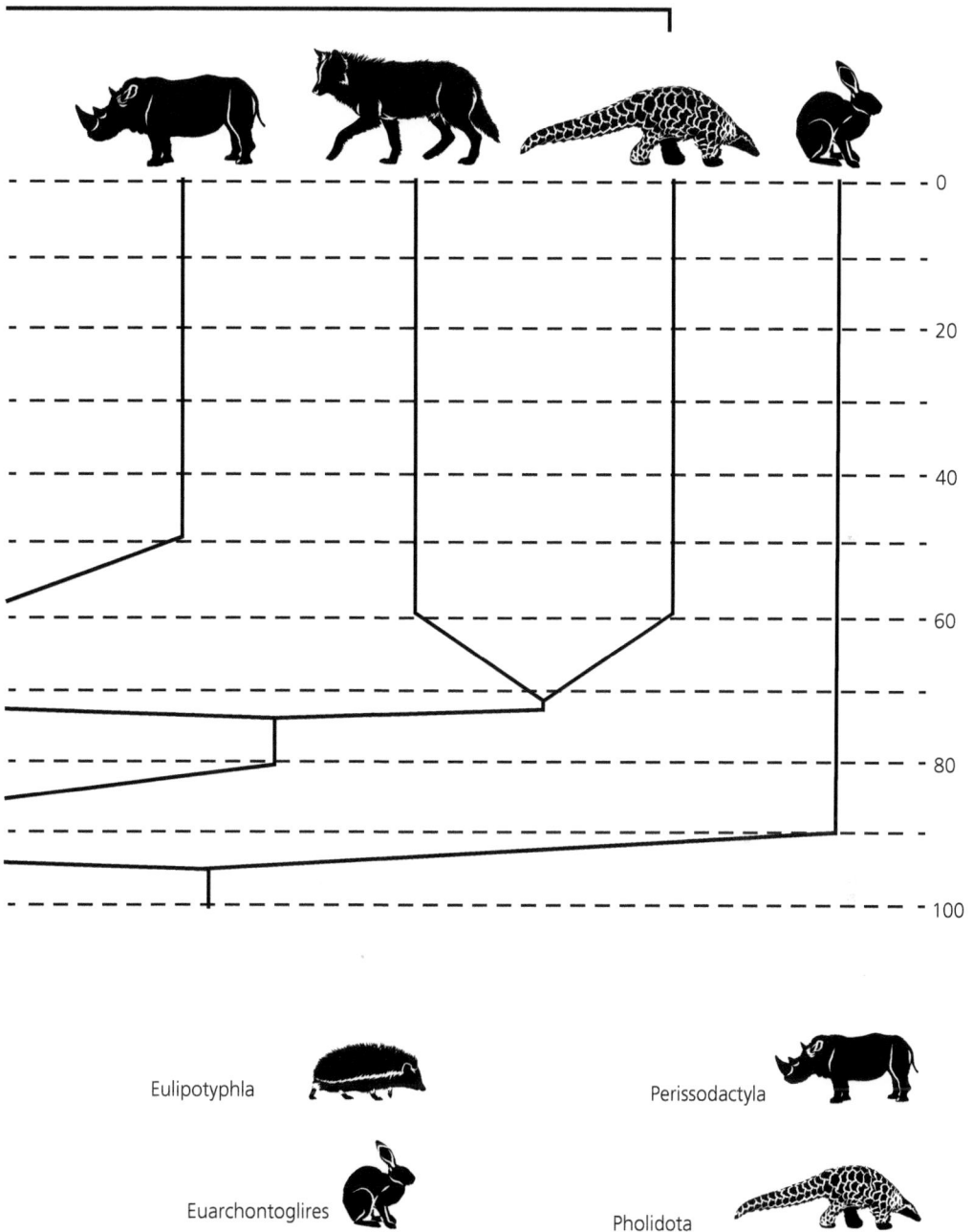

Eulipotyphla

Euarchontoglires

Perissodactyla

Pholidota

Figure 3.2 *continued*

about 64 Ma (Goin *et al.*, 1999). Australia saw a modest radiation of carnivorous metatherians across several orders, although this was a fraction of the radiation of carnivorans on the northern continents. The likely reasons for so few predaceous forms, particularly large-bodied ones, include the generally low plant productivity of Australian lands. As well, reptiles may have partially

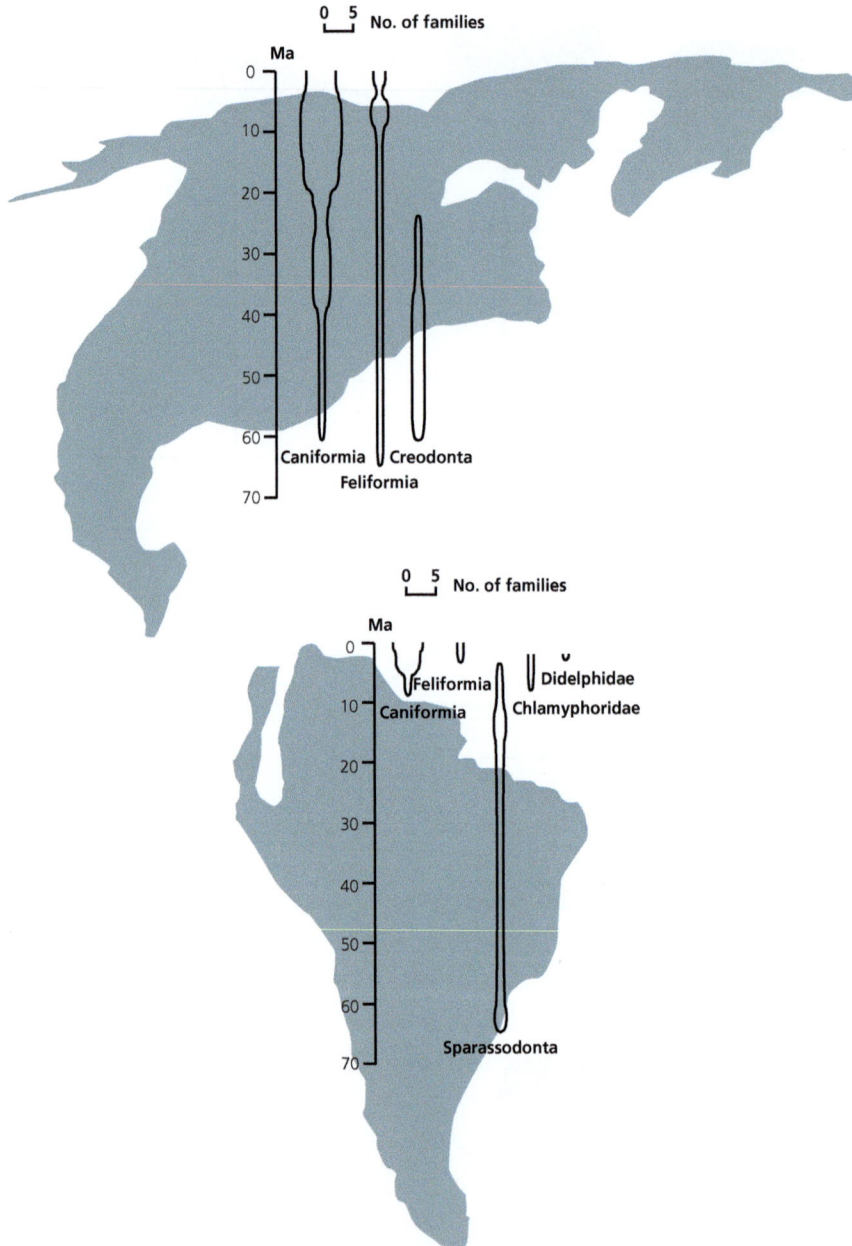

Figure 3.3 The numbers of families (in some cases of genera) of carnivorans or their ecological analogues through the Cenozoic, for New World (A) and Old World (B), continents showing continental coastlines as of around 40 Ma. The width of each diversity polygon is proportional to the number of fossil families in the suborder (or other taxon). Taxonomy and paleontological ranges follow McKenna and Bell (1997) for eutherians and New World metatherians, and Wroe (2003) for Australasian marsupials. Highly predaceous marsupials are those (Order Dasyuromorphia + Thylacoleonidae + Hypsiprymnodontidae) recognized by Wroe (2003). African families do not include the Eupleridae, isolated on Madagascar. For South America, Chlamyphoridae refers to *Macroeuphractus* and Didelphidae refers to *Thylophorops*. Gaps in the fossil record of less than 6,000,000 years in the pre-Pliocene fossil record are not depicted.

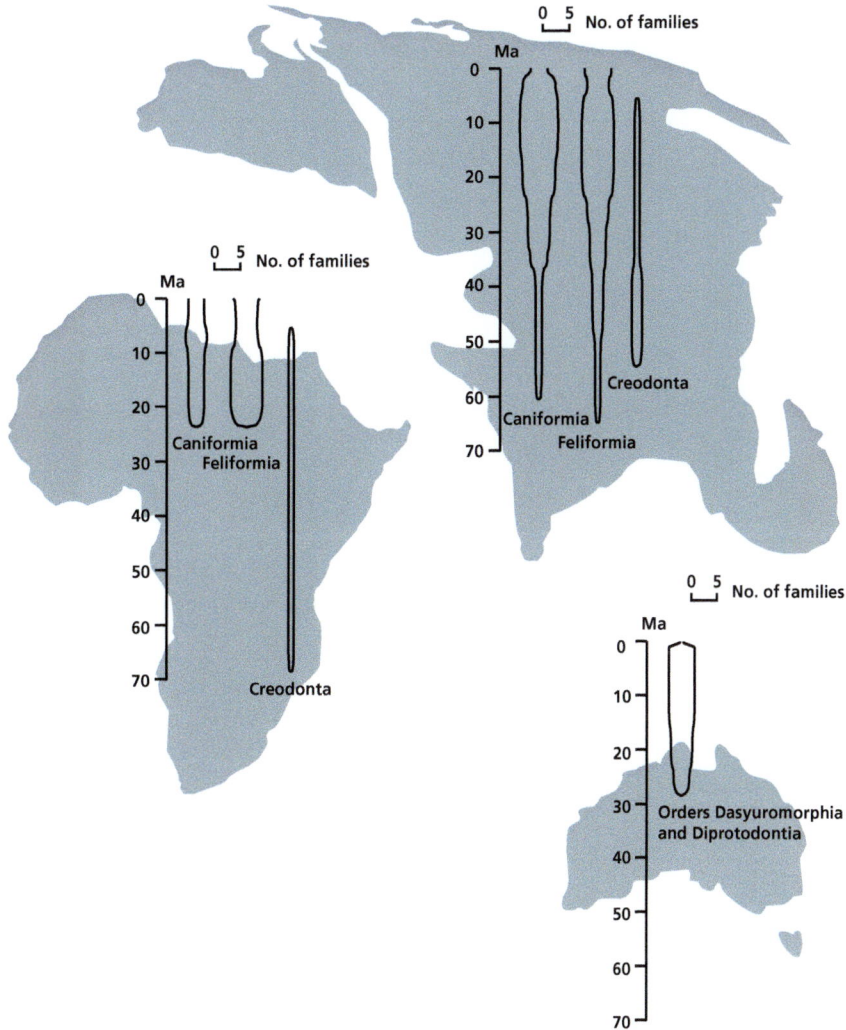

Figure 3.3 *continued*

filled the large predator niche, so that metatherians encountered few vacant niche spaces there (Wroe, 2003). By the middle of the Oligocene the Thylacinidae, Thylacoleonidae, Dasyuridae, and Hypsiprymnodontidae were leaving fossils, eventually including traces of "marsupial lions," canid- or felid-like thylacines, and smaller generalists and omnivores. These lineages persisted until the Late Pleistocene, when humans arrived via island chains 100,000–60,000 years ago, later bringing their companion dingos from Asia. Together, humans

and dogs drove most large mammalian herbivores and their predators to extinction (Miller *et al.*, 2005). The Tasmanian devil clings to existence on Tasmania—the largest-bodied extant marsupial morphologically and functionally analogous to any living carnivoran.

Laurasia (comprising North America and Eurasia) had drifted northward from the southern continents (Gondwana) about 200 Ma, and then separated, the two northern continents drifting apart on an east-west axis (Wildman *et al.*, 2007).

Figure 3.4 *Macroeuphractus outsie*, a Miocene–Pliocene armadillo of South America, as interpreted by Velizar Simeonovski. This predator had dental and bony adaptations to carnivory, but was limited by its heavy, armored body and slow cursation. It is inferred to have preyed on burrowing rodents excavated with its stout forelimbs. It would have been a weak competitor against newly arriving Carnivorans.
Painting: © Darin Croft, used by permission.

It was the western of these landmasses, now North America, where the Carnivora underwent their early radiations (Figure 3.5)—the earliest firmly dated carnivoran fossil, the earliest identifiable canid and ursid fossils, and the suspected geographic origin of mustelids were all North American (Rose, 2006). The ancestral lineage was formerly inferred to be Order Cimolesta, a small, morphologically generalized eutherian known from the Late Cretaceous and early Paleocene, now regarded as having no modern descendants. Molecular evidence recently showed Order Pholidota, the pangolins, as the closest extant lineage to the Carnivora, but a fossil of a plausible stem carnivoran-pangolin has not been reported. One Paleocene lineage from North America, Superfamily Miacoidea, was long regarded as basal or sibling to the Carnivora, but that possibility is less clear today. A group of miacoids, the Viverravidae (Figure 3.6), once considered early carnivorans, now are thought to be a separate branch, also with no modern descendants (Polly *et al.*, 2006). Another miacoid branch, the Miacidae (Figure 3.7), has clearer similarities to crown carnivorans, and may have been ancestral to Suborder Caniformia.

3.3.2 Early carnivoran radiations

Climatic events during the Eocene were plausible triggers for the radiations of some mammal lineages, including Carnivora. The Paleocene-Eocene Thermal Maximum began about 56 Ma and lasted for around 175,000 years. Believed to have resulted from major volcanic and uplifting events, it saw sea surface temperatures increased by 5–8°C. Coincidentally, the first appearances of the lineages that led to the Artiodactyla and Perissodactyla first appear in the fossil record at 55 Ma (Gingerich, 2006; Secord *et al.*, 2012). The appearances of these important prey of modern carnivorans could have resulted from dispersals rather than local differentiation, but this was also the time of the first appearances of the two major carnivoran lineages—Suborders Caniformia and Feliformia. Extant members of these two lines differ in several ways, conspicuously in the structure of the auditory bulla. In the caniforms, this structure is single-compartmented, comprising a single bone, whereas in the feliforms it is double-compartmented, of two bones. The two major carnivoran lineages underwent their earliest radiations on separate continents—the caniforms in North America and the feliforms in Eurasia—in the early

Eocene (Springer *et al.*, 2011). Ungulates and their main predators may have arisen in concert.

Earth's climate began a general cooling trend about 49 Ma, having varied randomly or warmed over the previous 17 million years. Along with cooling came the latitudinal zonation of Earth's climates and vegetation; no longer would lush trees and ectothermic vertebrates thrive above the Arctic Circle. This set the stage for the evolution of carnivoran ecomorphotypes (Section 2.6). It also affected carnivoran evolution via fluctuations in sea level, changing land connections, and the filtering of carnivoran dispersal. Dispersal filters affected caniform vs. feliform distributions differently. Most caniform families found their way to Eurasia and Africa, whereas of the Feliformia only the Felidae, Barbourofelidae, and one genus of Hyaenidae dispersed to North America. The arriving lineages fared differently as well, depended on their direction of travel. For example, the Canidae, Ursidae, Amphicyonidae, and Mustelidae experienced major radiations in North America and then dispersed to Eurasia, the Canidae 18 million years after the others. Once in Eurasia, all four families underwent stronger radiations than they had in North America, and fairly quickly, within 3–5 million years. By contrast, the Felidae, Barbourofelidae, Procyonidae, and Mephitidae dispersed to North America, and all but the procyonids underwent radiations less pronounced than in Eurasia. For some of these latter species, their rates of radiation were no greater in North America than in Eurasia at the same time. Explanations for these asymmetrical evolutionary bursts include climate shifts and extinctions of carnivoran-like mammals that opened niche spaces for lineage radiation. Hyaenodontid creodonts were in decline in Eurasia by about 38 Ma, and ursids and amphicyonids may have replaced them, or hastened their disappearance. Similarly, the Eurasian radiation of canids may have been strengthened by a synchronous decline of hyaenids (Pires *et al.*, 2015). As occurred multiple times in carnivoran evolution, competitive interactions

likely shaped the appearance of new lineages and the extinction of old ones.

Van Valkenburgh (1999) described three major phases in the macroevolution of carnivorous eutherians, each involving the replacement of one carnivorous fauna with another, seemingly similar, one in terms of adaptive features that had arisen convergently. In each instance, the mechanism underlying the replacement is unclear, whether through changed prey communities, arrival of superior competitors, or altered environments due to climate. For cat-like feliforms in North America, there is an apparent gap in the fossil record from about 23 Ma, when nimravids disappeared, until 17.5 Ma, when true cats appeared again, having migrated from Eurasia (Van Valkenburgh 1999). Again, the mechanisms underlying this "cat gap" are not understood but seem to be related to competition.

The Creodonta, a likely paraphyletic group, were important early competitors with the Carnivora (Figure 3.5). They were among the first dominant carnivorous land mammals of the Paleocene, and arguably the most numerically abundant ones—surpassing carnivorans—until about 25 Ma (Muizon and Lange-Badré 1997). Where creodonts originated remains uncertain, but one branch of the lineage (Oxyaenidae) appeared in Laurasia in the late Paleocene and in Eurasia during the Eocene. Family Hyaenodontidae, although more primitive, makes its initial appearance in Laurasia later, in the early Eocene, but in Eurasia during the late Cretaceous or early Paleocene (Gheerbrant *et al.*, 2006; Borths and Stevens, 2017). Hyaenodontids lived in Africa by about 68 Ma but were joined by carnivorans only 44 million years later. Ecological relationships between carnivorans and creodonts influenced the trajectories of both lineages. Like carnivorans, creodonts wielded carnassial cheek teeth (Section 1.1). Creodont carnassials occurred more posteriorly in the tooth row, comprising M^1/M_2 or M^2/M_3, rather than the carnivoran P^4/M_1. In some cases, multiple cheek tooth positions had shearing functions. By 10 Ma, creodonts had dis-

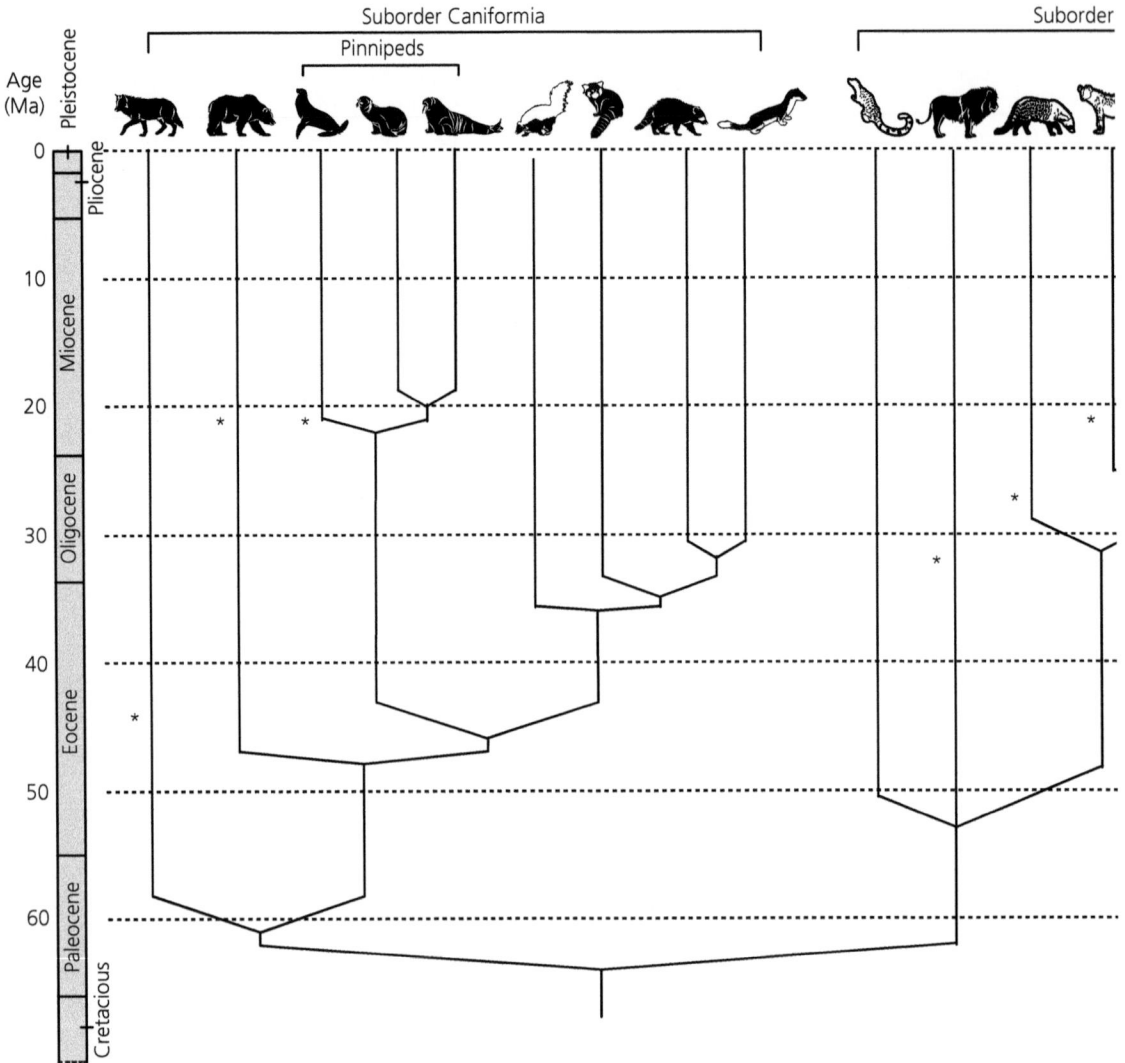

Figure 3.5 Reconstructed phylogeny of extant and two extinct (Barbourofelidae, Nimravidae) families of Carnivora, with outgroup Creodonta, based on branching patterns, divergence times, and other data in Nyakatura and Bininda-Emonds (2012). Asterisks denote earliest identified fossils of the crown taxon. The monophyletic origin of pinnipeds follows Arnason et al. (2006) and Nyakatura and Bininda-Emonds (2012). The joint origin of the nimravid-barbourofelid lineage follows Wang et al. (2020). Priodontidae, an extant sibling family of Felidae, is recognized in some modern classifications, but not shown here.

appeared from all continents, apparent losers to carnivorans in multiple continental competitions. It has been widely conjectured why creodonts would have been competitively inferior. The brains of the two lineages differed little in relative size (Radinsky, 1978), contrary to early understandings. Current explanations concern locomotor and dental constraints—creodonts possessed limb joints consistent with walking, but not well suited to climbing, leaping, or sprinting (Jenkins and Camazine, 1977), denying them several niches (Box 3.1). Their multiple carnassial pairs may have reduced their ability to process non-meat foods when prey were scarce (Van Valkenburgh, 1999, Figure 3).

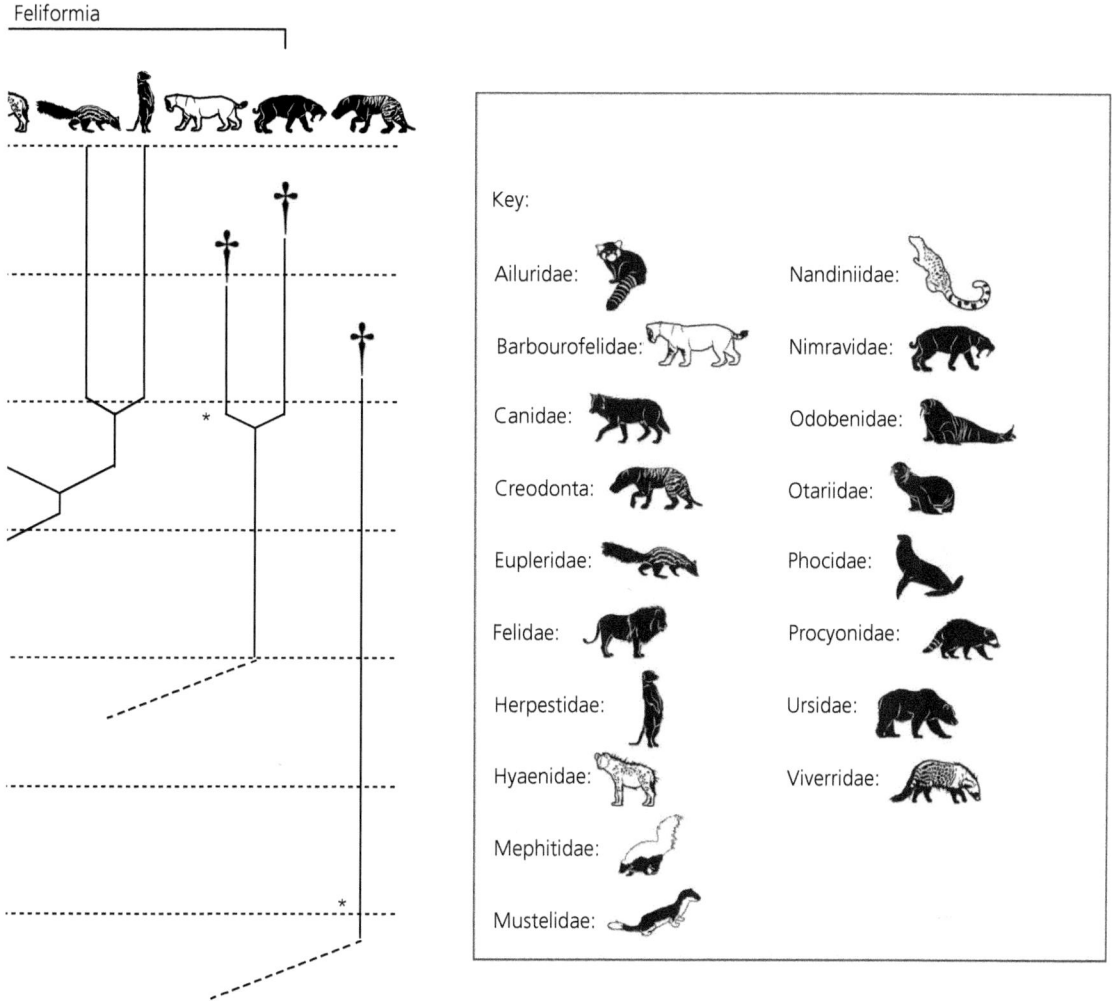

Figure 3.5 *continued*

All of the family-level diversification of Carnivora occurred in North America and Eurasia; South America, Africa, Australia, and Antarctica were completely isolated from the northern continents during the early Cenozoic. Exchanges between the northern continents occurred during the Eocene by two routes: one connecting northeastern North America to northwestern Europe and the other via successive land bridges across Beringia. As the climatic cooling progressed, both high-latitude routes increasingly functioned as dispersal filters, providing corridors for cold-adapted, but not more equatorial, species (McKenna and Bell, 1997). After the Eocene, only the Beringian connection appeared, and only intermittently (Martin, 1989).

Diversification of carnivorans had an iterative aspect, characterized by the successive appearance, diversification, domination, decline, and extinction of various family- and subfamily-level lineages, within a landmass (Van Valkenburgh, 1999). This progression was in some cases accompanied by increases in body size and specialization for carnivory, both of which predisposed large-bodied

Figure 3.6 *Viverravus* sp., a member of Viverravidae within superfamily Miacoidea, of the Fossil Lake Fauna (North America, Eocene), interpreted by Velizar Simeonovski. This lineage formerly was regarded as one of the earliest carnivorans, but one that left no modern descendants. Its inclusion in Carnivora is now uncertain (Wesley-Hunt and Flynn, 2005).
Artwork: by permission of U.S. National Park Service.

Figure 3.7 *Vulpavus* sp., a member of the Miacidae, of the Fossil Lake fauna (North America, Eocene), interpreted by Velizar Simeonovski. The lineage was adapted for climbing and had a reduced number of molars, consistent with a mostly carnivorous diet.
Artwork: by permission of U.S. National Park Service.

dietary specialists to reduced subsequent morphological diversification (Holliday and Steppan, 2004). Some evidence suggests that this leads to lineage extinction. This is observed in extant carnivorans—dietary specialization for large prey contributes to rarity (Van Valkenburgh *et al.*, 2004). Order of occupancy also has played an important role in car-

nivoran diversification. For example, all caniform families had evolved or arrived in North America by the peak of the GABI, in the Late Miocene. Of them, the Canidae, Mustelidae, Procyonidae and Ursidae migrated to an as-yet feliform-free South America, where they radiated into thirty-six species (vs. thirteen feliforms) today. When feliforms finally

Box 3.1 Fossil carnivorans

Fortunately for students of the evolution of carnivorans, they are better represented in the fossil record than some mammalian orders. This is partly because of the good preservation of carnivoran bones and teeth, which tend to be larger and sturdier than for smaller mammals, for example, rodents, bats, and insectivores. In some fossil sites, carnivorans are better represented in numbers of specimens or of taxa than their contemporaries (see Andrews *et al.*, 1979, Rose, 1981). This may occur because carnivorans are more likely than other mammals to shelter or overwinter in caves, where bones and teeth are protected from erosion and weathering, and where paleontologists often dig. An extreme example is the preservation of hundreds of thousands of skulls of Pleistocene cave bears of various species in European caves—the largest collection of carnivoran remains, living or extinct (Figure 3.8). The Drachenhöhle

in central Austria held the remains of tens of thousands of cave bears dating from 65,000–35,000 years ago. It and other caves have provided remarkable insights into mortality rates, mortality causes, and evolutionary change among cave bears (Kurtén, 1955, 1958). For a different reason, the La Brea Tar Pits in Los Angeles, California are a uniquely valuable site for Upper Pleistocene (less than 60,000 years ago) mammalian fossils. Of the sixty mammalian species recovered from La Brea, 36% are carnivoran (Stock, 1992), compared with the 5% of modern Californian terrestrial mammal species that are carnivorans. Of all fossil elements collected at La Brea, around 90% are carnivoran (Figure 3.9), whereas in modern mammal communities non-carnivoran individuals outnumber carnivorans by orders of magnitude. This particularly strong sampling bias may reflect the unusual context for fossil deposition at La Brea:

Figure 3.8 Skeleton of the Pleistocene cave-bear *Ursus spelaeus* found in Bear Cave, Chiscau, Romania. The large numbers of cave-bear bones found in European caves tended to be broken, scattered, and buried under sediments, but a few were found nearly intact and exposed, where the bear died.
Photo: Zátonyi Sándor/Wikimedia Commons.

Box 3.1 *Continued*

large mammals were attracted to an apparent water source, only to be trapped in viscous tar. Their struggles to escape and the odors of their carrion attracted predators and scavengers, which themselves became trapped (Stock, 1992). Enhancing their preservation, all skeletal materials became encased in dried tar. Natural trap caves are similar to La Brea in attracting predators and scavengers. For example, the Malapa, South African assemblage occurs in a cave that collected mammals via falls through a roof hole about 2 Ma. Of the eighteen mammalian species identified from Malapa, ten are carnivoran, identified from thirteen of the thirty-eight individual mammals found. By contrast, the mammalian assemblage of Southern Africa today includes only 13% carnivoran species, and an even smaller proportion of carnivoran individuals (Skinner and Chimimba, 2005; Val *et al.*, 2015). So, under various conditions, fossil mammal assemblages can overrepresent the diversity or abundance of the Carnivora, adding to our understandings of past species, if not representing the composition of paleocommunities.

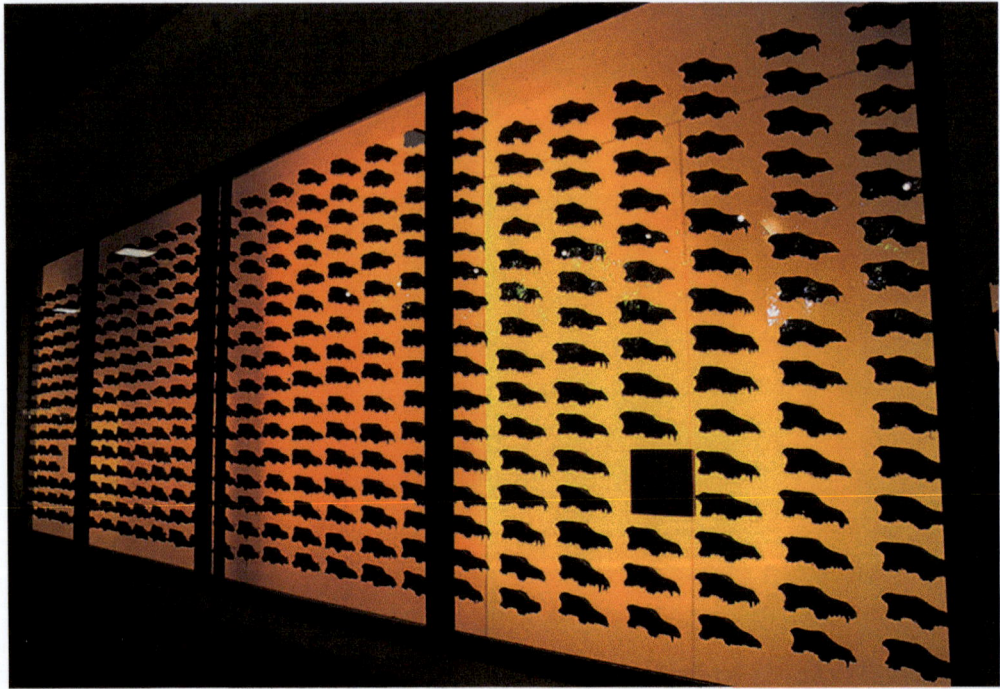

Figure 3.9 A display of over 400 dire wolf skulls in the Page Museum, Rancho La Brea tar pits, California, representing around one-tenth the number of specimens of the species found at La Brea. Dire wolves account for more fossil specimens than any other mammalian species, outnumbering specimens of *Canis lupus* by a factor of 100.
Photo: Pyry Matikainen/Wikimedia commons, CC by SA 2.5.

colonized South America around 2 Ma, only the Felidae did so; North American hyaenids went extinct in the early Pleistocene (Pedersen *et al.*, 2014). Therefore, a late-arriving and depauperate feliform fauna found insufficient niche space in South America for a strong radiation. Africa, by contrast, saw nearly synchronous arrivals of feliforms and caniforms from Eurasia about 24 Ma, followed by rapid diversification of both suborders. The feliforms in particular occupied niche spaces in forested and shrubland habitats, and today comprise fifty-seven African species (not including the Eupleridae on Madagascar), compared to nineteen caniform species (Wilson and Mittermeier, 2009). So, the order of colonization has influenced within-continent diversity below the level of family.

One notable feature of carnivoran radiations has been the marked family-level variation in morphological disparity, that is, variation in shape without regard to body size. Disparity across species within a family suggests the evenness with which a family (or other taxon above the species level) occupies the available adaptive space (Werdelin and Wesley-Hunt, 2010, 2014). Using the skull as a proxy for trophic adaptation, extant Felidae and Canidae exhibit the lowest levels of within-family disparity among carnivorans, Hyaenidae and Eupleridae the highest levels. The Hyaenidae are disparate by nature of including the termite-eating aardwolf with the vertebrate-eating members. The high disparity in the Eupleridae is consistent with its phylogenetic history: a single species, having rafted from the African mainland, radiated on an otherwise carnivoran-free Madagascar over a 20-million-year period. The modern descendants include cat-like, mongoose-like, and insectivorous species (Werdelin and Wesley Hunt, 2010). The Felidae exhibit the lowest disparity of all carnivoran families, whereas among canids, the low disparity finds exception in the insectivorous bat-eared fox. Otherwise, canids partition niche space largely based on body size (Werdelin and Wesley-Hunt, 2014). The disparity within Mustelidae is contributed mostly by Subfamily Lutrinae, the otters, absent which the mustelids are similar in their feeding apparatus, although not their post-cranial morphology.

Disparity can also be analyzed for geographic regions and shows how communities of carnivorans on the continents filled niche spaces via adaptation, colonization, and extinction. Werdelin and Wesley Hunt (2010, Table 8.3) found that the mean pairwise disparity of carnivoran species on the continents (considering Eurasia, Africa, North America, and South America) is remarkably consistent, ranging from a low of 0.581 in Eurasia to 0.595 in South America. The within-continent variances are likewise small, in part because the same species pairs occur on more than one continent. These results imply that carnivoran associations have evolved at the level of continents such that continents have similar amounts of morphological diversity across their carnivoran communities. This suggests strongly that competition structures communities at the continental level.

3.4 Body size in macroevolution

Carnivorous eutherians decreased in body size coincident with climatic warming early in the Eocene (Morlo *et al.*, 2010). Then, for about the next fifty million years, the largest carnivorans generally grew in size, as the climate cooled and various prey lineages, particularly ungulates, likewise became larger. This trend followed that of mammals generally—although fossils of mammals weighing 10–1000 g tended to predominate over the Cenozoic, larger mammals became asymptotically more common (Alroy, 1998). Few Paleocene mammals weighed more than 10 kg, but by the Miocene–Pliocene many did, and the largest-bodied mammals became relative giants: ground sloths of over 2000 kg, mammoths of over 5000 kg, and bovids of over 1000 kg. So, mammals in general provide support for the general pattern described by Edward D. Cope in the late nineteenth century (eponymously "Cope's rule"): animal lineages tended to evolve larger body size, followed by extinction (Box 3.2). Within several linages of the Caniformia (Canidae, Ursidae, and Amphicyonidae) body size increased from that of the last common ancestor. By contrast, within Superfamily Musteloidea (Families Mustelidae, Mephitidae, Ailuridae, Procyonidae) body size has generally decreased from the last common ancestor (Finarelli and Flynn, 2006). Exceptionally large-bodied species appeared in multiple lineages: viverrids the size of modern jackals, canids of 170 kg, mustelids of 200 kg, and hyenas of 190 kg (Turner and Antón, 1996; Geraads, 1997; Sorkin, 2008; Geraads *et al.*, 2011). These trends do not reflect the return of proto-pinnipeds to the marine environment in the early Miocene, about 21 Ma (Rybczynski *et al.*, 2009), in which large size strongly reflects adaption to the aquatic environment.

Box 3.2 The largest carnivoran

The largest-bodied land carnivoran known from fossils was *Arctotherium angustidens*, a short-faced bear (Subfamily Tremarctinae) of South America during the mid-Pleistocene (1–0.8 Ma). The largest fossil found to date is that of an old male estimated to have weighed 1600 kg (Figure 3.10). The tremarctines first appeared in North America, where they persisted from the Pliocene to the latest Pleistocene, during which time some colonized South America. There, the early

fossils were huge, decreasing in size and degree of carnivory over time. *A. angustidens* seems to have occupied an ecosystem rich in mammalian prey and carrion, perhaps reflecting the recent arrival of bears on a carnivoran-poor continent (Soibelzon and Schubert, 2011). The sole modern survivor of the tremarctines, the spectacled bear, is a modestly sized dweller of the northern Andes. It is scansorial and mostly herbivorous (Peyton 1980).

Figure 3.10 The largest known land carnivoran, the Pleistocene short-faced bear *Arctotherium angustidens* of South America (left), with a modern coastal brown bear (center), and a modern spectacled bear (right). The largest short-faced bear weighed as much as 1600 kg.

Key points

- Macroevolution—patterns of phylogeny above the species level—is inferred from fossil evidence and genomics. The two approaches are merged to provide trees with branching patterns informed by genetic studies, with dates of divergence calibrated to fossil ages.
- The earliest eutherian mammals diverged from metatherians (which include marsupials) about 160 Ma. Early eutherians co-occurred with dinosaurs, but were small, rare, and mostly nocturnal, presumably because predaceous dinosaurs were large, common, and mostly diurnal. The Carnivora arose around 70 Ma, depending on whether the Viverravidae are considered carnivoran or ancestral.
- The K–Pg event, during which a bolide impacted Earth's surface in what is now Mexico, caused the extinction of all large vertebrates, but a few

mammalian lineages—likely burrowers or cave dwellers—survived. Of several major continental centers of mammalian evolution, Laurasia (now North America) was the site of origin of the Carnivora.
- Other continents evolved predaceous mammals, but they tended to be constrained to forms that could not compete with carnivorans once the latter arose or arrived.
- The Paleocene–Eocene Thermal Maximum, which occurred 56 Ma and saw ocean surface temperatures rise by 5–8° C, coincided with the first appearances of lineages that led to Artiodactyls, Perissodactyls, and carnivoran suborders.
- The Caniformia radiated strongly in North America, dispersed to Eurasia, and radiated even more strongly there. By contrast, some families of Feliformia dispersed to North America, where they radiated more weakly than they had in Eurasia.

- The creodonts—morphologically similar to carnivorans—were important early competitors with carnivorans but went extinct on each continent where carnivorans arose or arrived. The implied competitive inferiority of creodonts may have resulted from their narrow range of body sizes, limited omnivory, and weak locomotor adaptations to climbing or running.

- Carnivorans are well represented in various fossil deposits, in some cases outnumbering non-carnivoran species or specimens by orders of magnitude. This facilitates studies of individual carnivoran species but misrepresents the composition of paleocommunities.

- The colonization of South America by carnivorans during the Great American Biotic Interchange shows how prior occupancy affects competitive interactions. The Canidae arrived in South America around 10 Ma and radiated into thirty-six species today. By contrast, the Felidae arrived only 2 Ma and are represented by only fourteen species.

- Body size aside, morphological variation (disparity) is remarkably similar across continents, suggesting that carnivoran communities evolved to a common standard of niche overlap.

- The rise of some carnivoran ecomorphs, particularly coursing forms, can be linked to a changing climate and the opening of prairies away from riparian zone edges.

References

Alroy, J. (1998) "Cope's rule and the dynamics of body mass evolution in North American fossil mammals," *Science*, 280, pp. 731–4.

Andrews, P., Lord, J.M. and Evans, E.M.N. (1979) "Patterns of ecological diversity in fossil and modern mammalian faunas," *Biological Journal of the Linnean Society*, 11, pp. 177–205.

Borths, M.R. and Stevens, N.J. (2017) "The first hyaenodont from the late Oligocene Nsungwe Formation of Tanzania: paleoecological insights into the Paleogene–Neogene carnivore transition," *PLoS ONE*, 12, p. e0185301.

Croft, D.A. (2006) "Do marsupials make good predators? Insights from predator–prey diversity ratios," *Evolutionary Biology Research*, 8, pp. 1193–214.

de Muizon, C. and Lange-Badré, B. (1997) "Carnivorous dental adaptations in tribosphenic mammals and phylogenetic reconstruction," *Lethaia*, 30, pp. 353–66.

Dornburg, A. *et al.* (2012) "Relaxed clocks and inferences of heterogeneous patterns of nucleotide substitution and divergence time estimates across whales and dolphins (Mammalia: Cetacea)," *Molecular Biology and Evolution*, 29, pp. 721–36.

Faurby, S. and Svenning, J.-C. (2016) "The asymmetry in the Great American Biotic Interchange in mammals is consistent with differential susceptibility to mammalian predation," *Global Ecology and Biogeography*, 25, pp. 1443–53.

Finarelli, J.A. and Flynn, J.J. (2006) "Ancestral state reconstruction of body size in the Caniformia (Carnivora, Mammalia): the effects of incorporating data from the fossil record," *Systematic Biology*, 55, pp. 301–13.

Forasiepi, A.M. *et al.* (2014) "Carnivorans at the Great American Biotic Interchange: new discoveries from the northern neotropics," *Naturwissenschaften*, 101, pp. 965–74.

Geraads, D. (1997) "Carnivores du Pliocène terminal de Ahl al Oughlam (Casablanca, Maroc)," *Geobios*, 30, pp. 127–64.

Geraads, D. *et al.* (2011) "*Enhydriodon dikikae*, sp. nov. (Carnivora: Mammalia), a gigantic otter from the Pliocene of Dikika, Lower Awash, Ethiopia," *Journal of Vertebrate Paleontology*, 31, pp. 447–53.

Gheerbrant, E. *et al.* (2006) "Early African hyaenodontid mammals and their bearing on the origin of the Creodonta," *Geological Magazine*, 143, pp. 475–89.

Gingerich, P. (2006) "Environment and evolution through the Paleocene–Eocene thermal maximum," *Trends in Ecology and Evolution*, 21, pp. 246–53.

Goin, F.J. *et al.* (1999) "New discoveries of 'opposumlike' marsupials from Antarctica (Seymour Island, Medial Eocene)," *Journal of Mammalian Evolution*, 6, pp. 335–65.

Goin, F.J. *et al.* (2009) "A new large didelphid of the genus *Thylophorops* (Mammalia: Didelphimorphia: Didelphidae), from the late Tertiary of the Pampean Region (Argentina)," *Zootaxa*, 2005, pp. 35–46.

Gura, T. (2000). "Bones, molecules . . . or both?" *Nature*, 406, pp. 230–3.

Holliday, J.A. and Steppan, S.J. (2004) "Evolution of hypercarnivory: the effect of specialization on morphological and taxonomic diversity," *Paleobiology*, 30, pp. 108–28.

Hu, Y. *et al.* (2005) "Large Mesozoic mammals fed on young dinosaurs," *Nature*, 433, pp. 149–52.

Jenkins, F.A., Jr. and Camazine, S.M. (1977) "Hip structure and locomotion in ambulatory and cursorial carnivores," *Journal of Zoology*, 181, pp. 351–70.

Kurtén, B. (1955) "Contribution to the history of a mutation during 1,000,000 years," *Evolution*, 9, pp. 107–18.

Kurtén, B. (1958) "Life and death of the Pleistocene cave bear," *Acta Zoologica Fennica*, 95, pp. 1–59.

Luo, Z.-X. *et al.* (2012). "Phylogenomic analysis resolves the interordinal relationships and rapid diversification of the Laurasiatherian mammals," *Systematic Biology*, 61, pp. 150–64.

Luo, Z.-X. *et al.* (2011) "A Jurassic eutherian mammal and divergence of marsupials and placentals," *Nature*, 476, pp. 442–5.

Martin, L.D. (1989) "Fossil history of the terrestrial Carnivora," in: Gittleman, J.L. (ed.) *Carnivore behavior, ecology, and evolution*. Boston: Springer, pp. 536–68.

McKenna, M.C. and Bell, S.K. (1997) *Classification of mammals above the species level*. New York: Columbia University Press.

Miller, G.H. *et al.* (2005). "Ecosystem collapse in Pleistocene Australia and a human role in megafaunal extinction," *Science*, 309, pp. 287–90.

Morlo, M., Gunnell, G.F. and Nagel, D. (2010) "Ecomorphological analysis of carnivore guilds in the Eocene through Miocene of Laurasia," in Goswami, A. and Friscia, A. (eds.) *Carnivoran evolution: new views on phylogeny, form, and function*. Cambridge: Cambridge University Press, pp. 269–310.

Nyakatura, K. and Bininda-Emonds, O.R.P. (2012) "Updating the evolutionary history of Carnivora (Mammalia): a new species-level supertree complete with divergence time estimates," *BMC Biology*, 10, p. 12.

Pedersen, R.Ø., Sandel, B. and Svenning, J.-C. (2014) "Macroecological evidence for competitive regional-scale interactions between the two major clades of mammal carnivores (Feliformia and Caniformia)," *PLoS ONE*, 9, p. e100553.

Peyton, B. (1980) "Ecology, distribution, and food habits of spectacled bears, *Tremarctos ornatus*, in Peru," *Journal of Mammalogy*, 61, pp. 639–52.

Pires, M.M., Silvestro, D. and Quental, T.B. (2015) "Continental faunal exchange and the asymmetrical radiation of carnivores," *Proceedings of the Royal Society B*, 282, p. 20151952.

Polly, P.D. *et al.* (2006) "Earliest known carnivoran auditory bulla and support for a recent origin of crown-group Carnivora (Eutheria, Mammalia)," *Palaeontology*, 49, pp. 1019–27.

Prevosti, F. J. and Soibelzon, L. H. (2012) "Evolution of the South American carnivores (Mammalia: Carnivora): a paleontological perspective," in Patterson, B.D. and Costa, L.P. (eds.) *Bones, clones, and biomes: the history and geography of recent neotropical mammals*. Chicago: University of Chicago Press, pp. 102–22.

Radinsky, L. (1978) "Evolution of brain size in carnivores and ungulates," *American Naturalist*, 112, pp. 815–31.

Robertson, D.S. *et al.* (2004). "Survival in the first hours of the Cenozoic," *Geological Society of America Bulletin*, 116, pp. 760–8.

Rose, K.D. (1981) "Composition and species diversity in Paleocene and Eocene mammal assemblages: an empirical study," *Journal of Vertebrate Paleontology*, 1, pp. 367–88.

Rose, K.D. (2006) *The beginning of the age of mammals*. Baltimore: Johns Hopkins University Press.

Rybczynski, N., Dawson, M.R. and Tedford, R.H. (2009) "A semi-aquatic Arctic mammalian carnivore from the Miocene epoch and origin of Pinnipedia," *Nature*, 458, pp. 1021–4.

Secord, R. *et al.* (2012) "Evolution of the earliest horses driven by climate change in the Paleocene–Eocene Thermal Maximum," *Science*, 335, pp. 959–62.

Skinner, J.D. and Chimimba, C.T. (2005) *The mammals of the southern African subregion*. 3rd edn. Cambridge: Cambridge University Press.

Soibelzon, L.H. and Schubert, B.W. (2011) "The largest known bear, *Arctotherium angustidens*, from the early Pleistocene Pampean region of Argentina: with a discussion of size and diet trends in bears," *Journal of Paleontology*, 85, pp. 69–75.

Sorkin, B. (2008) "A biomechanical constraint on body mass in terrestrial mammalian predators," *Lethaia*, 41, pp. 333–47.

Springer, M.S. *et al.* (2011) "The historical biogeography of Mammalia," *Philosophical Transactions of the Royal Society B*, 366, pp. 2478–502.

Stock, C. (1992). Rancho La Brea: a record of Pleistocene life in California. 7th edn. Revised by J.M. Harris. Science Series No. 37. Los Angeles: Natural History Museum of Los Angeles County.

Tarquini, S.D., Ladevèze, S. and Prevosti, F.J. (2022) "The multicausal twilight of South American native mammalian predators (Metatheria, Sparassodonta)," *Scientific Reports*, 12, p. 1224

Turner, A. and Antón, M. (1996) "The giant hyaena *Pachycrocuta brevirostris* (Mammalia, Carnivora, Hyaenidae)," *Geobios*, 29, pp. 455–68.

Val, A. *et al.* (2015) "Taphonomic analysis of the faunal assemblage associated with the hominins (*Australopithecus sediba*) from the early Pleistocene cave deposits of Malapa, South Africa," *PLoS ONE*, 10, p. e0126904.

Van Valkenburgh, B. (1999) "Major patterns in the history of carnivorous mammals," *Annual Review of Earth and Planetary Sciences*, 27, pp. 463–93.

Van Valkenburgh, B., Wang, X. and Damuth, J. (2004) "Cope's rule, hypercarnivory, and extinction in North American canids," *Science*, 306, pp. 101–4.

Vizcaíno, S.F. and de Iuliis, G. (2003) "Evidence for advanced carnivory in fossil armadillos (Mammalia: Xenarthra: Dasypodidae)," *Paleobiology*, 29, pp. 123–38.

Wang, X., White, S.C. and Guan, J. (2020) "A new genus and species of sabretooth, *Oriensmilus liupanensis* (Barbourofelinae, Nimravidae, Carnivora), from the middle Miocene of China suggests barbourofelines are nimravids, not felids," *Journal of Systematic Palaeontology*, 18, pp. 783–803.

Werdelin, L. and Wesley-Hunt, G.D. (2010). "The biogeography of carnivore ecomorphology," in Goswami, A. and Friscia, A. (eds.) *Carnivoran evolution: new views on phylogeny, form, and function*. Cambridge: Cambridge University Press, pp. 225–45.

Werdelin, L. and Wesley-Hunt, G.D. (2014) "Carnivoran ecomorphology: patterns below the family level," *Annales Zoologici Fennici*, 51, pp. 259–68.

Wesley-Hunt, G.D. and Flynn, J.J. (2005) "Phylogeny of the Carnivora: basal relationships among the Carnivoramorphans, and assessment of the position of 'Miacoidea' relative to Carnivora," *Journal of Systematic Palaeontology*, 3, pp. 1–28.

Wildman, D.E. *et al.* (2007) "Genomics, biogeography, and the diversification of placental mammals," *Proceedings of the National Academy of Sciences*, 104, pp. 14395–400.

Wilson, D.E. and Mittermeier, R.A. (eds.) (2009) *Handbook of the mammals of the world. Vol. 1. Carnivores*. Barcelona: Lynx Edicions.

Wroe, S. (2003). "Australian marsupial carnivores: recent advances in palaeontology," in Jones, M., Dickman, C., and Archer, M. (eds.) *Predators with pouches: the biology of carnivorous marsupials*. Collingwood: CSIRO Publishing, pp. 102–23.

Yoder, A.D. *et al.* (2003) "Single origin of Malagasy Carnivora from an African ancestor," *Nature*, 421, pp. 734–7.

CHAPTER 4

Physiological ecology

How does function at cellular and molecular levels relate to diet, ambient temperature, and ambient pressure? Carnivoran diets range from strictly carnivorous to those entirely of plant leaves and stems. Other carnivorans eat almost entirely insects or fruits. What do these specializations mean for digestive and metabolic processes? Some of our subjects live in the world's tallest mountain ranges, while others dive to thousands of meters below the ocean surface. A few live without free water for long periods, whereas others are in contact with water for most of their lives. How does this tremendous variety of niches occupied affect the range of homeostatic mechanisms: metabolism, respiration, kidney function, and reproduction? This chapter examines these mechanisms and how they are based on or relate to phylogeny or ecology in the Carnivora.

4.1 Digestion

Ecologists debate which mammalian order is most trophically diverse. The over 1,300 species of bats include frugivores, nectarivores, insectivores, sanguivores, and vertebrate predators (Jiao *et al.*, 2019). By comparison, the much less speciose Carnivora occupies most of those niches (except for sanguivory and nectarivory), but also includes two dietary specializations not found in bats: folivory and eating marine invertebrates. The crabeater seal is the most extreme carnivoran trophic specialist for aquatic or marine invertebrates, eating almost entirely (estimated 88%) Antarctic krill (Huckstadt *et al.*, 2012). In terms of digestion and assimilation, sanguivory differs from carnivory mostly in water content, both diets featuring high protein. Nectar differs modestly from fruit because both are concentrated sources of simple sugars. Carnivorans,

therefore, occupy trophic niches close to those of nectar- and blood-feeding bats, plus some that are very dissimilar from any found in bats. Clearly, carnivorans win the trophic diversity trophy by a comfortable margin.

In contrast to animal flesh, the diets of herbivores include digestible cell solutes, but also cellulose, lignin, hemicellulose, and other carbohydrates not directly digestible by any mammal. Herbivores eat frequently in response to appetite, have voluminous digestive tracts, and hold diverse gut microbial flora that produce cellulase and other enzymes that digest plant fibers. Consistent with less-frequent and regular eating than seen in herbivores, carnivorans have lost the genes coding for the INSL5 hormone and its receptor, which regulate appetite in response to gut contents (Hecker *et al.*, 2019). This may have resulted in the decoupling of killing behaviors from eating, which is consistent with the surplus-killing behaviors occasionally observed in carnivorans (Section 8.7). In addition to lacking large-capacity digestive tracts, carnivorans lack special structures to house microbial fermenters—compartmented stomachs or enlarged caecae. Indeed, the digestive tract of an herbivorous red panda differs little from that of a comparably sized felid (Nijboer and Dierenfeld, 2011). Not surprisingly, the microbial communities of carnivoran guts are about half as diverse as those of typical mammalian herbivores (Ley *et al.*, 2008). Further, the microbes in carnivoran guts are not suited for a plant-fiber diet; giant pandas have the least amount of cellulase and xylanase activity of any herbivorous mammal studied (Guo *et al.*, 2018). Kinkajous and binturongs, both strongly frugivorous, show some bacterial processing of soluble carbohydrates (Lambert *et al.*, 2014), but carnivorans feeding on

Carnivoran Ecology. Steven W. Buskirk, Oxford University Press. © Steven W. Buskirk (2023). DOI: 10.1093/oso/9780192863249.003.0004

leaves and stems absorb mostly cytoplasmic and nuclear contents, leaving cell walls undigested. For an herbivore, this trophic strategy is inherently less efficient than one employing fermentation.

Chitin, the major polysaccharide constituent of arthropod exoskeletons, is digestible via chitinases produced by extant insectivorous eutherians, as they were by early eutherians (Emerling *et al.*, 2018). The genes that code for chitinases (*CHIA1–5*) have been partially or wholly lost in most eutherian orders that came to specialize on non-insect foods, including extant ungulates, lagomorphs, fruit-eating bats, and carnivorans. In the case of carnivorans, *CHIA* genes were lost in various lineages, depending on their degree of omnivory. Canids, skunks, and meerkats show chitinase activity in the gastric mucosa, but stoats, domestic ferrets, stone martens, and domestic cats do not (Cornelius *et al.*, 1975; Tabata *et al.*, 2022).

4.1.1 Soluble carbohydrates

Dissolved carbohydrates are important components of fruits, plant leaves, and fungi, but mammals cannot absorb disaccharides or polysaccharides, which must be cleaved into monosaccharides such as glucose before absorption. The gene (*Treh*), which codes for trehalase, a digestive enzyme, is a useful marker for dietary diversity across mammals. Trehalose is a glucose dimer found in insect hemolymph and in fungi but it is absent from vertebrate tissues (Jiao *et al.*, 2019). Carnivorans that lack the intact gene for *Treh* cannot utilize important dissolved carbohydrates found in insects, which were important foods of basal carnivorans. Therefore, the functionality of *Treh* genes reveals how much and when diets of extant carnivorans diverged from the ancestral condition. Most canids, ursids, the striped hyena, mustelids, and the cheetah have intact *Treh* genes. By contrast, *Treh* function has been lost via mutation in the giant panda, domestic cat, American mink, sea otter, and all pinniped species studied (Figure 4.1) (Jiao *et al.*, 2019). None of the species shown to lack a functional *Treh* gene is a major insect feeder. Therefore, this pattern of loss of *Treh* function reflects specialization for vertebrate flesh, fruit, or plant leaves, leading to lapsed selection for functioning *Treh*.

It is therefore puzzling that fungi, good sources of trehalose, are uncommonly reported as foods in carnivoran diets, including species that eat insects and occupy moist habitats where fungi are common. This may be due to our traditionally limited means of understanding the dietary importance of carnivoran foods. Whereas fur, feathers, and bone leave strong signatures in carnivoran feces, fungi leave mostly spores to indicate their presence in the diet. Researchers who have found fungal sporocarps in feces of fishers in western North America have inferred that fungi may have been ingested secondarily in guts of fungus-eating rodents. Zielinski and colleagues (1999) considered this possibility upon finding that mammal remains and fungal spores were each present in 92% of fisher feces from the southern Sierra Nevada of California. In a revelatory use of stable isotope analyses, Smith and colleagues (2022) examined fisher diets from the same area and found that fungi dominated the summer diet (47% of metabolized energy vs. 14% for vertebrate foods) at one study site. The high value is comparable to that for some mycophagous rodents in mesic forests and would be possible only via functioning *Treh* genes. Future stable isotopic studies may show greater trophic importance of fungi to other carnivoran species in other areas, further expanding the known trophic range of the Carnivora.

Starch is a major energy source in human diets, but carnivorans encounter little starch in non-agricultural landscapes, mostly in underground plant parts and unripe fruits. Starch and glycogen break down to simpler sugars via amylase enzymes, but no studies have compared amylase coding regions of the genome across many carnivoran lineages. Felids and wolves show little amylase activity, but domestic dogs, having coevolved with humans during the rise of agriculture, have increased copy numbers of amylase genes (*AMY*) involved in starch digestion (Pajic *et al.*, 2019). Comparing wolves and domestic dogs, Axelsson and colleagues (2013) found that wolves had two copies of the *AMY2B* gene, whereas dogs of various breeds had 4–30 (mean = 15) copies, and showed, on average, twenty-eight times higher amylase gene expression than did wolves.

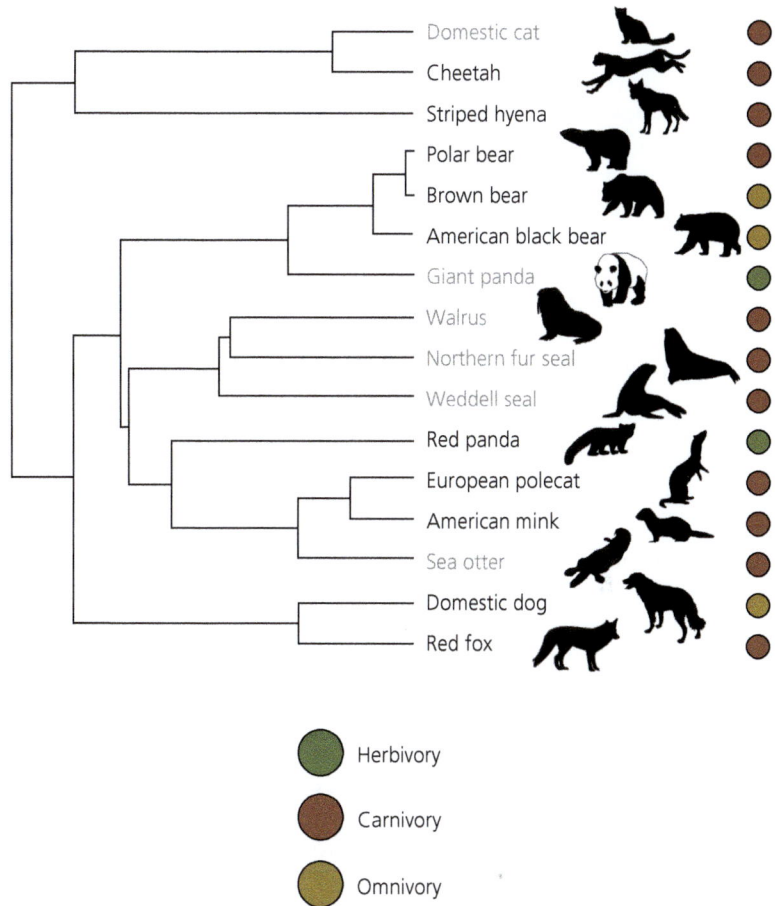

Figure 4.1 Ancestral carnivorans had functioning genes for the digestive enzyme trehalase, which splits the disaccharide trehalose. Trehalose is abundant in the bodies of insects, but absent from vertebrate bodies. This enzyme has disappeared inconsistently across the Carnivora as lineages adapted to non-insect diets and lost copies of their trehalase-coding genes (*Treh*). The phylogram shows functioning *Treh* (black font) and its absence (gray font) along with dietary niche. Most entirely carnivorous or entirely herbivorous species lack functioning *Treh*, but exceptions (e.g. cheetahs and European polecats) are apparent.
Modified from Jiao *et al.* (2019, Figure S2) to reflect omnivory in brown bear, American black bear, and domestic dog.

Except for dissolved carbohydrates, waxes, and some amino acids, most carnivorans eating plant leaves or stems derive little caloric benefit from these foods. This is true for the giant panda, a near-exclusive folivore, which shows assimilation efficiencies on a bamboo diet of around 20%, compared with 26% for red pandas fed bamboo leaves (Wei *et al.*, 2000), and around 65% for various non-ruminant mammals eating plant leaves and stems (Karasov and Martinez del Rio, 2007, p. 126). This low digestive efficiency suggests why the giant panda is so dependent on large tracts of intact bamboo forest; it requires high densities of bamboo available with little time spent searching. Of course, carnivorans consuming vertebrate bodies encounter indigestible materials as well—bone, hair, feathers, and keratinized skin, but in spite of this, the assimilation efficiency for whole vertebrates can be over 80%. Karasov ranked

the digestive efficiencies of various foods to non-ruminant mammals: nectar (highest, around 99% digestible), vertebrate soft tissue, seeds, insects, whole vertebrates, and plant leaves and stems (lowest, around 65% digestible, see Karasov and Martinez del Rio, 2007, p. 127). This range can be compared with the 20% value for giant pandas eating bamboo. Soluble carbohydrates may be readily digested, but they are uncommon, diluted, or transient in most terrestrial communities.

4.1.2 Gut passage

That carnivorans ingest less food of higher digestibility, carry less digesta in their guts, and retain food for briefer periods is now confirmed (De Cuyper *et al.*, 2020). Carnivoran gut contents (dry weight basis) and gut retention times are less than one-third those for non-carnivorans, but the two folivorous panda species are intermediate in their attributes. Some omnivorous carnivorans optimize nutrient uptake by adjusting food passage rate based on the meal ingested. Generally, berry diets generate the shortest retention times, animal carcasses somewhat longer, and plant leaves and stems variable lengths of time. For brown bears fed wild berries or mammal carcasses in Norway, retention times averaged six hours for berries vs. fifteen hours for carcasses (Elfström *et al.*, 2013). Wild ungulate soft tissue is around 90% digestible protein if eaten by a brown bear or American black bear vs. 18% for blueberries (dry weight basis, see Pritchard and Robbins, 1990). The digestion of concentrated protein in animal bodies requires longer times for gastric and small-intestinal peptidase enzymes to be produced and to act on substrate, and for chyme to be in contact with the intestinal mucosa. It is plausible that carnivorans that gorge on soft tissue can sense its passage through the pyloric sphincter, so that that assimilation efficiency is optimized. By contrast, the abundant simple sugars in berries can be absorbed quickly, but nitrogen deficit can result from a sustained diet of berries with so little protein. The kinkajou and binturong, mentioned earlier, can become nitrogen deprived on a pure-fruit diet (Lambert *et al.*, 2014).

Gut passage in pinnipeds has been little studied because of the challenges of observing semi-liquid feces in an aquatic environment. The digestive efficiencies of pinniped diets are high: 90% for cephalopods and 95% for herring in Steller sea lions (Rosen and Trites, 2000). Studies of captive southern elephant seals showed a mean passage time of thirteen hours, which is similar to that for raccoons and domestic cats although they have much shorter digestive tracts (Krockenberger and Bryden 1994). Comparable studies of captive leopard seals showed mean passage times 2–3 times greater than for elephant seals for reasons that were not clear (Hall-Aspland *et al.*, (2011).

4.2 Dietary requirements

4.2.1 Amino acids and fatty acids

Most study of the amino acid requirements of carnivorans has focused on domestic dogs and cats,

with some corroborating evidence from canids and felids held in zoos. Like other eutherians with simple stomachs, carnivorans require nine amino acids in the diet, called "essential" because they cannot be synthesized. In addition to these nine, some felids and species of *Mustela* require taurine and arginine. Taurine is a sulfur-containing amino acid consumed in bile-salt production and inefficiently synthesized from methionine and cysteine. Lacking sufficient taurine, domestic cats and some captive felids exhibit a range of immune-system and developmental pathologies (Hedberg *et al.*, 2007). Arginine is a conditionally essential amino acid, a precursor of ornithine, which binds with toxic ammonia produced from catabolizing amino acids to urea, the primary mammalian nitrogenous waste product. Arginine is so important to domestic cats in detoxifying ammonia that cats fed diets lacking it will die quickly (Morris and Rogers, 1978). More omnivorous carnivorans (e.g. domestic dogs) that depend less on amino acid catabolism require correspondingly less arginine (Baker and Czarnecki-Maulden, 1991).

Carnivorans, like other mammals, are unable to synthesize certain essential polyunsaturated long-chain fatty acids, notably omega-6 linoleic acid, which is a precursor of several other n-6 fatty acids with important roles in immunity, nervous system function, and signaling (Nakamura and Nara, 2003; Malcicka *et al.*, 2018). Therefore, fatty acids such as alpha linolenic acid and linoleic acid must be ingested, and only a few natural sources are abundant and particularly rich. Oily plant seeds contain these fatty acids, which some carnivorans living away from agriculture find in various tree nuts, such as chestnut, hazelnut, and beechnut. Stone pines, in Subgenus *Strobus*, produce large seeds in cones that do not open at maturity. Seeds of these five pine species commonly contain >50%, and in some cases 75% lipid, and fatty acid fractions tend to be dominated by linoleic and oleic acids (Lanner and Gilbert, 1994), making them particularly valuable to bears, sables, and martens that eat pine nuts (Huygens and Hayashi, 2001; Kirkpatrick and Pekins, 2002). Marine fish likewise contain abundant polyunsaturated fatty acids of high nutrient value (Section 6.1) and are eaten by bears and other carnivorans near spawning sites and other sources of fish carcasses.

4.2.2 Macronutrients

It has been long understood that, comparing herbivorous and carnivorous mammals, herbivores ate nutritionally incomplete foods, switching among items to balance the diet. Omnivorous feeders could switch among plant parts and plant species, perhaps between plant and animal foods, and to abiotic mineral sources as needed. By contrast, carnivorous species were thought to eat inherently complete and balanced diets. After all, carnivores were composed of the same constituent amino acids and lipids as their prey. The logical inference was that highly carnivorous species must be limited by food abundance rather than food quality—one prey carcass or soft tissue should be as nutritious as another. We now recognize that diets of carnivorous carnivorans are not uniformly complete; captive animals in cafeteria trials show evidence of balancing diets among the macronutrients—protein, fat, and carbohydrate—if forced to eat an unbalanced diet first. For example, American minks previously fed a diet dominated by calories from protein with little fat compensated by preferring food with more fat and less protein. The reciprocal experiment showed similar switching. Most remarkably, a previous diet low in carbohydrates also produced a compensatory response—minks preferred a diet high in carbohydrate after a carbohydrate fast (Jensen *et al.*, 2014). Domestic cats tended to balance caloric intake of macronutrients from cafeteria offerings at 52:46:2 (% protein:fat:carbohydrate), whereas farmed minks tended to maintain a ratio of 32:51:17. These results could reflect selection in the farmed mink lineage for a carbohydrate-rich diet—they have been fed grains since their first domestication in the mid-nineteenth century, much as early domestic dogs were during the agricultural revolution (cat: Jensen *et al.*, 2014; mink: Mayntz *et al.*, 2009). Interestingly, domestic cats balance macronutrients via a sense for the macronutrients per se; they do not choose foods on the basis of odors or texture, which might be expected to provide sensory cues to macronutrient composition (Hewson-Hughes *et al.*, 2016). Macronutrient balancing has not been shown for wild carnivorans, but needs for specific nutrients, rather than simple availability, may motivate their selection of prey species, or the body parts they consume (Kohl *et al.*, 2015).

Certain pinnipeds differ from all other mammals in the lactose composition of their milk, and therefore the diets of nursing young. Otariid seals and walruses, uniquely among mammals, produce milk with no detectable lactose, whereas phocid seals produce trace amounts of milk lactose. These lineages differ in their lactation strategies—phocids lactate continuously for a few days to a few weeks, until weaning. By contrast, otariids and the walrus nurse young for months to over one year, with intermittent foraging trips of days or weeks. During these non-nursing periods, milk production ceases, but milk remains in mammary ducts. Storing lactose-free milk for days at a time may reduce bacterial souring, providing the selective basis for this trait (Reich and Arnould, 2007).

4.3 Metabolism and growth

4.3.1 Metabolism

Resting metabolic rates of carnivorans are about 25% higher than for eutherians generally, controlling for body mass. The more carnivorous the carnivoran, the faster its metabolism. Felids have resting metabolic rates around 30% higher than mammals generally, whereas the striped hyena (a vertebrate scavenger and invertebrate eater) and the aardwolf (a termitivore) have rates near standard mammalian values. Likewise, the tayra, an omnivorous mustelid of the New World tropics, metabolizes more slowly than predicted from body mass. Frugivorous carnivorans from various families metabolize at lower resting rates than mammals generally, the differences being greatest for large-bodied species. For example, the 3-kg kinkajou and 14-kg binturong have resting rates lower than for comparably sized carnivorans (McNab, 2002); the kinkajou needs to shiver to arouse itself from daily sleep (Lambert *et al.*, 2014). The large folivorous giant panda has an exceptionally low metabolic rate (Nie *et al.*, 2015), but the rate of the trophically similar red panda is comparable to values predicted from body mass (Fei *et al.*, 2017). Eating plants requires less energy expenditure, and returns less energy, than preying on vertebrates. The higher energy density of vertebrate foods (relative to insects or fruits) facilitates and requires higher metabolic rates among vertebrate predators, and the increased year-round availability of vertebrates (compared to more seasonal fruits and ectothermic insects) allows and requires them to maintain high activity levels, with only moderate circannual variation (McNab, 1992, 2000, 2002, 2008).

Resting metabolism is measured on inactive animals within a range of ambient temperatures that require no expenditure of additional energy to stay cool or warm. This thermoneutral resting metabolic rate represents the animal's minimal energy expenditure at normal body temperatures. Within the thermoneutral range, modifying thermal conductance slightly is possible via changes in posture and in the thickness of the fur layer, the latter being adjusted via the piloerector muscles. Below the lower boundary (the lower critical temperature) of this range, endotherms typically raise their metabolic rates by increasing their physical activity, by shivering, or, for some, by non-shivering thermogenesis. Above the range of thermoneutrality, carnivorans redistribute blood flow to the head and extremities, extend their tongues and pant to dissipate heat through water evaporation.

Carnivorans exhibit some of the most remarkable adaptations to extreme cold of any endotherm. Most notably, the arctic fox, a 3-kg predator of coastal tundra and sea-ice environments, has a thermoneutral zone in winter extending to at least −40°C before shivering (Scholander et al., 1950). As ambient temperature approaches −40°C, metabolism slows, so that the fox becomes mildly hypothermic rather than expending extra energy to maintain a warm body core (Prestrud, 1991). By contrast, red foxes in Alaska, weighing around twice as much, have much higher rates of heat loss, and a lower critical temperature of −13°C. They increase metabolic rate twofold relative to thermoneutral values at −50°C (Irving et al., 1955)—clearly not an arctic adaptation.

Field metabolic rate is a more ecologically relevant measure of energy consumption than metabolic rate measured while resting in a chamber. It expresses the energetic costs of free-living, including locomotion, foraging, digestion, thermoregulation, and other functions. It has been measured for a few carnivoran species and shows the highest body mass exponent (0.87) of any endotherm group examined (Nagy et al., 1999). This disproportionate increase in energetic costs with increasing body size means that carnivorans, more than any other taxon examined, pay higher energetic costs for being large-bodied animals. The allocation of energy to locomotion vs. thermoregulation has been estimated for several carnivoran species, based on daily distance travelled, activity times, and field metabolic rates. Carnivorans use an estimated 5% of daily energy expenditure for locomotion, around five times the value predicted for non-carnivoran mammals. This allocation increases with body size: a 100-g carnivoran is predicted to allocate 3% of its total energy expenditure for locomotion, whereas the value for a 100-kg carnivoran is 14% (Garland, 1983). Field metabolic rate of carnivorans also has been related to sex, rate of travel, habitat type, and substrate. Male Pacific martens expended more mass-specific energy than females, and both sexes moved faster and more erratically, expending more energy, in open than in forested areas. This is consistent with the well-documented aversion by boreal forest martens to leaving areas with overhead cover; they tend to move through open areas in ways that reduce their exposure to predators. Martens also spent more energy to move through deep snow than shallow snow (Martin et al., 2019).

For several reasons, the metabolic physiology of marine carnivorans is especially challenging to study. First, training seals or otters to rest in metabolic chambers or tolerate a respiratory mask is difficult. Second, wild pinnipeds resting on the water surface may not be at thermoneutrality, may be absorbing nutrients from their most recent feeding dive, or may exchange gases at elevated rates that reflect their most recent anerobic exertion (Williams et al., 2001). As a result, it is possible that measured metabolic rates (via gas exchange) of aquatic carnivorans resting on the water surface are higher than for terrestrial carnivorans. They are 60% higher in the case of the Weddell seal, twofold higher for the California sea lion, and 2.8-fold higher for the sea otter (California sea lion, Liao, 1990; Weddell seal and sea otter, Williams et al., 2001).

What allows the sea otter to metabolize at such a high rate remained unclear until the discovery that it employs non-shivering thermogenesis in skeletal muscle (Wright et al., 2021). Non-shivering thermogenesis in brown adipose tissue has been understood for decades to enable neonatal mammals and rodents emerging from torpor to generate heat from adipose cells that house mitochondria. The same process occurs in skeletal muscle of the sea otter, but because muscle makes up a much larger proportion of body mass than does brown fat, it exerts stronger, more sustained effects on thermogenesis. The mechanism underlying this thermogenesis is mitochondrial proton leaking, in which electron flow from reduced substrate to oxygen is coupled to hydrogen ion pumping from the mitochondrial matrix. This creates an electrochemical gradient across the mitochondrial inner membrane. The gradient drives protons back into the matrix via adenosine triphosphate (ATP) synthase, which phosphorylates adenosine diphosphate to ATP. The process is not perfectly efficient, and some hydrogen ions leak back into the matrix without coupling to ATP production (Brand *et al.*, 1994). This leakage produces heat and increases metabolic rate via non-oxidative pathways. The process is inducible in the sea otter and other marine carnivores, the former depending on proton leak to generate sufficient heat to compensate for its small body size and lack of subcutaneous fat. More remarkably, although some muscle attributes (e.g. myoglobin concentration) are only partially developed in neonate sea otters, their mitochondrial proton leak system performs as well as that of adults. Leak metabolism has been measured for the northern elephant seal (Wright *et al.*, 2020) and other cold-climate mammals, but likely is not critical for thermoregulation in species with adaptive features that the sea otter lacks. Only shrews weighing 5–6 g have resting metabolic rates comparable to that of the sea otter. Mitochondrial proton leak metabolism has potentially important ecological implications; it may account for the strength of per-capita and per-kg effects exerted by some carnivoran predators on communities (Section 9.2.1).

4.3.2 Growth

Postnatal growth is a trait with strong ecological implications. Neonates that grow rapidly reduce certain risks to themselves (predation, becoming separated, hypothermia, and starvation) as well as to their mothers. Mammals exhibit a nearly hundredfold range in growth rates, even controlling for body size, and carnivorans are arguably the most variable mammalian order, including some of the slowest- and fastest-growing mammals. Some pinnipeds with adult weights of 180 kg grow by 7 kg/day until weaning, a rate fiftyfold greater than that of the brown bear (Case, 1978; Bowen *et al.*, 1985). Canids also grow quickly by mammalian standards, up to several hundred grams/day, whereas most other terrestrial carnivorans have intermediate rates. Rapid growth in pinnipeds is consistent with immobile neonates being vulnerable to predation on land until they can swim and experiencing high thermal losses once in cold water. Growing quickly reduces both threats. Pinniped milk is rich in fat and protein; typical lipid values are 30–50%, and protein composition is 8–12% (Schulz and Bowen, 2004), which is more similar to cetacean values than those for fissipeds. Neonatal pinnipeds are weaned quickly at ages of a few days to a few weeks. The hooded seal has the shortest time to weaning—four days—of any mammal of any body size (Bowen *et al.*, 1985), and rapid growth continues through the first year. By contrast, the slow somatic growth of bears is a correlate of their slow life histories, which include winter torpor, long inter-birth intervals and low rates of population increase. Growth is constrained by the low nutrient density of most ursid diets. Growth by neonates may also be influenced by litter size. For example, in captivity triplet brown bear cubs gain around 70 g/day, whereas twins grow twice as fast (Robbins *et al.*, 2012a).

4.4 Body temperature and torpor

4.4.1 Body temperature

Body core temperature varies across endotherm lineages, with birds maintaining slightly warmer body cores than do mammals. Among mammals, carnivorans have the lowest body temperatures, strict herbivores the highest, and omnivores intermediate temperatures, although the differences are small. This pattern is opposite to that suggested by resting metabolic rates and becomes more puzzling when we consider that, within the Carnivora, species that feed on vertebrate prey have higher body temperatures than those that feed on invertebrates (Clarke and O'Connor, 2014). Among herbivorous mammals, the highest body temperatures are observed in species with the highest proportion of cellulose in the diet (folivores), suggesting that the needs of microbial fermenters determine the body temperatures of their folivorous hosts. This is plausible, although the differences between the core temperatures of leaf eaters, vertebrate eaters, and invertebrate eaters are small (1–2°C). Still, a difference of 1.5°C produces a 25% difference in microbial metabolic rate (assuming microbial respiratory $Q_{10} = 2$), which could affect gut passage time and assimilation efficiency for the herbivore.

Many carnivorans thermoregulate behaviorally, selecting activity times and resting microenvironments to minimize thermal costs. In cold environments, this is especially apparent in small-bodied carnivorans. Ermines and least weasels in the boreal and tundra zones live almost entirely in the subnivean zone in winter (Figure 4.2A), where temperatures can be 40°C warmer than at the snow-air interface. The somewhat larger boreal forest martens (*Martes* spp.) of coniferous forests are too large to traverse their home ranges beneath the snow, but travel on the snow surface, accessing the subnivean space to find prey (Buskirk *et al.*, 1989). They also rest beneath the snow surface in cold weather, warming protected microsites to thermoneutrality with waste heat from recent locomotion (Taylor and Buskirk, 1994). At the opposite end of the body size and climate spectra, lions and other large carnivorans in hot environments seek shaded sites to reduce costs of cooling (Figure 4.2B) (Hayward and Hayward, 2006).

4.4.2 Torpor

Endotherms display various states of reduced physical activity, metabolic rate, and responsiveness. These states include shallow torpor, characterized by sleep from which the animal can be readily aroused, to deeper multi-day states of hypometabolism and non-responsiveness. No carnivoran achieves the deep torporous states of some wintering rodents or bats (Barnes, 1989). However, several lineages undergo shallow torpor. American black bears in winter dens exhibit body core temperatures of around 34°C, 5–7°C lower than active values, with no circadian pattern, and are easily aroused (Harlow *et al.*, 2004). Both American badgers and European badgers undergo torpor during subterranean inactivity in winter; body temperatures of European badgers fall to 28–34°C, and those of American badgers to 29°C, accompanied by a 50% reduction in heart rate. American badgers can remain underground for over seventy consecutive days (European badger, Fowler and Racey, 1988; American badger, Harlow, 1981). The aardwolf of eastern and southern Africa enters shallow torpor when ambient temperatures fall below 9°C and its preferred *Trinervitermes* termites become unavailable. Body core temperature lowers to around 34°C and animals share dens to further reduce thermal losses (Anderson, 2004). Striped skunks are capable of the deepest torpor documented among carnivorans and achieve body core temperatures as low as 26°C during daily bouts. The depth of their torpor depends in part on whether the skunks rest singly or in groups, with solitary skunks reaching lower temperatures (mean = 27°C) during longer bouts than group-resting animals (mean = 31°C), which experienced briefer bouts of hypothermia (Hwang *et al.*, 2007).

Although carnivorans show limited employment of torpor, ursids have been studied closely for their unique adaptations to winter denning—they do not walk, eat, drink, defecate, or urinate for up to 150 days, yet they avoid toxic levels of circulating urea and maintain muscle mass and tone, unlike humans would under comparable conditions. For this reason, bears are considered "metabolic magicians" by human nephrologists (Stenvinkel *et al.*, 2013). The primary mechanism underlying this ability is high reliance on non-nitrogenous energy stores—fat. Also, bears recycle almost 100% of their urea into amino acids via microbial action in the gut (Barboza *et al.*, 1997). These amino acids are reassembled into muscle protein, and retention of muscle strength may be due to a combination of shivering and voluntary contractions (Harlow *et al.*, 2004).

4.5 Energy storage and fasting

Carnivorans in seasonal environments have evolved diverse strategies of energy storage and

(a)

(b)

Figure 4.2 Behavioral thermoregulation is key to survival and cold and hot climates. A: An ermine emerges from a subnivean space, where it forages and rests at temperatures warmer than at the snow surface. B: A pride of lions crowds into the shade of an acacia in Tanzania. Shade is an important microhabitat for African lions on hot days.
Photos: A. John Bakkila/iStock; B. Oskanov/ CanStock.

mobilization. Circannual variability in food availability results from production (e.g. seasonal production of leaves or mast) or accessibility (e.g. hibernation of prey, migration of prey, or depth or hardness of snow). Seasonal energetic strategies are also constrained by body size, body form, and metabolic plasticity. Larger mammals, even disregarding marine species, have higher proportions of body fat and can fast for longer periods than small ones (Lindstedt and Boyce, 1985). Further, carnivorans in open or marine environments might not be constrained by carrying somatic reserves, as would small-bodied species that forage in tight spaces. A polar bear that adds subcutaneous fat loses little mobility on land and enjoys greater buoyancy in water, whereas a least weasel that stores fat is constrained in tight spaces. The result is that some carnivorans store energy reserves for seasons of food scarcity, whereas others cannot. Most large mammals in seasonal environments balance energy budgets over periods of about a year, but small carnivorans do so over brief periods, as short as a few days. For some annual balancers, a brief season of food abundance requires them to store energy—either inside or outside their bodies—quickly. Most northern latitude bears become hyperphagic during late summer and fall, and conserve energy by denning. Brown bears have late-autumn body fat depots that are 30–50% of body mass, declining over winter. Among females in Scandinavia, the reduction is 39%, most of it representing catabolism of fat (Swenson *et al.*, 2007). American black bears are obese in autumn, but 92% of total energy expenditure while hibernating results from fat oxidation (Harlow *et al.*, 2002). Of course, fasting bears require some protein for nitrogen and water balance, and often use labile protein reserves such as smooth muscle, rather than somatic muscle, to maintain nitrogen and water balance (Lohuis *et al.*, 2007). Polar bears differ from other northern bears in foregoing torpor except for pregnant females, and in that their main foods—seals—are least accessible during the warm ice-free season (Thiemann *et al.*, 2006). Their peak fat reserves, at least during the ice conditions of the 1980–90s, provided a fasting endurance of about 300 days for pregnant females, which is among the longest regular fasting endurances of any mammal species (Robbins *et al.*, 2012b).

The opposite strategy is also found among carnivorans; giant pandas rely almost entirely on low-energy, low-protein bamboo leaves, and cannot accumulate somatic reserves to allow extended fasting. Therefore, they remain active, feeding on bamboo throughout winter (Schaller *et al.*, 1985). Most small-bodied carnivorans resemble the giant panda in this regard and require access to food all year. They have limited fat reserves and catabolize skeletal muscle protein at the risk of losing locomotor function. North American martens are such a species and have winter body fat reserves that are only 1.2–8.8% of body mass. An animal with the mid-range of these values, catabolizing only lipid, is predicted to exhaust all metabolizable fat reserves, not including lipids in cell membranes, in about 90 hours (Buskirk and Harlow, 1989). Even employing shallow torpor while inactive, these martens cannot forego foraging for long. The most energetically challenged carnivorans—ermines and least weasels living in arctic environments—have even shorter fasting endurance times: a 30-g least weasel, assuming 2.5% body fat and with the standard allometry for fasting endurance (Lindstedt and Boyce, 1985), is expected to survive without food for fewer than forty-eight hours, at which time fat would be irreversibly lost from smooth muscle and

other organs. The arctic fox is an outlier in this pattern, having fat reserves of 3–38% of body mass. A 3.5-kg arctic fox with 22% body fat is estimated to have a fasting endurance of thirty days (Prestrud, 1991) with no apparent loss of function.

Only a few mammalian species store energy reserves outside the body. Many carnivorans opportunistically use carrion sources—killed by them or scavenged—over extended periods in cold climates. They carry and cache carcass parts around their home ranges (Box 4.1), although this does not necessarily provide long-term storage because it can degrade or be usurped (Macdonald, 1976, Table 3). Ermines cache rodent and shrew carcasses in winter nests; these caches can contain dozens of uneaten carcasses by spring.

Likewise, an arctic fox has been observed to cache over 100 sea bird carcasses in a single belowground site in summer, for use over winter (Murie, 1959, p. 297; Prestrud, 1991). Perhaps most remarkably, the neotropical tayra caches mature (fully grown, ready to ripen) but unripe fruits of multiple species, collecting fruits before they are attractive to competitors, and placing them in sites less likely to be discovered by usurpers (Figure 4.4). This caching of foods that are inedible when harvested, and hidden until optimal, raises the possibility of prospective cognition in tayras (Soley and Alvarado-Diaz, 2011).

4.6 Osmoregulation and kidney function

Marine pinnipeds must maintain tissue homeostasis in a hyperosmotic environment, so they must produce urine with osmolality greater than seawater. This suggests special urine-concentrating mechanisms, such as those seen in desert rodents, and should be a particular feature in fasting animals, because they are not ingesting preformed water in prey. Pinnipeds, like cetaceans, exhibit kidney morphology consistent with urine concentrating ability, including a relatively thick medulla. However, the urine osmolality of pinnipeds—and the ratios of urine-to-plasma osmolality among pinniped species—differ little from those of the North American river otter and domestic dog (much of this discussion is based on Ortiz, 2001). This paradox may be the result of marine pinnipeds having poor urine concentrating abilities, mediated mostly

Box 4.1 Interment of the dead

Frehner and colleagues (2017) described belowground storage of large mammal carcasses by American badgers in winter. Two badgers were remotely photographed burying domestic cow carcasses, one of them entirely; pelage patterns showed that only one badger and no other scavengers fed on the carcass at each site, for up to 52 days (Figure 4.3).

(a)

(b)

Figure 4.3 American badgers reduce competition for carrion by burying entire carcasses of large mammals. A: A badger excavates around a domestic cow carcass in Utah, US that it had discovered two days earlier. B: One day later, the entire carcass is buried. The badger dug a new burrow entrance nearby and spent the next several days underground.
Photos: © Evan Buechley.

Figure 4.4 A tayra picks fruit from a Panama rubber tree in Costa Rica. The tayra has been documented to store semi-ripe fruit where it cannot be usurped by other frugivores, and then retrieve the fruit when fully ripened.
Photo: © Max Waugh.

via osmoregulator hormones and rate of urine production. Fasting post-weaning seal pups experience particular challenges of osmoregulation and water loss. The fasting period lasts 2–3 months in some species, during which pups selectively oxidize fat, which produces more metabolic water than does protein. However, due to the high energy density of fat and concomitantly high metabolic rate, high respiratory evaporative water loss on exposed rookery sites can be problematic. They minimize this water loss via multiple mechanisms, including countercurrent exchange of water vapor in the nasal cavities.

4.7 Detoxification and self-medication

Like other mammals, carnivorans encounter a wide range of naturally occurring toxic substances, as well as some of human origin. These include the plant defensive compounds that deter herbivory: phenolics, turpenoids, and nitrogenous compounds. There is a general trend for carnivorous (including insectivorous and piscivorous) eutherian lineages to have lost the *NR1I3* gene involved in detoxifying plant toxins (Wagner *et al.*, 2022). Most carnivorans also face toxins that herbivores encounter less often: toxic prey body parts, toxic sprays, and venoms. Resistance to envenomation has arisen in a number of marsupial and eutherian lineages, including several carnivoran lines (Voss and Jansa, 2012). The Herpestidae, Mephitidae, and Mustelidae include multiple species that readily eat snakes, are commonly envenomated, and show some venom resistance. The Indian gray mongoose and Egyptian mongoose neutralize the hemorrhagic activity of some venoms via serum factors analogous to those found in the common opossum, another venom-resistant species (Qi *et al.*, 1994). By contrast, honey badgers deactivate neurotoxic venoms via muscular nicotinic cholinergic receptors that do not bind to the toxins that target them. Honey badgers have an extraordinary reaction to cobra envenomation; they become comatose for up to several hours, then regain consciousness and mobility with no apparent further effects. Thus, snake-eating carnivorans and their venomous prey have coevolved at the molecular level, which allows some carnivorans regularly to prey on venomous species (Barchan *et al.*, 1992; Drabeck *et al.*, 2015).

One last class of toxins that scavenging carnivorans are especially likely to encounter comprises the products of putrefaction in carcasses. Especially in warm climates, large carcasses have high concentrations of putrescine, cadaverine, and other polyamines produced by enzymes (intrinsic to the decomposing flesh or from microbes) acting on animal flesh. Microbes evolved these compounds as a means of reducing exploitation competition with vertebrate consumers, particularly in warm climates (Janzen, 1977); they represent a form of interference competition by microbes against vertebrate scavengers. At lower concentrations, they volatilize and attract vertebrate scavengers, so that they have curious, opposing functions—facilitating and deterring vertebrate competition. In addition, cadaveric anaerobic bacteria produce botulinum toxins of several types, some highly toxic to mammals. The coyote carries serum antibodies to several botulinum toxins, but at lower levels than does the turkey vulture (Ohishi *et al.*, 1979). How scavenging carnivorans cope with various toxins (which may involve the action of gut microbes, as it does for vultures; see Mendoza *et al.*, 2018) is a fertile area for further study.

Some mammals use natural materials to manipulate their own behaviors or alter the body's response to invading organisms. This self-medication is best documented for primates but observed in various mammalian orders. Carnivorans are second only to primates in the frequency of reports; the phenomenon seems to be most frequent in large-bodied species with large brains (Neco *et al.*, 2019).

4.8 Reproduction

4.8.1 General patterns

Even accounting for body size, carnivorans vary strongly in most reproductive attributes, including age at first reproduction, total gestation length, litter size, and lifetime reproductive output. With abundant food resources, 3-kg arctic foxes can give birth at one year and yearly afterward, and produce litters of up to fifteen kits (Audet *et al.*, 2002). A 100-kg female brown bear without good access to animal protein might produce its first young at age five, produce two cubs/litter, and give birth at four-year intervals. This would be among the lowest reproductive rates of any land mammal (Pasitchniak-Arts, 1993). Comparing species, an arctic fox with access to abundant food might produce more neonates by age three than a brown bear living in unproductive habitat over a twenty-five-year lifetime. This kind of variation strongly constrains life history strategies (Section 10.4).

Neonatal terrestrial carnivorans are altricial—lightly furred, closed-eyed, and requiring extended maternal care. Virtually all terrestrial carnivorans give birth in protected structures, which can include subterranean burrows, cavities in tree boles, or other aboveground sites protected by woody or rock structures. Pinnipeds and the sea otter, by contrast, give birth to well-furred neonates in highly exposed sites. Pinnipeds do so on beaches, rocks, or sea ice, whereas the sea otter does so in water (Estes, 1980). Neonatal pinnipeds are fully furred and open their eyes quickly, and some catabolize brown adipose tissue for postpartum thermoregulation. Nevertheless, neonate pinnipeds are highly vulnerable to trampling, infanticide, and predation (Bininda-Emonds and Gittleman, 2000). Countering these risks, some species gain weight rapidly, consuming large volumes of milk with high fat content to gain mobility and independence (Baker *et al.*, 1995).

4.8.2 Embryonic diapause

Order Carnivora, in particular Family Mustelidae, is remarkable for the frequency of embryonic diapause (delayed implantation)—the suspension of growth at the blastocyst stage, with the embryo floating free in the uterine lumen for up to several months (Figure 4.5). Over half of all mammalian delayed implanters are carnivorans, of which most are mustelids. Across the Carnivora, implantation and resumption of embryonic development respond to environmental temperature, photoperiod, nutritional condition, or combinations of these factors (Lopes *et al.*, 2004). The specific mechanism underlying implantation and resumed development is the presence of estrogen-related polyamines, in the absence of which implantation does not occur (Fenelon *et al.*, 2016). Diapause is adaptive by decoupling the time of mating from that of parturition, so that both events can occur at optimal seasons (Thom *et al.*, 2004). Using the example of the brown bear and ermine, both species mate in mid-summer, and in brown bears, diapause lasts 20–25 weeks, with young born in winter dens in January or February. The ermine, by contrast, undergoes nearly forty-three weeks of diapause, giving birth in the northern hemisphere in April or May. The distribution of diapause across the Carnivora is idiosyncratic; for example, one North American spotted skunk species exhibits it, but another, living at similar latitudes, does not. The North American river otter delays implantation, but the Eurasian otter, a close ecological analogue in the same subfamily, does not. Geographic latitude is useful predictor of the occurrence and duration of diapause; species of Carnivora living farther from the equator

Puma - induced ovulation, no diapause

Ermine - induced ovulation, diapause

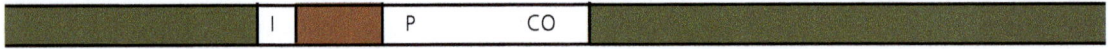

Least weasel - induced ovulation, no diapause

American black bear - induced ovulation, diapause

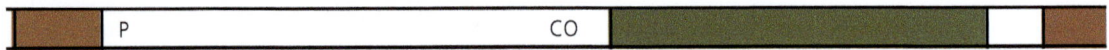

California sea lion - spontaneous ovulation, diapause

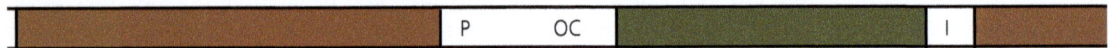

Red fox - spontaneous ovulation, no diapause

1 Jan.   31 Dec.

Embryonic diapause   C: Copulation

Active gestation   I: Implantation

                                                   O: Ovulation

                                                   P: Parturition

Figure 4.5 Reproductive cycles of carnivorans, illustrating the possible permutations of spontaneous vs. induced ovulation, and embryonic diapause vs. its absence. Each horizontal bar represents one year, with seasonality of the north temperate zone. The puma and ermine are semi-seasonal breeders; the timing of events in their cycles is variable.
Reproductive data are from Mammalian Species accounts and Greig *et al.* (2007).

are more likely to employ it, and their delays tend to be more prolonged (Heldstab *et al.*, 2018). Thus, diapause and seasonality appear to be related.

Once thought to have originated multiple times in the carnivoran lineage, embryonic diapause is now believed to have arisen once, among the early

Caniformia, after the Canidae had diverged. Under this scenario, which assumes the fewest origin-loss events, it was secondarily lost in various lineages, under selective pressures that are not understood (Lindenfors *et al.*, 2003). It is less common among the smallest-bodied species. For them, life expectancy at birth is brief. As a result, delaying sexual maturation until late in the first year of life, combined with delaying implantation for most of another year, is too limiting to lifetime reproductive output. Least weasels and several other *Mustela* spp. have lost diapause, and least weasels can even give birth during the summer of their own birth. Female ermines illustrate how delaying implantation constrains a short-lived species; they first conceive at only a few weeks of age, while only partially grown, and give birth at age one. This requires adult males to impregnate immature females in the dens where the latter are developing (Sandell, 1984). Diapause affects population ecology by delaying numerical responses to prey fluctuations. This lagged numerical response can contribute to carnivoran–prey population cycles (Section 8.6), because if an ermine gives birth nine months after it conceives during the low phase of the prey population cycle, its numerical response will reflect conditions during the previous biological year.

4.8.3 Induced ovulation

The species-specific occurrence of another reproductive mechanism—induced ovulation—is easier to rationalize. In it, ovulation occurs in response to contact with or proximity to a male conspecific, rather than to an abiotic cue. In some carnivorans, the sensory threshold for stimulating ovulation is low; the mere proximity of a caged male domestic cat triggers some ovulations (Binder *et al.*, 2019). In many carnivorans, brief copulation is sufficient to trigger the surge of luteinizing hormone and elevated progesterone, which cause rupture of the ovarian follicle. In some carnivorans, however, a high stimulation threshold must be crossed for ovulation. These females ovulate only in response to repeated or aggressive, prolonged copulations, which may be accompanied by biting and loud vocalizations (Figure 4.6). Across the Carnivora, few species have been examined for induced ovulation, because of the need for experimental manipulation of captive animals. Induced ovulation is known to occur in some felids, some mustelids, and some ursids. Among the Felidae, there are known examples of all ovulatory patterns—spontaneous, low-threshold induced, and high-threshold induced (Jorge-Neto *et al.*, 2020). The baculum, present in the Carnivora

Figure 4.6 Copulating lions. In many high-threshold induced ovulators, mating is prolonged or repeated and accompanied by loud vocalizations and the male biting the female's head or neck. Photo: J.W. Elzenga/iStock.

and three other mammalian orders, was once suspected of playing a role in high-threshold induced ovulation. It stiffens the penis and plausibly could prolong copulation or increase vaginal stimulation. Lariviere and Ferguson (2002) found, however, that bacular length did not vary with duration of copulation or differ between spontaneous and induced ovulators.

The adaptive significance of induced ovulation for low-density carnivorans seems intuitive—rather than females ovulating and then seeking mates that may be seldom encountered, they ovulate only after encountering a male. Induced ovulators tend to have large home ranges and multi-male mating systems (Lariviere and Ferguson, 2003; Soulsbury, 2010). Males of polyandrous species benefit from limiting time invested after copulation, excepting some attendance near the female to assure that fertilization has been monopolized. Then, males can search for other receptive females. Females benefit from insemination coinciding with ovulation, with the additional potential for post-copulatory sexual selection. This selection results from females terminating pregnancies of pairings that do not provide specific behavioral or biochemical cues. There is a tendency for induced ovulation to occur in less seasonal environments than is the case for spontaneous ovulation (Heldstab *et al.*, 2018).

4.9 Scaling physiology to populations

Several of these organismal and molecular processes have obvious consequences for populations and communities. For example, the biomass on the landscape of highly carnivorous mammals relative to that of their prey is constrained by the second law of thermodynamics, a corollary of which states that energy transfers, whether trophic, radiative, or kinetic, are not perfectly efficient. Because trophic levels denote energy transfers, the further an organism or population is from the system's original source of biosynthesis, the less mass and chemical energy it will contain. So, predators of herbivores are much rarer than herbivores; mammalian carnivores weighing over 10 kg occur at population densities less than 5% those of mammalian herbivores of the same body size (Silva *et al.*, 2001). At

100 g body weight, the densities are more similar. The relative rarity of carnivorans, especially those of large size, is a recurring theme in community ecology.

As an extension of this principle, individual carnivorans require much more area to subsist than do herbivores. The home range—the area traversed by a mobile animal in its day-to-day living—is a measure of individual energy requirements. Carnivorans have home ranges around twenty-five times larger than those of mammalian herbivores, a multiplier that is unchanged if one substitutes non-flying carnivorous mammals for carnivorans (Tamburello *et al.*, 2015). A related spatial attribute—dispersal distance—is the distance traveled by young animals seeking to establish home ranges. Dispersal distance is proportional to home-range diameter (Bowman *et al.*, 2002), so that carnivorans travel much more widely—on a daily and lifelong basis—than do herbivores. This contributes to the high field metabolic rates of carnivorans compared to herbivorous mammals.

Body size also is pervasively important in predatory relationships, whether we consider invertebrates or vertebrates, ectotherms or endotherms. Across pairs of metazoan predators and their respective prey, predators are larger in about 90% of pairs. Considering only vertebrate endotherms, predators weigh (on average) about 1.4 times as much as their prey (Cohen *et al.*, 1993). Among only carnivorans, this relationship varies with body size: small carnivorans (body mass < 15–20 kg) tend to prey on invertebrates and vertebrates a small fraction of their own size. Foxes that weigh less than 10 kg, for example, feed on insects, small rodents, and lagomorphs. Carnivorans that weigh more than 20 kg prey on a much wider range of vertebrate body sizes, most of which are near or greater than their own sizes. They also experience much higher rates of energy expenditure during prey capture, and higher probability of injury from subduing prey. For example, a 40-kg wolf cooperating with pack members routinely tests, harasses, injures, and kills moose that weigh 15 times as much (Carbone *et al.*, 2007). Even within species, body size is important in predation behaviors. Male wolves are larger than females and are the primary large ungulate hunters. Smaller-bodied male wolves are most effective at tasks that require speed and agility (prey selection and isolation), whereas larger males are better suited to strength-related tasks (subduing prey) (MacNulty *et al.*, 2009). For marine carnivorans—the pinnipeds—these trends do not hold. Pinnipeds show much higher predator–prey body-size ratios than do fissipeds. No pinniped kills prey larger than itself, and pinnipeds occupy higher trophic levels (1.3 trophic levels higher) than terrestrial carnivorans (Tucker and Rogers, 2014). Thus, the tendency for larger carnivorans to prey on larger prey is highly contextual.

Key points

- Carnivora is the most trophically diverse mammalian order, consuming diets ranging from entirely meat to entirely plant leaves and stems. Some species specialize on fruits, insects, or aquatic invertebrates.
- Folivorous carnivorans lack strong adaptations for digesting the cellulose in cell walls: large or compartmented guts, diverse gut microbial floras, or associated cellulose-digesting enzymes.
- Carnivorans have lost the gene coding for the hormone that links appetite to gut fullness, perhaps contributing to surplus killing behaviors
- The genes for enzymes that digest chitin and trehalose, the latter a disaccharide found in insect bodies and fungi, follow phylogenetic lines. Their presence or absence in extant species tracks divergence from an early insectivorous diet. Carnivorans require some amino acids beyond the standard mammalian essential ones in species that are highly carnivorous.
- Domestic carnivorans—dogs and the American mink—have biochemical adaptation to the starches that humans have introduced to their diets with domestication.
- Carnivorans metabolize at high rates, and the more carnivorous, the higher the rate. By mammalian standards, they show only modest metabolic plasticity and do not enter deep torpor.
- The group shows great variation in fasting endurance, from only a few days to more than six months.
- Delayed implantation is especially common in the Carnivora relative to other mammals and

allows mating and parturition to occur at optimal seasons. Induced ovulation also is common inasmuch as it allows mating and ovulation to be synchronized, which is adaptive for a lineage in which potential mates are solitary much of the time and seldom meet at the optimal season for active gestation to begin.
- Physiological traits, particularly body size, trophic level, and metabolic rate, explain key features of carnivoran ecology.

References

Anderson, M.D. (2004) "Aardwolf adaptations: a review," *Transactions of the Royal Society of South Africa*, 59, pp. 99–104.

Audet, A.M., Robbins, C.B. and Larivière, S. (2002) "*Alopex lagopus*," *Mammalian Species*, 713, pp. 1–10.

Axelsson, E., *et al.* (2013) "The genomic signature of dog domestication reveals adaptation to a starch-rich diet," *Nature*, 495, p. 11837.

Baker, D.H. and Czarnecki-Maulden, G.L. (1991) "Comparative nutrition of cats and dogs," *Annual Review of Nutrition*, 11, pp. 239–63.

Baker, S.R., Barrette, C. and Hammill, M.O. (1995) "Mass transfer during lactation of an ice-breeding pinniped, the grey seal (*Halichoerus grypus*), in Nova Scotia, Canada," *Journal of Zoology*, 236, pp. 531–42.

Barboza, P.S., Farley, S.D. and Robbins, C.T. (1997) "Whole-body urea cycling and protein turnover during hyperphagia and dormancy in growing bears (*Ursus americanus* and *U. arctos*)," *Canadian Journal of Zoology*, 75, pp. 2129–36.

Barchan, D., *et al.* (1992) "How the mongoose can fight the snake: the binding site of the mongoose acetylcholine receptor," *Proceedings of the National Academy of Sciences*, 89, pp. 7717–21.

Barnes, B.M. (1989) "Freeze avoidance in a mammal: body temperatures below 0 °C in an Arctic hibernator," *Science*, 244, pp. 1593–5.

Binder, C. *et al.* (2019) "Spontaneous ovulation in cats—uterine findings and correlations with animal weight and age," *Animal Reproduction Science*, 209, pp. 106167.

Bininda-Emonds, O.R. and Gittleman, J.L. (2000) "Are pinnipeds functionally different from fissiped carnivores? The importance of phylogenetic comparative analyses," *Evolution*, 54, pp. 1011–23.

Bowen, W.D., Oftedal, O.T. and Boness, D.J. (1985) "Birth to weaning in 4 days: remarkable growth in the hooded seal, *Cystophora cristata*," *Canadian Journal of Zoology*, 63, pp. 2841–6.

Bowman, J., Jaeger, J.A.G. and Fahrig, L. (2002) "Dispersal distance of mammals is proportional to home range size," *Ecology*, 83, pp. 2049–55.

Brand, M.D. et al., (1994) "The causes and functions of mitochondrial proton leak," *Biochimica et Biophysica Acta*, 1187, pp. 132–9.

Buskirk, S.W. et al. (1989) "Winter resting site ecology of marten in the Central Rocky Mountains," *Journal of Wildlife Management*, 53, pp. 191–6.

Buskirk, S.W., and Harlow, H.J. (1989) "Body-fat dynamics of the American marten (*Martes americana*) in winter," *Journal of Mammalogy*, 70, pp. 191–3.

Carbone, C., Teacher, A. and Rowcliffe, J.M. (2007) "The costs of carnivory," *PLoS Biology*, 5, p. e22.

Case, T.J. (1978) "On the evolution and adaptive significance of postnatal growth rates in the terrestrial vertebrates," *Quarterly Review of Biology*, 53, pp. 243–82.

Clarke, A. and O'Connor, M.I. (2014) "Diet and body temperature in mammals and birds," *Global Ecology and Biogeography*, 23, pp. 1000–8.

Cohen, J.E. et al. (1993) "Body sizes of animal predators and animal prey in food webs," *Journal of Animal Ecology*, 62, pp. 67–78.

Cornelius, C., Dandrifosse, G. and Jeuniaux, C. (1975) "Biosynthesis of chitinases by mammals of the order Carnivora," *Biochemical Systematics and Ecology*, 3, pp. 121–2.

De Cuyper, A., et al. (2020) "The uneven weight distribution between predators and prey: comparing gut fill between terrestrial herbivores and carnivores," *Comparative Biochemistry and Physiology Part A: Molecular & Integrative Physiology*, 243, p. 110683.

Drabeck, D.H., Dean, A.M. and Jansa, S.A. (2015) "Why the honey badger don't care: convergent evolution of venom-targeted nicotinic acetylcholine receptors in mammals that survive venomous snake bites," *Toxicon*, 99, pp. 68–72.

Elfström, M. et al. (2013) "Gut retention time in captive brown bears *Ursus arctos*," *Wildlife Biology*, 19, pp. 317–24.

Emerling, C.A., Delsuc, F. and Nachman, M.W. (2018) "Chitinase genes (CHIAs) provide genomic footprints of a post-Cretaceous dietary radiation in placental mammals," *Science Advances*, 4, p. eaar6478.

Estes, J.A. (1980) "*Enhydra lutris*," *Mammalian Species*, 133, pp. 1–8.

Fei, Y. et al. (2017) "Metabolic rate of the red panda, *Ailurus fulgens*, a dietary bamboo specialist," *PLoS ONE*, 12, p. e0173274.

Fenelon, J.C. et al. (2016) "Regulation of diapause in carnivores," *Reproduction in Domestic Animals*, 52(Suppl. 2), pp. 12–7.

Fowler, P.A., and Racey, P.A. (1988) "Overwintering strategies of the badger, *Meles meles*, at 57° N," *Journal of Zoology*, 214, pp. 635–51.

Frehner, E.H. et al. (2017) "Subterranean caching of domestic cow (*Bos taurus*) carcasses by American badgers (*Taxidea taxus*) in the Great Basin Desert, Utah," *Western North American Naturalist*, 77, pp. 124–9.

Garland, T., Jr. (1983) "Scaling the ecological cost of transport to body mass in terrestrial mammals," *American Naturalist*, 121, pp. 571–87.

Greig, D.J. et al. (2007) "Seasonal changes in circulating progesterone and estrogen concentrations in the California sea lion (*Zalophus californianus*)," *Journal of Mammalogy*, 88, pp. 67–72.

Guo, W., et al. (2018) "Metagenomic study suggests that the gut microbiota of the giant panda (*Ailuropoda melanoleuca*) may not be specialized for fiber fermentation," *Frontiers in Microbiology*, 9, p. 229.

Hall-Aspland, S. et al. (2011) "Food transit times in captive leopard seals (*Hydrurga leptonyx*)," *Polar Biology*, 34, pp. 95–9.

Harlow, H.J. (1981) "Torpor and other physiological adaptations of the badger (*Taxidea taxus*) to cold environments," *Physiological Zoology*, 54, pp. 267–75.

Harlow, H.J. et al. (2004) "Body surface temperature of hibernating black bears may be related to periodic muscle activity," *Journal of Mammalogy*, 85, pp. 414–9.

Harlow, H.J. et al. (2002) "Body mass and lipid changes by hibernating reproductive and nonreproductive black bears (*Ursus americanus*)," *Journal of Mammalogy*, 83, pp. 1020–5.

Hayward, M.W. and Hayward, G.J. (2006) "Activity patterns of reintroduced lion *Panthera leo* and spotted hyaena *Crocuta crocuta* in the Addo Elephant National Park, South Africa," *African Journal of Ecology*, 45, pp. 135–41.

Hedberg, G.E., Dierenfeld, E.S. and Rogers, Q.R. (2007) "Taurine and zoo felids: considerations of dietary and biological tissue concentrations," *Zoo Biology*, 26, pp. 517–31.

Hecker, N., Sharma, V., and Hiller, M. (2019) "Convergent gene losses illuminate metabolic and physiological changes in herbivores and carnivores," *Proceedings of the National Academy of Sciences*, 116, pp. 3036–41.

Heldstab, S.A. et al. (2018) "Geographical origin, delayed implantation, and induced ovulation explain reproductive seasonality in the Carnivora," *Journal of Biological Rhythms*, 33, pp. 402–19.

Hewson-Hughes, A. K. et al. (2016) "Balancing macronutrient intake in a mammalian carnivore: disentangling the influences of flavor and nutrition," *Royal Society Open Science*, 3, p. 160081.

Hückstädt, L.A. *et al.* (2012) "Diet of a specialist in a changing environment: the crabeater seal along the western Antarctic peninsula," *Marine Ecology Progress Series* 455, pp. 287–301.

Huygens, O.C. and Hayashi, H. (2001) "Use of stone pine seeds and oak acorns by Asiatic black bears in central Japan," *Ursus*, 12, pp. 47–50.

Hwang, Y.T., Larivière, S. and Messier, F. (2007) "Energetic consequences and ecological significance of heterothermy and social thermoregulation in striped skunks (*Mephitis mephitis*)," *Physiological and Biochemical Zoology*, 80, pp. 138–45.

Irving, L., Krogh, H. and Monson, M. (1955) "The metabolism of some Alaskan animals in winter and summer," *Physiological Zoology*, 28, 173–85.

Janzen, D.H. (1977) "Why fruits rot, seeds mold, and meat spoils," *American Naturalist*, 111, pp. 691–713.

Jensen, K. *et al.* (2014) "Nutrient-specific compensatory feeding in a mammalian carnivore, the mink, *Neovison vison*," *British Journal of Nutrition*, 112, pp. 1226–33.

Jiao, H. *et al.* (2019) "Trehalase gene as a molecular signature of dietary diversification in mammals," *Molecular Biology and Evolution*, 36, pp. 2171–83.

Jorge-Neto, P.N., *et al.* (2020) "Can jaguar (*Panthera onca*) ovulate without copulation?", *Theriogenology*, 147, pp. 57–61.

Karasov, W.H. and Martinez del Rio, C. (2007) *Physiological ecology: how animals process energy, nutrients, and toxins*. Princeton: Princeton University Press.

Kirkpatrick, R.L. and Pekins, P.J. (2002) "Nutritional value of acorns for wildlife," in McShea, W.J. and Healy, W.M. (eds.) *Oak forest ecosystems: ecology and management for wildlife*. Baltimore: The Johns Hopkins University Press, pp. 173–81.

Kohl, K.D., Coogan, S.C.P. and Raubenheimer, D. (2015) "Do wild carnivores forage for prey or for nutrients?", *Bioessays*, 37, pp. 701–9.

Krockenberger, M.B. and Bryden M.M. (1994) "Rate of passage of digesta through the alimentary tract of southern elephant seals (*Mirounga leonina*) (Carnivora: Phocidae)," *Journal of Zoology*, 234, pp. 229–37.

Lambert, J.E. *et al.* (2014) "Binturong (*Arctictis binturong*) and kinkajou (*Potos flavus*) digestive strategy: implications for interpreting frugivory in Carnivora and Primates," *PLoS ONE*, 9, p. e105415.

Lanner, R.M. and Gilbert, B.K. (1994) "Nutritive value of whitebark pine seeds, and the question of their variable dormancy," in Schmidt, W.C. and Holtmeier, F.-K. (eds.) *Proceedings—international workshop on subalpine stone pines and their environment: the status of our knowledge*. USDA Forest Service, General Technical Report, INT–309. Washington, DC:, pp. 206–11.

Lariviere, S. and Ferguson, S.H. (2002) "On the evolution of the mammalian baculum: vaginal friction, prolonged intromission or induced ovulation?", *Mammal Review*, 32, pp. 283–94.

Lariviere, S. and Ferguson, S.H. (2003) "Evolution of induced ovulation in North American carnivores," *Journal of Mammalogy*, 84, pp. 937–47.

Ley, R.E., *et al.* (2008) "Evolution of mammals and their gut microbes," *Science*, 320, pp. 1647–51.

Liao, J. A. (1990) *An investigation of the effect of water temperature on the metabolic rate of the California sea lion* Zalophus californianus. Master's thesis. University of California, Santa Cruz.

Lindenfors, P., Dalen, L. and Angerbjörn, A. (2003) "The monophyletic origin of delayed implantation in carnivores and its implications," *Evolution*, 57, pp. 1952–6.

Lindstedt, S.L. and M.S. Boyce. (1985) "Seasonality, fasting endurance, and body size in mammals," *American Naturalist*, 125, pp. 873–8.

Lohuis, T.D. *et al.* (2007) "Hibernating bears conserve muscle strength and maintain fatigue resistance," *Physiological and Biochemical Zoology*, 80, pp. 257–69.

Lopes, F.L., Desmarais, J.A. and Murphy, B.D. (2004) "Embryonic diapause and its regulation," *Reproduction*, 128, pp. 669–78.

Macdonald, D.W. (1976) "Food caching by red foxes and some other carnivores," *Zeitschrift für Tierpsychologie*, 42, pp. 170–85.

MacNulty, D.R. *et al.* (2009) "Body size and predatory performance in wolves: is bigger better?", *Journal of Animal Ecology*, 78, pp. 532–9.

Malcicka, M., Visser. B. and Ellers. J. (2018) "An evolutionary perspective on linoleic acid synthesis in animals," *Evolutionary Biology*, 45, pp. 15–26.

Martin, M.E., Moriarty, K.M. and Pauli, J.N. (2019) "Forest structure and snow depth alter the movement patterns and subsequent expenditures of a forest carnivore, the Pacific marten," *Oikos*, 129, pp. 356–66.

Mayntz, D., *et al.* (2009) "Balancing of protein and lipid intake by a mammalian carnivore, the mink, *Mustela vison*," *Animal Behaviour*, 77, pp. 349–55.

McNab, B.K. (1992) "A statistical analysis of mammalian rates of metabolism," *Functional Ecology*, 6, pp. 672–9.

McNab, B.K. (2000) "The standard energetics of mammalian carnivores: Felidae and Hyaenidae," *Canadian Journal of Zoology*, 78, pp. 2227–39.

McNab, B.K. (2002) *The physiological ecology of vertebrates: a view from energetics*. Ithaca: Cornell University Press.

McNab, B.K. (2008) "An analysis of the factors that influence the level and scaling of mammalian BMR," *Comparative Biochemistry and Physiology A*, 151, pp. 5–28.

Mendoza, M.L.Z., *et al.* (2018) "Protective role of the vulture facial skin and gut microbiomes aid adaptation to scavenging," *Acta Veterinaria Scandinavica*, 60, p. 61.

Morris, J.G. and Rogers, Q.R. (1978) "Arginine: an essential amino acid for the cat," *Journal of Nutrition*, 108, pp. 1944–53.

Murie, O.J. (1959) "Fauna of the Aleutian Islands and Alaska Peninsula," North American Fauna, 61, pp. 1–364.

Nagy, K.A., Girard, I.A. and Brown, T.K. (1999) "Energetics of free-ranging mammals, reptiles, and birds," *Annual Review of Nutrition*, 19, pp. 247–77.

Nakamura, M.T. and Nara, T.Y. (2003) "Essential fatty acid synthesis and its regulation in mammals," *Prostaglandins, Leukotrienes and Essential Fatty Acids*, 68, pp. 145–50.

Neco, L.C. *et al.* (2019) "The evolution of self-medication behaviour in mammals," *Biological Journal of the Linnean Society*, 128, pp. 373–8.

Nie, Y., *et al.* (2015) "Exceptionally low daily energy expenditure in the bamboo-eating giant panda," *Science*, 349, pp. 171–4.

Nijboer, J. and Dierenfeld, E.S. (2011) "Red panda nutrition: how to feed a vegetarian carnivore," in Glatston, A.R. (ed.) *Red panda: biology and conservation of the first panda*. London: Academic Press, pp. 257–70.

Ohishi, E. *et al.* (1979) "Antibodies to *Clostridium botulinum* toxins in free-living birds and mammals," *Journal of Wildlife Diseases*, 15, pp. 3–9.

Ortiz, R.M. (2001) "Osmoregulation in marine mammals," *Journal of Experimental Biology*, 204, 1831–44.

Pajic, P., *et al.* (2019) "Independent amylase gene copy number bursts correlate with dietary preferences in mammals," *eLife*, 8, p. e44628.

Pasitschniak-Arts, M. (1993) "*Ursus arctos*," *Mammalian Species*, 439, pp. 1–10.

Prestrud, P. (1991) "Adaptations by the Arctic fox (*Alopex lagopus*) to the polar winter," *Arctic*, 44, pp. 132–8.

Pritchard, G.T. and Robbins, C.T. (1990) "Digestive and metabolic efficiencies of grizzly and black bears," *Canadian Journal of Zoology*, 68, pp. 1645–51.

Qi, Z.-Q. *et al.* (1994) "Characterization of the antihemorrhagic factors of mongoose (*Herpestes edwardsii*)," *Toxicon*, 32, pp. 1459–69.

Reich, C.M. and Arnould, J.P.Y. (2007) "Evolution of Pinnipedia lactation strategies: a potential role for α-lactalbumin?", *Biology Letters*, 3, pp. 546–9.

Robbins, C.T. *et al.* (2012a) "Maternal condition determines birth date and growth of newborn bear cubs," *Journal of Mammalogy*, 93, pp. 540–6.

Robbins, C.T. *et al.* (2012b) "Hibernation and seasonal fasting in bears: the energetic costs and consequences for polar bears," *Journal of Mammalogy*, 93, pp. 1493–503.

Rosen, D.A.S. and Trites, A.W. (2000) "Digestive efficiency and dry-matter digestibility in steller sea lions fed herring, pollock, squid, and salmon," *Canadian Journal of Zoology*, 78, pp. 234–9.

Sandell, M. (1984) "To have or not to have delayed implantation: the example of the weasel and the stoat," *Oikos*, 42, pp. 123–6.

Schaller, G.B. *et al.* (1985) *The giant pandas of Wolong*. Chicago: University of Chicago Press.

Scholander, P.F. *et al.* (1950) "Body insulation of some arctic and tropical mammals and birds," *Biological Bulletin*, 99, pp. 225–36.

Schulz, T.M. and Bowen, W.D. (2004) "Pinniped lactation strategies: evaluation of data on maternal and offspring life history traits," *Marine Mammal Science*, 20, pp. 86–114.

Silva, M., Brimacombe, M. and Downing, J.A. (2001) "Effects of body mass, climate, geography, and census area on population density of terrestrial mammals," *Global Ecology and Biogeography*, 10, pp. 469–85.

Smith, G.B, Tucker, J.M. and Pauli, J.N. (2022) "Habitat and drought influence the diet of an unexpected mycophagist: fishers in the Sierra Nevada, California," *Journal of Mammalogy*, 103, pp. 328–38.

Soley, F.C. and Alvarado-Díaz, I. (2011) "Prospective thinking in a mustelid? *Eira barbara* (Carnivora) cache unripe fruits to consume them once ripened," *Naturwissenschaften*, 98, pp. 693–8.

Soulsbury, C.D. (2010) "Ovulation mode modifies paternity monopolization in mammals," *Biology Letters*, 6, pp. 39–41.

Stenvinkel, P., Jani, A.H. and Johnson, R.J. (2013) "Hibernating bears (Ursidae): metabolic magicians of definite interest for the nephrologist," *Kidney International*, 83, pp. 207–12.

Swenson, J.E. *et al.* (2007) "Brown bear body mass and growth in northern and southern Europe," *Oecologia*, 153, 37–47.

Tabata, E., *et al.* (2022) "Noninsect-based diet leads to structural and functional changes of acidic chitinase in Carnivora," *Molecular Biology and Evolution*, 39, p. msab331.

Tamburello, N., Côté, I.M. and Dulvy, N.K. (2015) "Energy and the scaling of animal space use," *American Naturalist*, 186, pp. 196–211.

Taylor, S.L. and Buskirk, S.W. (1994) "Forest microenvironments and resting energetics of the American marten *Martes americana*," *Ecography*, 17, pp. 249–56.

Thiemann, G.W., Iverson, S.J. and Stirling, I. (2006) "Seasonal, sexual and anatomical variability in the adipose tissue of Polar bears (*Ursus maritimus*)," *Journal of Zoology*, 269, pp. 65–76.

Thom, M.D., Johnson, D.D.P. and Macdonald, D.W. (2004) "The evolution and maintenance of delayed implantation in the Mustelidae (Mammalia: Carnivora)," *Evolution*, 58, pp. 75–83.

Tucker, M.A. and Rogers, T.L. (2014) "Examining predator–prey body size, trophic level and body mass across marine and terrestrial mammals," *Proceedings of the Royal Society B: Biological Sciences*, 281, p. 20142103.

Voss, R.S. and Jansa, S.A. (2012) "Snake-venom resistance as a mammalian trophic adaptation: lessons from didelphid marsupials," *Biological Reviews*, 87, pp. 822–37.

Wagner, F., *et al.* (2022) "Reconstruction of evolutionary changes in fat and toxin consumption reveals associations with gene losses in mammals: a case study for the lipase inhibitor *PNLIPRP1* and the xenobiotic receptor *NR1I3*," *Journal of Evolutionary Biology*, 35, pp. 225–39.

Wei, F. *et al.* (2000) "Seasonal energy utilization in bamboo by the red panda (*Ailurus fulgens*)," *Zoo Biology*, 19, pp. 27–33.

Williams, T.M. *et al.* (2001) "A killer appetite: metabolic consequences of carnivory in marine mammals," *Comparative Biochemistry and Physiology A*, 129, pp. 785–96.

Wright, T.R., *et al.* (2020) "Changes in northern elephant seal skeletal muscle following thirty days of fasting and reduced activity," *Frontiers in Physiology*, 11, p. 564555.

Wright, T.R. *et al.* (2021) "Skeletal muscle thermogenesis enables aquatic life in the smallest marine mammal," *Science*, 373, pp. 223–5.

Zielinski, W.J. *et al.* (1999) "Diet of fishers (*Martes pennanti*) at the southernmost extent of their range," *Journal of Mammalogy*, 80, pp. 961–71.

CHAPTER 5

Sensory biology and neuroanatomy

How do the sensory and central nervous systems of carnivorans resemble or differ from those of other mammalian orders? Does a predatory trophic niche select for greater brain size? Do predators have senses tuned to their predatory styles, and does predation require stronger cognition than avoiding predation? Some neural and sensory attributes of carnivorans are distinctive, whereas others are highly conserved across mammals.

5.1 The senses

5.1.1 Tactile

Various regions of the integument provide tactile inputs that permit situational awareness, coordination of movement, and locating prey. These include the pads of the feet that contribute to lateral balance control as terrestrial species walk (Park *et al.*, 2019) and the use of vibrissae to navigate in the dark, as shown by Schmidberger (1932) for domestic cats. Sea otters respond rapidly to the texture of objects sensed by pads of their forefeet, both above and below water (Strobel *et al.*, 2018). These tactile functions occur in most terrestrial small mammals but seem to be exaggerated in some carnivorans. The tactile sense of the forelimb digits is especially well developed in the northern raccoon, which uses its lightly furred forefeet to feel for prey in turbid water and low light (Oddie *et al.*, 2015). Perhaps the most remarkable display of the tactile sense is by pinnipeds, which use vibrissae to locate underwater prey without inputs from vision, olfaction, or physical contact. Both otariids and phocids use hydrodynamic information—water currents—to locate swimming fish via the turbulence wakes behind their prey (Hanke *et al.*, 2013). Also remarkable is

the ability of harbor seals to locate stationary flatfish buried under sand by swimming above them and sensing the water currents ejected by opercular pumping of water across the gills (Niesterok *et al.*, 2017). This system of sensing small variations in water pressure is analogous to the lateral line system of fish.

5.1.2 Chemosense

The chemical sense in mammals can be divided into three functional and morphological components. Non-volatile molecules (e.g. salts, sugars, and amino acids) contact taste receptors on the tongue. Volatile molecules pass over the nasal epithelia and contact nasal olfactory receptors (ORs). Mixtures of volatile and non-volatile molecules in gaseous solution or carried on airborne droplets provide information about conspecifics via the vomeronasal organ (VNO) system.

5.1.2.1 Taste

Mammals sense five basic non-volatile chemical qualities via taste buds: sweetness, umami, bitterness, saltiness, and sourness. Omnivorous land carnivorans retain most of these modalities, but the more carnivorous lineages have lost some or most taste senses. Carnivorans are less sensitive to saltiness than are herbivorous mammals, which is consistent with the ample availability of sodium and potassium in animal flesh relative to plant leaves and stems (Bradshaw, 2006). In addition, the domestic cat, Asian small-clawed otter, spotted hyena, fossa, banded linsang, harbor seal, and two species of otariids—all obligate flesh-eaters—have lost their senses of sweetness (mono- and disaccharides) via mutations in the *Tas1r2* gene. The protein product

Carnivoran Ecology. Steven W. Buskirk, Oxford University Press. © Steven W. Buskirk (2023). DOI: 10.1093/oso/9780192863249.003.0005

of this gene forms a dimer with another protein to detect simple sugars. By contrast, the aardwolf, North American river otter, giant panda, spectacled bear, northern raccoon, and red wolf have functional *Tas1r2* genes; most of these species include some plant material in their diets. The coding errors that cause dysfunction of the sweetness genes differ across lineages, showing that the sweetness sense has been lost multiple times over carnivoran evolution (Li *et al.*, 2009; Jiang *et al.*, 2012).

The umami taste gene encodes a protein, TAS1R1, which likewise forms a protein dimer enabling detection of glutamic acid and other amino acids. Protein-rich foods are best sensed via umami, so strict carnivores and omnivores have strong umami senses: the polar bear, domestic dog, maned wolf, arctic fox, and domestic cat all have intact *TAS1R1* genes. By contrast, the giant panda and red panda independently lost the umami taste sense, via one nucleotide deletion error in the case of the giant panda, and by three insertion-deletion errors in the red panda (Hu *et al.*, 2017). The error in the giant panda genome occurred about 4 Ma, early in its bamboo specialization (Zhao *et al.*, 2010). Pinnipeds are also unable to sense umami, as are all cetaceans, which is puzzling inasmuch as pinnipeds are strict flesh-eaters. It has been suggested that pinnipeds lost this—and perhaps all taste senses—as a result of their typical swallowing of prey whole, without mastication, and the absence of protein-poor foods in the open ocean. They have no need to select for protein content. Alternatively, the high salt concentrations of marine environments may overwhelm or interfere with umami or other taste modalities (Sato and Wolsan, 2012; Jiang *et al.*, 2012; Zhu *et al.*, 2014).

Because the bitter taste sense is hypothesized to be important to detect plant defensive compounds and other toxins, the highly predaceous felids were once suspected to lack bitter taste receptors altogether. Recent work has shown, however, that cats have intact bitter taste genes with sensitivities to a range of molecules, much as do humans. This has forced a re-examination of the dietary sources of bitter tastes that might have selective importance to carnivorans (Lei *et al.*, 2015). Clearly, several carnivoran taxa have lost some taste modalities, reflecting the relaxation of selection for some taste senses in some dietary specialist lineages.

5.1.2.2 Olfaction

The olfactory sense in Carnivora is more complex than that for taste at morphological, histological, transductional, and genetic levels. The OR and VNO, with their associated structures, provide partially overlapping information about the environment. Chemicals used in animal communication—semiochemicals—are ubiquitous and well developed across Class Mammalia, except for Cetacea. In Carnivora, the locations, structures, and chemical compositions of the products of scent glands are diverse, and the ecological functions of these chemicals are crucial. However, our understandings of the specific functions of each chemical and how it is deposited are very incomplete. We also lack key knowledge about the receptor organs, which are generally assumed to be the nasal OR or VNO.

Olfactory transmitters can be classified by their location on the body, by the kinds of chemicals they produce, their intended target species (inter- or intraspecific), or by how the context of deposition affects the signal. The clear ecological advantage of chemical signals is that they persist after the signaler leaves the area—an asset when the signal relates to territorial defense. One disadvantage of chemical signaling is that species other than the intended target gain useful information by eavesdropping, posing risks to the signaler. Ferrero and colleagues (2011) showed that a blend of carnivoran urines contained one specific amine compound that triggered genetically coded aversion circuits in rodent brains. Mammalian scent glands tend to be modified skin glands—typically either derived sweat glands that secrete water-soluble chemicals, or sebaceous glands that produce lipid-based ones. Mixed aqueous-oily suspensions also occur, and bacterial action can modify semiochemicals. Some carnivoran scents are easily detected by humans, and may be either pleasant or repulsive, whereas others are imperceptible to the human nose. The locations of carnivoran scent glands include the sweat glands on the feet of domestic dogs and wolves, which were recognized for their scent function early in the nineteenth century (Gurlt, 1835). The secretions from these glands are mostly glycoconjugates (Meyer and Tsukise, 1995) and tend to be perceived as unoffensive or pleasant to humans.

Martens and wolverines have skin glands along the abdominal midline, the unidentified products of which are deposited by rubbing the belly across objects while walking (Hall, 1926). North American martens, sables, fishers, North American otters, and wolverines have glands on the plantar pads of the hind feet. In cold climates in species with these glands, the glands take the form of inflorescences in winter, with tentacle-like projections extending from a central skin pore. On the unfurred feet of the North American river otter, the glands are fused columns of papillae (Buskirk et al., 1986). The products of these glands have not been described.

Urine is an important scent matrix for many carnivoran species. The signaling molecules are highly variable across species, as are the functions and modes of deposition. Some scents are attractive to heterospecifics, whereas others repel them. For several species, the signaler raises a leg or the hind quarters to elevate the urine deposit, and the combination of chemical and context is commonly thought to communicate species, sex, social rank, or other information (Vogt et al., 2016). Feces also carry information and is deposited in ways that suggest their importance for signaling, but it is unclear how much of the signal in feces originates from the anal sacs. These glands, unique to the Carnivora, are perhaps the best known of carnivoran scent transmitters and show great functional variation across taxa. Of course, among the Mephitidae the glands are large and can spray their products some distance in defense, whereas other species deposit the products of the glands on feces via ducts that empty at the anal sphincter. The repellent odor of mephitid—and some mustelid—anal glands is the result of sulfur-containing thiatanes and dithiolanes (Crump, 1980). However, the anal glands of boreal forest martens produce few or no sulfur-containing compounds, instead producing benzaldehyde. While bilateral anal sacs are unique to the carnivora, the odorant-binding proteins responsible for their function are common to a wide range of mammals, a homology extending to before the divergence of placental from other mammals (Janssenswillen et al., 2021). European otters produce mostly proteins, mucopolysaccharides, and lipids in their anal glands (Brinck et al., 1983; Burger et al., 2005). Anal gland secretions are accompanied by several deposition behaviors,

including "hand-stands" that have been reported in martens, mongooses, spotted skunks, and giant pandas (e.g. Mykytowycz, 1972; Sharpe, 2015). Apparently unique among the Carnivora, the European badger possesses a subcaudal gland. Located on the midline between the base of the tail and anus, the gland opening is a horizontal slit 30–50 mm wide, covering a pouch holding up to several grams of odoriferous sebaceous material (Kruuk et al., 1984). The badger scent-marks its territory via rump-dragging behaviors, the secretions being spread across other clan members. African lion manes have been proposed to serve a chemical signaling role; distinct head-rubbing behaviors are common, and volatile fatty acids have been isolated from the bases of mane hairs (Poddar-Sarkar et al., 2008).

Nasal ORs are imbedded in the olfactory epithelium, mounted on thin bony sheets (conchae) that extend medially from the ethmoid bones. Anterior to this epithelial area, the maxillae also feature turbinate portions covered with epithelium, which moisturize inhaled air before it passes over the olfactory epithelium. ORs send nerve impulses to the main olfactory bulbs (MOBs) of the brain. Nasal olfactory acuity is generally considered to reflect the surface area of the olfactory portion of the conchae, the density of ORs on the epithelium, and the number of functional OR genes. Surface area of the nasal epithelium is determined by the amount of folding of the conchae, combined with the length of the rostrum. Felids have shorter rostra than long-snouted lineages (e.g. canids, ursids and hyaenids), and have been long suspected of having weaker OR acuity. Van Valkenburgh and colleagues (2014) examined possible compensating mechanisms that might equalize olfactory surface area across major carnivoran lineages but found no evidence for them. Indeed, felids seem to be more visual, and canids and ursids more olfactory in their sensory acuities. The number of functioning OR genes is important because they encode the molecules that are the bases of transduction. Domestic dogs have around 800 functional OR genes, compared with about 700 for felids, 300–400 for various primates, 400 for humans, and 800–1200 for herbivorous rodents and ungulates (Malnic et al., 2004; Niimura et al., 2014 for dogs and herbivores, Montague et al.,

2014 for felids). Each OR protein can detect more than one odorant molecule, with some odorants detected by multiple ORs working in combination, and processing occurring in the MOB and olfactory cortex.

The consequence of these differences in odorant detection is that terrestrial carnivorans—at least the long-snouted lineages—have OR olfactory acuity comparable or superior to most of their herbivorous prey, and vastly superior to humans. Canidae and Hyaenidae have particularly large olfactory bulbs, and domestic dogs, because they are tractable research subjects, show most clearly the extraordinary carnivoran olfactory sense. Some dogs can detect organic odorant molecules in air samples at concentrations 100–10,000,000 times lower than can humans (Marshall *et al.*, 1981; Krestel *et al.*, 1984). Some breeds are able to consistently intercept a human trail and correctly infer the direction of travel after sampling only a few human footprints, without visual cues (Thesen *et al.*, 1993; Wells and Hepper, 2003). Trained bloodhound dogs, in multiple cases, have tracked the scent of criminal suspects that had traveled in vehicles for multiple kilometers. The dogs had done this multiple days after the scent was deposited, and in busy pedestrian and vehicular traffic (Stockham *et al.*, 2004). However, other carnivorans have modest olfactory abilities. Semi-aquatic otters have especially small MOBs, consistent with scarce chemical cues in aquatic environments (Gittleman, 1991). Pinnipeds show no evidence of using olfaction in aquatic foraging, but some otariids use olfaction on land to detect threats, and otariid mothers can identify their pups via scent, although the VNO may account for that behavior (Gittleman, 1991; Pitcher *et al.*, 2010; Marriott *et al.*, 2013). Loss of the olfactory sense in aquatic carnivorans is apparent at the molecular level as well. Both the sea otter and giant otter show loss of function of OR genes over the 8–9 million years since their divergence from the terrestrial mustelid lineage (Beichman *et al.*, 2019).

The VNO system is responsible for an ancient chemosense, recognizable in modern amphibians, reptiles, and mammals of several orders. The VNO itself lies bilateral to the nasal septum, between the nasal and oral cavities, and is isolated from the nasal cavity by hyaline cartilage but connected to it by narrow ducts. In most carnivorans, but not rodents, it connects to the oral cavity via paired nasopalatine ducts, which pass through the incisive foramina between the premaxillae and maxillae (Figure 5.1). Non-volatile and airborne molecules enter the VNO via nasal or oral routes, and contact the epithelium, aided by flehmen behavior. This distinctive behavior involves raising the upper lip, elevating the snout, and inhaling shallowly and repeatedly, to bring odorants into contact with the VNO receptors (Verberne, 1976). Nerve impulses from the VNO travel to an accessory olfactory bulb within the MOB. The molecules detected by the VNO include peptides that bind larger molecules involved in the major histocompatibility complex—the important system responsible for recognizing family members and individual animals, in addition to antigens (Leinders-Zufall *et al.*, 2004). This system also detects molecules associated with reproductive condition, particularly estrus.

The VNO system functions highly variably across carnivoran families (Kelliher *et al.*, 2001). Indeed, the two known VNO receptor gene families, *V1R* and *V2R*, are among the most variable gene families across mammals. Chimpanzees have no functional *V1R* genes, whereas the platypus has 270 and the dog has eight (Keller and Vosshall, 2008). The VNO and associated accessory olfactory bulb are present in domestic ferrets but have not shown expected molecular responses to stimulation with conspecific odors (Woodley *et al.*, 2004). Both sexes of the domestic ferret respond to conspecific anal gland odors, but neural processing occurs in the MOB, rather than the accessory olfactory bulb (Woodley and Baum, 2004). This suggests processing of reproductive odors via the OR, rather than the VNO, which is consistent with the absence of flehmen behaviors in the ferret. By contrast, sea otters exhibit flehmen behaviors (Island *et al.*, 2017), as do maned wolves (Coelho *et al.*, 2012). Most felid species have intact VNO structures and genome regions and exhibit well-developed flehmen behaviors (Hecker *et al.*, 2019). The Felidae seem to have traded off nasal OR acuity against VNO acuity; they hunt by sight, with reduced reliance on the OR, but communicate with conspecifics with an acute VNO system (Montague *et al.*, 2014).

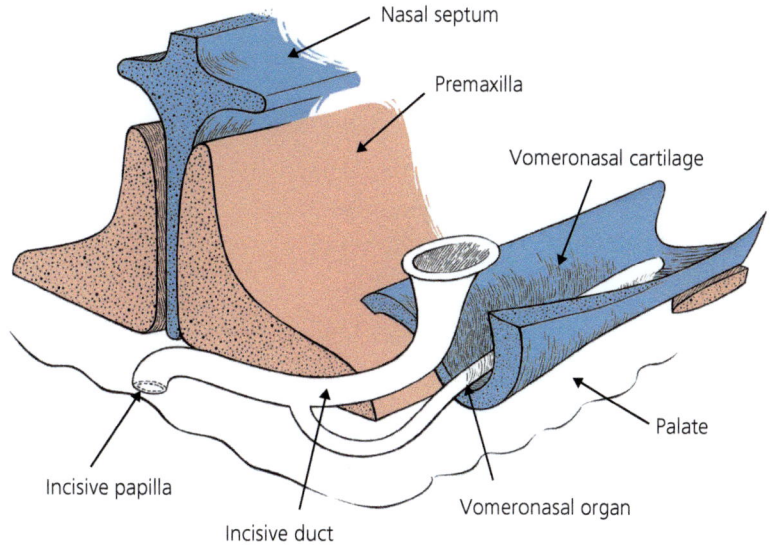

Figure 5.1 The vomeronasal organ of the domestic dog, with associated cartilaginous (blue) and bony (red) structures. The organ lies at the end of a branch of the incisive duct, which connects the oral and nasal cavities in carnivorans. The remote position and narrow ducting of the organ explains why flehmen behaviors are needed to bring odorant molecules into contact with the VNO receptors.
Modified from Evans and de Lahunta (2013).

5.1.3 Hearing

Mammals have a wider range of hearing frequencies than do other vertebrates, and carnivorans have some of the widest ranges of any mammal (Vater and Kössl, 2011). This reflects the modification of the cochlea and associated structures to add high-frequency sounds to the early mammalian hearing range. Across species, body size is associated with frequency range; large-bodied species are adapted to detect low-frequency sound, and vice versa. In this way the morphological diversity of carnivorans helps explain their collective hearing frequency range. The smallest-bodied carnivoran, the least weasel, is able to detect ultrasound in the range of 160 kHz, whereas the northern elephant seal, when under water, can detect frequencies down to 100 Hz, representing a 1600-fold range. Some terrestrial species are auditory generalists; the domestic cat has one of the broadest hearing ranges of any mammal studied: 48 Hz–85 kHz (Heffner and Heffner, 1985). Pinnipeds face special problems in vocal–auditory communication because of poor sound transmission across the air–water interface. Therefore, they require dissimilar hearing systems for the two media. Otariids are best adapted to hearing in air, shallow-diving phocids (e.g. harbor seal) hear better in water than terrestrial carnivorans (Möhl, 1968), and elephant seals (deep divers) hear best

underwater, at the expense of hearing in air (Kastak and Schusterman, 1998, 1999). The sea otter is even less well adapted to underwater hearing than sea lions (Ghoul and Reichmuth, 2014). Scientists have long hypothesized that pinnipeds should be able to echolocate in water, but repeated efforts have failed to demonstrate such an ability; the mechanics of hearing in air apparently constrain the morphological transformation that biosonar would require (Schusterman et al., 2000). The auditory acuity of fissipeds has been measured experimentally for few (if any) species. However, it is clear that some species have superb hearing, as evidenced by red foxes localizing small rodent sounds under deep snow, which is an excellent acoustic insulator (Figure 5.2).

5.1.4 Electromagnetic radiation

The electromagnetic spectrum spans frequencies from gamma rays with wavelengths less than 10^{-11} m through the range visible to humans (380–740 nanometers [nm]), to radio waves with wavelengths of thousands of kilometers. Various organs sense radiation and provide input to the brain, and several qualities are used to characterize the visual sense: resolution (acuity), sensitivity under low light, spectral quality (colors sensed), and stereopsis. By these criteria, the visual sense of most carnivorans is

Figure 5.2 A red fox captures small mammals under deep snow in Yellowstone National Park by sound localization. This pouncing behavior is used in snow as deep as one meter, although success rates in relation to snow depth have not been reported.
Photos: U.S. National Park Service by N. Herbert.

comparable to that of other diurnal or crepuscular non-primates, but better than most potential prey species. The resolution of an eye is a function of the retinal image size, which reflects the size of the retina and the density of ganglion cells on the retina. Large-bodied mammals tend toward greater visual acuity than small ones. When phylogeny and mass-specific retinal size are controlled, carnivorans display greater acuity than do artiodactyls, rodents, or bats (Veilleux and Kirk, 2014).

Most carnivorans have dichromatic color vision, with dogs having peak sensitivities in the 429-nm (blue-red) and 555-nm (green) wavelength regions. Domestic cats have short-wave peak sensitivities at about 450 nm (blue), and other carnivorans exhibit either monochromatic vision, or use information from rods and cones to produce a dichromatic effect (Jacobs, 1993). This is possible because nocturnal animals have higher ratios of rods to cones than do diurnal ones, and the spectral sensitivities of the receptor types can be compared. Marine carnivorans face particular visual challenges because of the differing refractive indices of water vs. air, and the low ambient light found at the great depths where some phocids forage. The spectral attributes of rod pigments of pinnipeds—except the elephant seal—are similar to those of the sea otter and land carnivorans, with peak sensitivities at 500 nm (blue-green). The elephant seal exhibits a spectral shift in its rod vision, with spectral peak sensitivity at around 487 nm (blue), consistent with the light environment at the great depths where they feed. Pinniped

middle-long wavelength cone pigments are similar to those of dogs and cats, having peak sensitivities at 545–560 nm (green). All pinnipeds seem to lack short-wavelength cones, suggesting either monochromatic vision or that they discern color via the rod-cone combination (Peichl *et al.*, 2001; Levenson *et al.*, 2006). Vision in the infrared region (7–1,000,000 nm) among carnivorans has not been widely explored but has been demonstrated for the domestic ferret (Newbold and King, 2009), which can detect wavelengths in the region of 870–920 nm. This has implications for the practical use of infrared wavelengths to detect or view wild carnivorans—those wavelengths are not necessarily invisible to them (Box 5.1).

Pinnipeds are also remarkable among carnivorans for their large retinas, which enhance sensitivity; among phocids, deep divers tend to have the largest eyeballs, but such a trend is not observed in otariids. This is consistent with the greater importance of vision to phocids that pursue fish in clear pelagic water as opposed to otariids searching for bottom-dwelling fish and invertebrates in turbid water (Debey and Pyenson, 2013). The importance to pinnipeds of visual acuity in air is suggested by the ability of harbor seals to navigate by identifying particular simulated stars based on their brightness in a swimming planetarium (Mauck *et al.*, 2008). By fixing on particular stars, they should be able to navigate across long oceanic distances by dead reckoning, in much the same way that Polynesian seafarers have done. Stereopsis, the perception of

Box 5.1 Weak thermal radiation

Another modality in which carnivorans surpass other mammals is detection of weak thermal radiation. Glaser and Kroger (2017) showed that zoo carnivorans had rhinaria— the naked, exterior parts of the nose— cooler than those of ungulates, rodents, or lagomorphs. This would make carnivoran rhinaria particularly sensitive to weak infrared

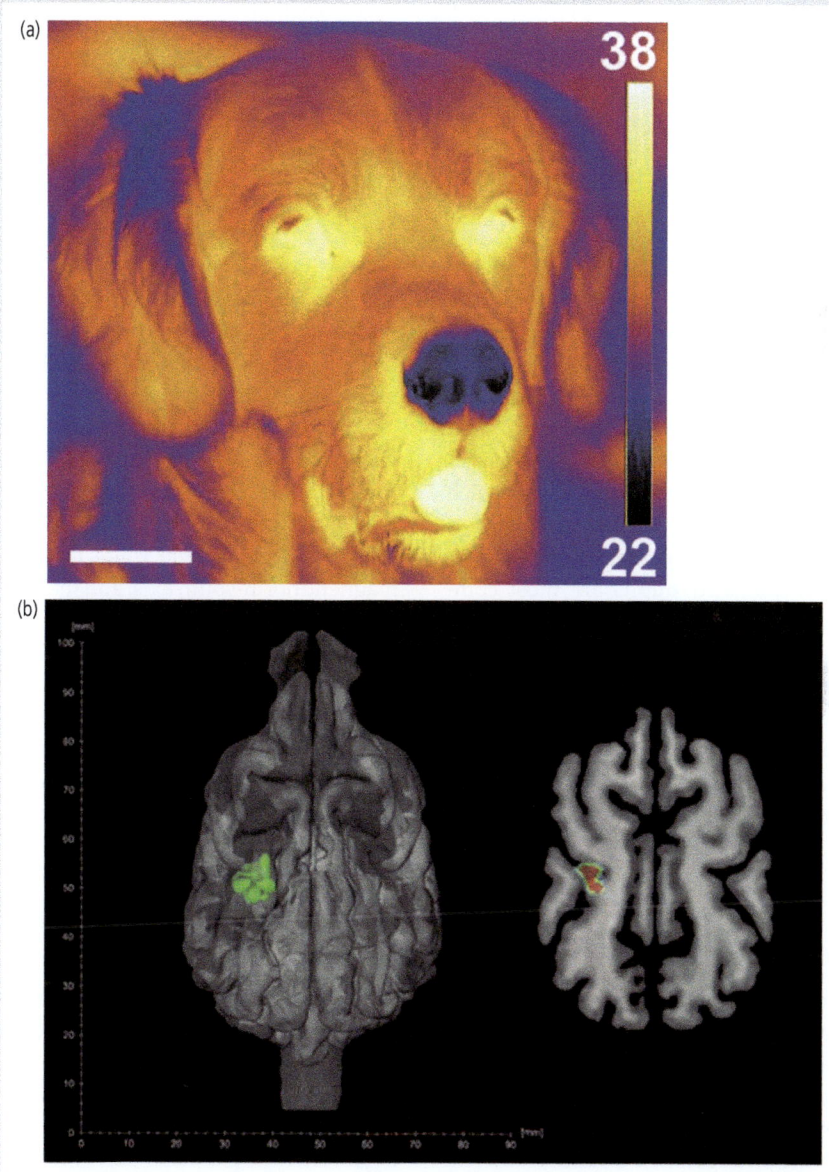

Figure 5.4 The domestic dog has a sense for weak thermal radiation, which allows it to localize heat sources when within a few meters. A: Thermograph of a dog's face in the shade at 27°C, showing high surface temperatures of the eyes and tongue, but low temperatures on the tip of the rhinarium. Temperature scale is in °C; scale bar is 50 mm. B: MRI-detected activity of the left somatosensory association complex of a dog in response to the presence of an object with a surface temperature 11°C warmer than ambient.

Photos: Bálint *et al*. (2020, Figures. 1, 2) CC.

Box 5.1 *Continued*

---

radiation. Bálint *et al.* (2020) then showed that domestic dogs can detect temperature differences between objects at ambient vs. endotherm surface temperatures from as far away as 1.6 meters. The retina cannot hold the transducer for weak thermal radiation because photons in the near-infrared region have too little energy to activate retinal pigments, especially after traversing the aqueous structures of the eye. So, the sensory transducer appears to be in the tip of the nose—the rhinarium itself. In ungulates, this patch of skin is near the temperature of other facial skin, but in

dogs it is colder when ambient temperatures are above 15 °C (Figure 5.4A). Bálint and colleagues (2020) also found pronounced activation of the left parietal somatosensory association complex of the brain in response to presenting a warm object a 1–2 meters from a dog (Figure 5.4,B). Other vertebrates, including pit vipers and common vampire bats, have analogous but far more acute infrared senses, with better localization abilities than have dogs. Dogs likely use this sense to detect motionless, cryptic endotherm prey, or nests recently abandoned by birds, a short distance away.

---

depth via binocular vision, confers several advantages. It facilitates estimating distance and breaking camouflage patterns. It also obviates the need for the observer to move to generate motion parallax. With their short rostra and forward-facing eyes, cats have such overlap, and are generally assumed to enjoy strong stereopsis. This ability is enhanced by having pupils shaped like vertical slits, which creates astigmatic depth of field, so that objects nearer and farther than the focal object are also in focus. Vertically elongated pupils are most common in ambush predators. Also, carnivorans with eyes near the ground (e.g. foxes and small felids) are more affected by vegetation obstructing their fields of view and are correspondingly more likely to have vertically elongated pupils than taller mammals (Figure 5.3) (Banks *et al.*, 2015). The pupils become enlarged and more circular when animals are excited.

5.1.5 Geomagnetic

Geomagnetic orientation is important among various vertebrates (Cervený *et al.*, 2011) and evidence supports the existence of a magnetic sense in various mammalian species. The red fox tends toward a northeast orientation when using a pounce to capture prey beneath vegetation or snow. Such an orientation leads to higher success rates than for pounces made in other directions (Begall *et al.*, 2013). Similarly, free-ranging domestic dogs show a tendency to run initially along a geomagnetic

north-south axis when trying to find their owners. Experimental dogs were allowed to explore off-leash in a forest and their movements recorded using GPS collars and cameras. When they were more than 100 meters from the owner, the owner hid and the animal's path in homing to the owner was recorded. A dog typically found its outbound trail and followed it back to the owner, but in a third of trials it navigated back using a novel route. Those using the novel route initially ran along a magnetic north-south axis, and then set a course that took them to their owners faster than if they followed their own trail. Dogs that initially oriented themselves north-south were more likely to set a novel course, rather than to follow their own trail (Benediktová *et al.*, 2020). These findings strongly suggest the existence of a magnetic sense, although the mechanisms underlying such a sense are not known.

5.2 Brain morphology

Biologists have taken interest in the relationship between form and function in mammalian brains for over a century. Scores of studies have tested the enticing premise that some aspect of brain gross morphology is related to functional traits (e.g. olfactory acuity, sociality, or success at solving mechanical puzzles). Carnivorans have received close attention in these studies, perhaps because of the presumed high levels of processing required to subsist on dispersed packets of resources spread over large home ranges. This interest in the link

Figure 5.3 Vertically elongated pupils are more common in ambushing carnivorans than other species, and more common in short than tall animals. Tall animals, such the tiger (A) and wolf (B), experience less visual obstruction from vegetation than short animals, such as the domestic cat (C) and red fox (D). Having vertically elongated pupils allows short animals to focus more sharply on vertically elongated obstructions, for example, grass stems, which may be closer and more distant than the focal object. This allows short animals to estimate distances to focal objects more accurately, thereby enhancing ambushing ability.

Photos: A: subinpumsom; B: fouroaks; C: Jeanninebryan; D: cybernesco, all CanStock.

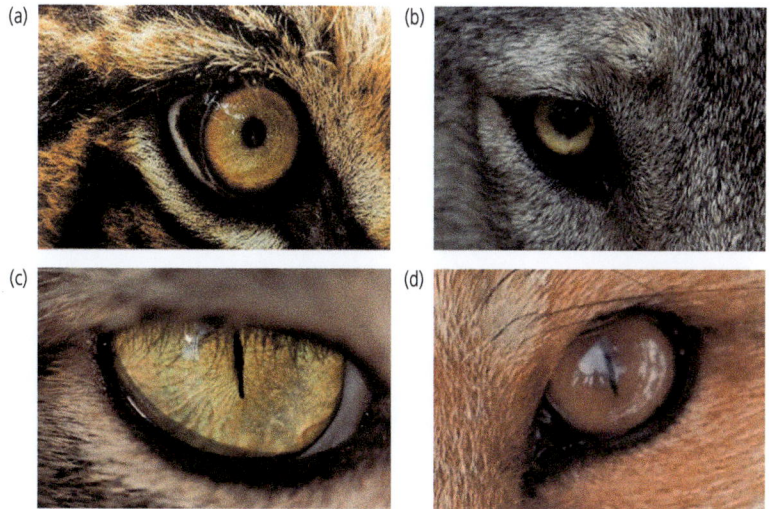

between neuroanatomy and function has only increased in recent years (Healy and Rowe, 2007). Brain morphology is characterized several ways: absolute brain size, brain size relative to body size, and size of specific brain regions relative to brain size. Reports show a number of significant correlations, although mechanistic explanations for the correlations often are lacking.

5.2.1 Brain size

Brain size (encephalization) has been especially attractive as a morphological trait to study because it can be measured on live animals via computed tomography, on cleaned skulls by filling and measuring the cranium with small pellets, and on fossils by virtual reconstruction. The size of the brain seems a promising correlate of ecological traits, firstly because brain size is so variable across mammalian species (Burger *et al.*, 2019), and secondly, because of the high energetic cost of maintaining brain tissue. Brain tissue requires around ten times the maintenance cost of non-brain mammalian tissues (Herculano-Houzel, 2011). This suggests trade-offs between opposing selective pressures associated with brain size; in other words, maintaining a large brain is costly. Several studies have shown that the brains of carnivorans, scaled to body size, are the third largest among mammalian orders—smaller than only those of primates and cetaceans

(e.g. Burger *et al.*, 2019). Within the Carnivora, we see the largest brains relative to body size in the pinnipeds, mirroring the large brains observed in cetaceans (Kruska, 2005, Figure 5.5). The viverrids, herpestids, and hyaenids tend to have brains smaller than expected for their body sizes, whereas canids and felids have larger brains. Musteloids have highly variable brain sizes, ranging from smaller than expected in mephitids to quite large in otters, the latter pattern demonstrating the convergent evolution of large brains in semi-aquatic and marine lineages. Arboreal and scansorial mustelids—tayras and boreal forest martens—also have large relative brain sizes (Kruska, 2005). Carnivoran brain size scales to the 0.75 power of body mass in interspecific comparisons, which is the same coefficient as for mammals generally; relative brain size decreases as body size increases (Burger *et al.*, 2019).

Other traits hypothesized to relate to absolute or relative brain size in carnivorans include sociality, extent of parental care, cognitive ability, and domestication. The social brain hypothesis stated that social species should have larger brains to cope with the amount of information they store and process about social rank, kin relations, and related factors (Dunbar, 1998). Across all mammalian lineages, many correlations and anecdotal accounts have variously supported or refuted such a relationship. However, within the Carnivora

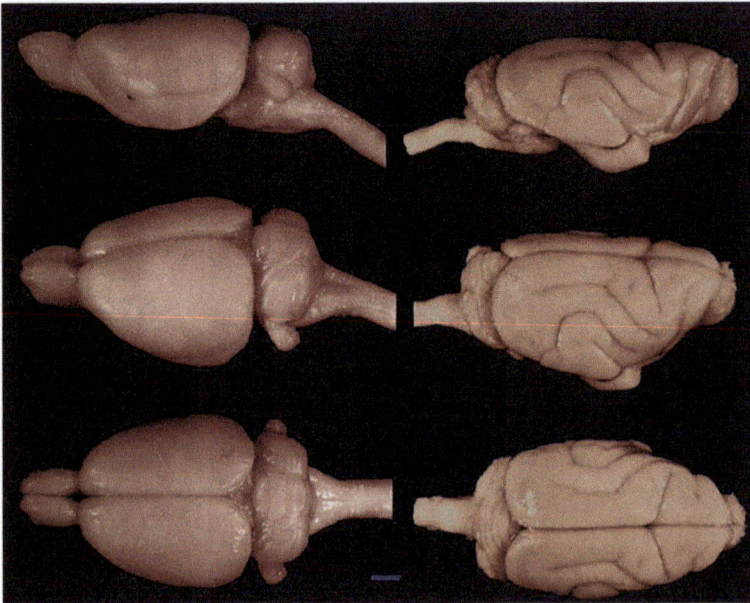

Figure 5.5 The brain of the brown rat (left) and ermine (right), to the approximately same scale. The neocortex of the ermine is proportionally larger, with a more fissured surface.
Photo: modified from University of Wisconsin Brain Collection online, accessed December 17, 2021.

the premise is not supported by data (Finarelli and Flynn, 2010; Benson-Amram *et al.*, 2016; Sakai *et al.*, 2016). Parental and alloparental care could relate to brain size because of the complex progression of behaviors required as neonates mature; this hypothesis is similar to the social brain hypothesis. Gittleman (1991) showed that females of carnivoran species that provide sole parental care have larger brains than do species in which care is bi-parental or communal. Carnivorans are second only to primates in providing allomaternal care to neonates—behaviors such as nursing, provisioning, and protecting shown by non-maternal animals. Within the Carnivora, these behaviors are especially strongly correlated with relatively large brains (Isler and van Schaik, 2012), again interpreted as reflecting the memory and processing required to provide neonatal care. Another hypothesized cognitive trait that correlates with brain size is ability to solve problems contrived by experimenters. Benson-Amram and colleagues (2016) presented carnivorans of thirty-nine species with puzzle boxes scaled to body size that required visual evaluation, dexterity, and repeated attempts to gain entry and reward. They found that species most likely to solve the puzzle were large-brained, large-bodied, and manually dexterous. Those species also spent long times trying to open the box. Only relative brain size was a significant predictor of success in general linear models.

Lastly, domestication has been proposed to affect relative brain size because of the profound changes that this process produces in procuring food, competing with other carnivorans, and most other ecological relationships. Several patterns are clear. First, domesticated forms commonly show greater variability of brain size, at any body size, compared with the ancestral wild form (much of this discussion is based on Kruska, 2005). For example, domestic dogs within and outside the body-size range of the ancestral wolf have more variable brain sizes than do wolves. Second, domesticated carnivorans have smaller brains than the corresponding wild form—20% smaller for the domestic American mink, 29% smaller for the domestic European ferret, 28% for the domestic cat, and 29% for the domestic dog. These effect sizes are higher than for non-carnivoran mammals; the loss of brain volume depends on relative brain size before domestication. Third, most of the reduction in brain size occurs in the telencephalon, which is mostly neocortex. The hippocampus, a part of the limbic system responsible for memory, particularly spatial memory, is reduced to varying degrees with carnivoran domestication, being conserved relative to other brain regions in the domestic American mink, but much reduced in the domestic dog.

5.2.2 Brain regional size

The large variation in mammalian (and carnivoran) proportional brain size is due not to uniform variation in the sizes of all brain regions; some brain regions are more variable in size than others. Specifically, variation in the size of the neocortex accounts for much of the interspecific variation in carnivoran brain size. Consider a Norway rat weighing 214 g, which has a brain weight of about 2 g, whereas a slightly smaller ermine weighing 185 g has a brain weighing 4.5 g. In the rat, the neocortex accounts for only 30% of total brain weight, whereas in the ermine it accounts for 45% (Figure 5.5). Further, the ermine brain features much more complex fissuring of the neocortical surface, and a more obvious layering of the gray matter (Kruska, 2005). Some brain regions have very specific functions. For example, the inferior colliculus, a region involved in auditory processing, is several times larger than expected in bats, whereas the superior colliculus, important in vision, is forty times smaller than expected in the Middle East blind mole rat (Lewitus, 2018). Using this approach, it is apparent that carnivorans have relatively large limbic as well as neocortical regions of the brain, in contrast to bats and primates, both of which show strong trade-offs between the sizes of these regions. The olfactory regions of the carnivoran brain are comparable in size to those of ungulates and some xenarthrans; however, those of pinnipeds are much reduced (Gittleman, 1991; Reep *et al.*, 2007; Kavoi and Jameela, 2011). The terrestrial carnivoran hippocampus is likewise comparably large to that of ungulates, smaller than that of xenarthrans, and much larger than that of pinnipeds.

Perhaps the most compelling evidence that relates neuroanatomy to specific behaviors in carnivorans takes advantage of the enormous neuroanatomical and behavioral variation among domestic dog breeds. Hecht and colleagues (2019) estimated brain regional sizes of thirty-three dog breeds from imaging studies. These sizes were compared to breed-specific behavioral tendencies, such as that for social interaction (as seen in companion breeds), olfaction and gustation (such as in hounds), and drive and reward (as found in retrievers and the border collie). They searched for these relationships in a phylogenetic context, recognizing that much variation among dog breeds has been applied through strong selection over brief generational time. Indeed, the size of brain regions correlates with functions they presumably control. Hounds have large brain regions for olfaction and gustation. Border collies and retrievers have large centers that control drive and reward, and Yorkshire terriers and miniature schnauzers have well-developed centers for social interaction.

Key points

- Sensory modalities are more diverse across the Carnivora than across most other mammalian orders, primarily because of the great ecological radiation that occurred in carnivorans.
- Major sensory derivations occur in the pinnipeds, in which hydrodynamic sensation is highly developed, the umami taste sense is lost, olfaction is greatly reduced, and vision in low light is enhanced in deep pelagic divers.
- Other sensory functions are modified according to trophic niche. Carnivorans are less sensitive to salt than herbivores, and some strict flesh-eaters have lost their sweetness sense. Both panda species have lost functioning genes for the umami taste sense.
- Carnivoran scent transmitters are diverse and poorly understood. They include urine, feces, abdominal skin secretions, the products of various foot glands, and others.
- The olfactory receptors comprise the nasal receptors on the epithelia of the nasal conchae, and the vomeronasal receptors in the vomeronasal organ.
- Nasal olfactory acuity is a function of the surface area of nasal epithelia, the density of receptors on the epithelia, and number of functional olfactory receptor genes. Long-snouted terrestrial carnivorans have superior olfactory acuity to short-snouted types.
- Vomeronasal olfactory acuity varies among carnivorans along phylogenetic lines, but with

unclear adaptive significance. Vomeronasal sensing is used mostly in intraspecific communication.

- Domestic dogs, the only carnivoran so far tested, can detect weak thermal radiation via the cold skin on the rhinarium. This newly recognized sense likely helps to localize stationary prey at short distances.
- Among carnivorans, aquatic and marine species have the largest brains for their body sizes, felids and canids intermediate-sized brains, and viverrids, herpestids, and hyaenids smaller relative brain sizes. Tree-climbing species tend to have large relative brain sizes.
- There is no evidence to support the social brain hypothesis—larger brains in more social species—across the Carnivora.
- Carnivorans have relatively large limbic and neocortical regions of the brain. Olfactory regions are comparable in size to those of ungulates and xenarthrans, excepting the small olfactory regions of pinniped brains.
- Advanced imaging shows that the volumes of functional regions of the brains of domestic dog breeds are correlated with behaviors that have been under strong human selection for a few centuries.

References

Bálint, A. et al. (2020) "Dogs can sense weak thermal radiation," Scientific Reports, 10, p. 3736.

Banks, M.S. et al. (2015) "Why do animal eyes have pupils of different shapes?" Science Advances, 1, p. e1500391.

Begall, S. et al. (2013) "Magnetic alignment in mammals and other animals," Mammalian Biology, 78, pp. 10–20.

Beichman, A.C. et al. (2019) "Aquatic adaptation and depleted diversity: a deep dive into the genomes of the sea otter and giant otter," Molecular Biology and Evolution, 36, pp. 2631–55.

Benediktová, K. et al. (2020) "Magnetic alignment enhances homing efficiency of hunting dogs," eLife, 9, p. e55080.

Benson-Amram, S. et al. (2016) "Brain size predicts problem-solving ability in mammalian carnivores," Proceedings of the National Academy of Sciences of the United States of America, 113, pp. 2532–7.

Bradshaw, J.W.S. (2006) "The evolutionary basis for the feeding behavior of domestic dogs (Canis familiaris) and cats (Felis catus)," Journal of Nutrition, 136, pp. S1927–31.

Brinck, C., Erlinge S. and Sandell, M. (1983) "Anal sac secretions in mustelids," Journal of Chemical Ecology, 9, pp. 727–45.

Burger, B.V. (2005) "Mammalian semiochemicals," Topics in Current Chemistry, 240, pp. 231–78.

Burger, J.R. et al. (2019) "The allometry of brain size in mammals," Journal of Mammalogy, 100, pp. 276–83.

Buskirk, S.W., Maderson, P.F.A. and O'Connor, R.M. (1986) "Plantar glands in North American Mustelidae," in Duvall, D., Müller-Schwarze, D. and Silverstein, R.M. (eds.) Chemical signals in vertebrates, Vol. 4. New York: Plenum Press, pp. 617–22.

Cervený, J. et al. (2011) "Directional preference may enhance hunting accuracy in foraging foxes," Biology Letters, 7, pp. 355–7.

Coelho, M.C., Schetini de Azevedo, C. and Young, R.J. (2012) "Behavioral responses of maned wolves (Chrysocyon brachyurus, Canidae) to different categories of environmental enrichment stimuli and their implications for successful reintroduction," Zoo Biology, 31, pp. 453–69.

Crump, D.R. (1980) "Thietanes and dithiolanes from the anal gland of the stoat (Mustela erminea)," Journal of Chemical Ecology, 6, pp. 341–7.

Debey, L.B. and Pyenson, N.D. (2013) "Osteological correlates and phylogenetic analysis of deep diving in living and extinct pinnipeds: what good are big eyes?" Marine Mammal Science, 29, pp. 48–83.

Dunbar, R.M. (1998) "The social brain hypothesis," Evolutionary Anthropology, 6, pp. 178–90.

Evans, H.E. and de Lahunta, A. (2013) Miller's anatomy of the dog. 4th edn. St. Louis: Elsevier.

Ferrero, D.M. et al. (2011) "Detection and avoidance of a carnivore odor by prey," Proceedings of the National Academy of Sciences of the United States of America, 108, pp. 11235–40.

Finarelli, J.A. and Flynn, J.J. (2010) "Brain-size evolution and sociality in Carnivora," Proceedings of the National Academy of Sciences of the United States of America, 106, pp. 9345–9.

Ghoul, A. and Reichmuth, C. (2014) "Hearing in the sea otter (Enhydra lutris): auditory profiles for an amphibious marine carnivore," Journal of Comparative Physiology A, 200, pp. 967–81.

Gittleman, J.L. (1991) "Carnivore olfactory bulb size: allometry, phylogeny and ecology," Journal of Zoology, 225, pp. 253–72.

Gläser, N. and Kröger, R.H.H. (2017) "Variation in rhinarium temperature indicates sensory specializations in placental mammals," Journal of Thermal Biology, 67, pp. 30–4.

Gurlt, E.F. (1835) "Vergleichende Untersuchungen über die Haut des Menschen und der Haus-Säugethiere, besonders in Beziehung auf die Absonderungs-Organe des Haut-Talges und des Schweisses," *Archive der Anatomie und Physiologie*, 8, pp. 399–418.

Hall, E.R. (1926) "The abdominal skin glad of Martes," *Journal of Mammalogy*, 7, pp. 227–9.

Hanke, W. *et al.* (2013) "Hydrodynamic perception in true seals (Phocidae) and eared seals (Otariidae)," *Journal of Comparative Physiology A*, 199, pp. 421–40.

Healy, S.D. and Rowe, C. (2007) "A critique of comparative studies of brain size," *Proceedings of the Royal Society B: Biological Sciences*, 274, pp. 453–64.

Hecht, E.E. *et al.* (2019) "Significant neuroanatomical variation among domestic dog breeds," *Journal of Neuroscience*, 39, pp. 7748–58.

Hecker, N. *et al.* (2019) "Convergent vomeronasal system reduction in mammals coincides with convergent losses of calcium signalling and odorant-degrading genes," *Molecular Ecology*, 28, pp. 3656–68.

Heffner, R.S. and Heffner, H.E. (1985) "Hearing range of the domestic cat," *Hearing Research*, 19, pp. 85–8.

Herculano-Houzel, S. (2011) "Scaling of brain metabolism with a fixed energy budget per neuron: implications for neuronal activity, plasticity and evolution," *PLoS ONE*, 6, p. e1751.

Hu, Y. *et al.* (2017) "Comparative genomics reveals convergent evolution between the bamboo-eating giant and red pandas," *Proceedings of the National Academy of Sciences of the United States of America*, 114, pp. 1081–6.

Island, H.D., Wengeler, J. and Claussenius-Kalman, H. (2017) "The flehmen response and pseudosuckling in a captive, juvenile southern sea otter (*Enhydra lutris nereis*)," *Zoo Biology*, 36, pp. 30–9.

Isler, K. and van Schaik, C.P. (2012) "Allomaternal care, life history and brain size evolution in mammals," *Journal of Human Evolution*, 63, pp. 52–63.

Jacobs, G.H. (1993) "The distribution and nature of colour vision among the mammals," *Biological Reviews*, 68, pp. 413–71.

Janssenwillen, S. *et al.* (2021) "Odorant-binding proteins in canine anal sac glands indicate an evolutionarily conserved role in mammalian chemical communication," *BMC Ecology and Evolution*, 21, p. 182.

Jiang, P. *et al.* (2012) "Major taste loss in carnivorous mammals," *Proceedings of the National Academy of Sciences*, 109, pp. 4956–61.

Kastak, D. and Schusterman, R.J. (1998) "Low-frequency amphibious hearing in pinnipeds: methods, measurements, noise, and ecology," *Journal of the Acoustical Society of America*, 103, pp. 2216–28.

Kastak, D. and Schusterman, R.J. (1999) "In-air and underwater hearing sensitivity of a northern elephant seal (*Mirounga angustirostris*)," *Canadian Journal of Zoology*, 77, pp. 1751–8.

Kavoi, B.M. and Jameela, H. (2011) "Comparative morphometry of the olfactory bulb, tract and stria in the human, dog and goat," *International Journal of Morphology*, 29, pp. 939–46.

Keller, A. and Vosshall, L.B. (2008) "Better smelling through genetics: mammalian odor perception," *Current Opinion in Neurobiology*, 18, pp. 364–9.

Kelliher, K.R., Baum, M.J. and Meredith, M. (2001) "The ferret's vomeronasal organ and accessory olfactory bulb: effect of hormone manipulation in adult males and females," *Anatomical Record*, 263, pp. 280–8.

Krestel, D. *et al.* (1984) "Behavioral determination of olfactory thresholds to amyl acetate in dogs," *Neuroscience and Biobehavioral Reviews*, 8, pp. 169–74.

Kruska, D.C.T. (2005) "On the evolutionary significance of encephalization in some eutherian mammals: effects of adaptive radiation, domestication, and feralization," *Brain, Behavior and Evolution*, 65, pp. 73–108.

Kruuk, H., Gorman, M. and Leitch, A. (1984) "Scent-marking with the subcaudal gland by the European badger, *Meles meles* L.," *Animal Behaviour*, 32, pp. 899–907.

Lei, W. *et al.* (2015) "Functional analyses of bitter taste receptors in domestic cats (*Felis catus*)," *PLoS ONE*, 10, p. e0139670.

Leinders-Zufall, T. *et al.* (2004) "MHC Class I peptides as chemosensory signals in the vomeronasal organ," *Science*, 306, pp. 1033–7.

Levenson, D.H. *et al.* (2006) "Visual pigments of marine carnivores: pinnipeds, polar bear, and sea otter," *Journal of Comparative Physiology A*, 192, pp. 833–43.

Lewitus, E. (2018) "Inferring evolutionary process from neuroanatomical data," *Frontiers in Neuroanatomy*, 12, p. 54.

Li, X. *et al.* (2009) "Analyses of sweet receptor gene (*Tas1r2*) and preference for sweet stimuli in species of Carnivora," *Journal of Heredity*, 100, pp. S90–100.

Malnic, B., Godfrey, P.A. and Buck, L.B. (2004) "The human olfactory receptor gene family," *Proceedings of the National Academy of Sciences of the United States of America*, 101, pp. 2584–9.

Marriott, S. *et al.* (2013) "Somatosensation, echolocation, and underwater sniffing: adaptations allow mammals without traditional olfactory capabilities to forage for food underwater," *Zoological Science*, 30, pp. 69–75.

Marshall, D.A., Blumer, L. and Moulton, D.G. (1981) "Odor detection curves for *n*-pentanoic acid in dogs and humans," *Chemical Senses*, 6, pp. 445–53.

Mauck, B. *et al.* (2008) "Harbour seals (*Phoca vitulina*) can steer by the stars," *Animal Cognition*, 11, pp. 715–8.

Meyer, W. and Tsukise, A. (1995) "Lectin histochemistry of snout skin and foot pads in the wolf and the domesticated dog (Mammalia: Canidae)," *Annals of Anatomy*, 177, pp. 39–49.

Mohl, B. (1968) "Auditory sensitivity of the common seal in air and water," *Journal of Auditory Research*, 8, pp. 27–38.

Montague, M.J. et al. (2014) "Comparative analysis of the domestic cat genome reveals genetic signatures underlying feline biology and domestication," *Proceedings of the National Academy of Sciences of the United States of America*, 111, pp. 17230–5.

Mykytowycz, R. (1972) "The behavioural role of the mammalian skin glands," *Die Naturwissenschaften*, 59, pp. 133–9.

Newbold, H.G. and King, C.M. (2009) "Can a predator see 'invisible' light? Infrared vision in ferrets (*Mustela furo*)," *Wildlife Research*, 36, pp. 309–18.

Niesterok, B. et al. (2017) "Hydrodynamic detection and localization of artificial flatfish breathing currents by harbour seals (*Phoca vitulina*)," *Journal of Experimental Biology*, 220, pp. 174–85.

Niimura, Y., Matsui, A. and Touhara, K. (2014) "Extreme expansion of the olfactory receptor gene repertoire in African elephants and evolutionary dynamics of orthologous gene groups in 13 placental mammals," *Genome Research*, 24, pp. 1485–96.

Oddie, M.A.Y., Coombes, S.M. and Davy, C.M. (2015) "Investigation of cues used by predators to detect snapping turtle (*Chelydra serpentina*) nests," *Canadian Journal of Zoology*, 93, pp. 299–305.

Park, H. et al. (2019) "Cutaneous sensory feedback from paw pads affects lateral balance control during split-belt locomotion in the cat," *Journal of Experimental Biology*, 222, p. jeb198648.

Peichl, L., Behrmann, G. and Kröger, R.H.H. (2001) "For whales and seals the ocean is not blue: a visual pigment loss in marine mammals," *European Journal of Neuroscience*, 13, pp. 1520–8.

Pitcher, B.J. et al. (2010) "Social olfaction in marine mammals: wild female Australian sea lions can identify their pup's scent," *Biology Letters*, 7, pp. 60–2.

Poddar-Sarkar, M. et al. (2008) "Putative pheromones of lion mane and its ultrastructure," in Hurst, J.L. et al. (eds.) *Chemical signals in vertebrates*, Vol. 11. New York: Springer, pp. 61–7.

Reep, R.L., Finlay, B.L. and Darlington, R.B. (2007) "The limbic system in mammalian brain evolution," *Brain, Behavior and Evolution*, 70, pp. 57–70.

Sakai, S.T. et al. (2016) "Big cat coalitions: a comparative analysis of regional brain volumes in Felidae," *Frontiers in Neuroanatomy*, 10, p. 99.

Sato, J.J. and Wolsan, M. (2012) "Loss or major reduction of umami taste sensation in pinnipeds," *Naturwissenschaften*, 99, pp. 655–9.

Schmidberger, G. (1932) "Uber die Bedeutung der Schnurrhaare bei Katzen," *Zeitschrift fur vergleichende Physiologie*, 17, pp. 387–407.

Schusterman, R.J. et al. (2000) "Why pinnipeds don't echolocate," *Journal of the Acoustical Society of America*, 107, pp. 2256–64.

Sharpe, L.L. (2015) "Handstand scent marking: height matters to dwarf mongooses," *Animal Behavior*, 105, pp. 173–9.

Stockham, R.A., Slavin, D.L. and Kift, W. (2004) "Specialized use of human scent in criminal investigations," *Forensic Science Communications* [online], 6(3). Available at: https://archives.fbi.gov/archives/about-us/lab/forensic-science-communications/fsc/july2004/research/2004_03_research03.htm (Accessed: 2 Oct 2021).

Strobel, S.M. et al. (2018) "Active touch in sea otters: in-air and underwater texture discrimination thresholds and behavioral strategies for paws and vibrissae," *Journal of Experimental Biology*, 221, p. jeb181347.

Thesen, A., Steen, J.B. and Doving, K.B. (1993) "Behaviour of dogs during olfactory tracking," *Journal of Experimental Biology*, 180, pp. 247–51.

Van Valkenburgh, B. et al. (2014) "Respiratory and olfactory turbinals in feliform and caniform carnivorans: the influence of snout length," *Anatomical Record*, 297, pp. 2065–79.

Vater, M. and Kössl, M. (2011) "Comparative aspects of cochlear functional organization in mammals," *Hearing Research*, 273, pp. 89–99.

Veilleux, C.C. and Kirk, E.C. (2014) "Visual acuity in mammals: effects of eye size and ecology," *Brain, Behavior and Evolution*, 83, pp. 43–53.

Verberne, G. (1976) "Chemocommunication among domestic cats, mediated by the olfactory and vomeronasal senses. II. The relation between the function of Jacobson's organ (vomeronasal organ) and flehmen behavior," *Zeitschrift für Tierpsychologie*, 42, pp. 113–28.

Vogt, K. et al. (2016) "Chemical composition of Eurasian lynx urine conveys information on reproductive state, individual identity, and urine age," *Chemoecology*, 26, pp. 205–17.

Wells, D.L. and Hepper, P.G. (2003) "Directional tracking in the domestic dog, Canis familiaris," *Applied Animal Behavior Science*, 84, pp. 297–305.

Woodley, S.K. and Baum, M.J. (2004) "Differential activation of glomeruli in the ferret's main olfactory bulb by anal scent gland odours from males and females: an early step in mate identification," *European Journal of Neuroscience*, 20, pp. 1025–32.

Woodley, S.K. *et al.* (2004) "Effects of vomeronasal organ removal on olfactory sex discrimination and odor preferences of female ferrets," *Chemical Senses*, 29, pp. 659–69.

Zhao, H. *et al.* (2010) "Pseudogenization of the umami taste receptor gene *Tas1r1* in the giant panda coincided with its dietary switch to bamboo," *Molecular Biology and Evolution*, 27, pp. 2669–73.

Zhu, K. *et al.* (2014) "The loss of taste genes in cetaceans," *BMC Evolutionary Biology*, 14, p. 218.

CHAPTER 6

Community ecology

Community ecology examines interactions among co-occurring species. It considers trophic interactions among plants, herbivores, predators, pathogens, and scavengers, as well as non-trophic relationships within trophic levels and relationships that are not trophic in nature. Non-trophic relationships include nutrient cycling, nutrient transport, pollination, dispersal of plant propagules, and direct effects on soils. Some community interactions are strong enough to affect ecosystem processes. For example, if herbivory, predation, or competition is strong enough to affect nutrient fluxes, primary production, standing crop, or multispecies processes, it may be considered important at an ecosystem scale.

Our understandings of the community ecology of the Carnivora have expanded rapidly since 1980. Two of the most frequently studied topics in ecology are predation and competition, and carnivorans are conspicuous participants in both processes (Chesson and Kuang, 2008). I organize the community interactions of carnivorans into several categories, the first of which includes nutrient cycling and transport, direct effects on soil, community aspects of disease ecology, facilitation of access to carrion, and direct carnivoran–plant interactions, all covered in this chapter. Interactions with non-prey animals are treated in Chapter 7. Carnivoran–prey interactions are covered in Chapter 8, and cascades are discussed in Chapter 9. Population ecology and microevolution, particularly as related to community processes, are treated in Chapter 10. Lastly, some of the most important community interactions of carnivorans are with humans, described in Chapters 11 and 12. The concepts of both Grinnellian and Eltonian niche are raised, at times implicitly, in the following chapters. Most commonly,

we characterize niches in Grinnellian terms—the habitats and other resources a species requires, and the species' adaptations to them. In this and in the following chapters, I emphasize Eltonian aspects of carnivoran niches—ways in which carnivorans influence the communities of which they are members.

6.1 Nutrient cycling and transport

Perhaps the most readily recognized but underappreciated ecosystem service of carnivorans is their processing of animal (mostly herbivore) bodies to products that re-enter air, water, and soil: exhaled CO_2, bio-available nitrogen and phosphorus in urine, and partially digested shell, bone, and soft tissue. This is a highly redundant service, provided by all metazoan predators and scavengers, as well as by fungal and microbial consumers (Schmitz et al., 2010), and carnivorans are relatively minor contributors at the ecosystem level. However, the functions performed by carnivorans are spatially non-random and contribute to landscape pattern. In some cases, they transport nutrients and metabolites across ecological boundaries, for example, from water to land, where they are concentrated in soil, rather than diluted. The importance of animal-mediated transport is now recognized for several reasons. First, whereas passive transport typically follows energy gradients—such as downstream water flow—animal-aided transport can run counter to such gradients and be comparable in magnitude (McInturf et al., 2019). Second, carnivoran-aided transport can concentrate nutrients so that they have strong influences.

The best-documented examples of carnivoran-mediated nutrient transport involve bears and

Carnivoran Ecology. Steven W. Buskirk, Oxford University Press. © Steven W. Buskirk (2023). DOI: 10.1093/oso/9780192863249.003.0006

Pacific salmon. Brown bears and black bears depend heavily on spawning salmon in coastal areas of the North Pacific Rim. Brown bears on Kodiak Island, Alaska have been shown to consume a mean of 1100 kg of salmon per bear over a median of forty-two days per year spent on salmon-spawning streams. The salmon represented a mean of 64% (and as much as 90% for some bears) of the annual assimilated diet (Deacy et al., 2018). In consuming so much salmon flesh, bears transport marine-derived nutrients into terrestrial systems. They deposit feces, urine, and partially eaten salmon (Figure 6.1A), and carcasses are further scavenged, including by other carnivorans: minks, martens, pumas, wolves, and coyotes (Schindler et al., 2003). Brown bears transport an estimated 40–70% of salmon of various species from streambeds to the adjacent forest floor (Quinn et al., 2009). Without such transport, most of the salmon-borne nutrients left in streambeds are flushed below ground and out of the terrestrial community (Figure 6.1B).

The nutrients transported include nitrogen, phosphorus, and carbon, the first two limiting to plants in the wet environments of many coastal spawning streams. Bear-transported salmon provide an estimated 14–60% of riparian nitrogen budgets at some North American sites (Helfield and Naiman, 2001; Hocking and Reynolds, 2012). Among nine study streams along the northwest coast of North America examined over a fifty-year period for which records were available, five showed correlations between the number of spawning salmon in one year and the subsequent growth of coniferous trees, as measured by annual rings. The variation in growth of conifers between the lowest and highest spawning years was dramatic—a factor of from two to over ten (Drake and Naiman, 2007). In southwestern Alaska, researchers experimentally showed that brown bears transported salmon carcasses to sites where soil ammonium (NH_4^+) concentration was elevated threefold relative to controls (Holtgrieve et al., 2009). Interestingly, some studies demonstrate that the effects of salmon-derived nutrient deposition on soils and plants are transient; no effect on woody plants could be detected after one year had elapsed since a focal salmon run. Salmon carcasses also are rich sources of polyunsaturated fatty acids,

which are difficult for mammals to synthesize, and provide conditionally essential nutrients to terrestrial mammals that consume marine fish (Section 4.2.1).

The ursid–salmon interaction is striking for its locally strong, yet geographically widespread effects on vegetation. In pre-industrial times, those effects were important along spawning streams from California to Japan. Most salmon spawning migrations are shorter than 100 km, but some extend 3000 km inland, so that bears can be effective nutrient transport agents far from the ocean. Bears also contribute to maintaining spawning habitats for salmon: bear-transported carcasses fertilize large riparian conifers, increasing growth rates nearly threefold over reference areas. Large trees provide shade, sediment filtration, and coarse woody debris to the spawning stream. Coarse woody debris in turn maintains spawning beds and juvenile fish habitat by forming small dams and plunge pools (Helfield and Naiman, 2001). This ecological function of bears in maintaining spawning habitat, appreciated only recently, drastically alters our understanding of causation in the community of bears, salmon, and vegetation. In early twentieth-century Alaska, a prevailing view saw brown bears as competitors with humans for salmon. Now, we understand that bears are important agents for the fertilization of riparian soils near spawning streams, contributing to tree growth and sustainable spawning habitat.

Marine carnivorans provide similar transport functions by nature of consuming large quantities of marine fish, but pinnipeds mostly urinate and defecate in water, and have limited mobility on land. Galapagos sea lions and gray seals transport marine-derived nutrients to land in their guts or bodies, nourishing plants when they urinate, defecate, or die and decompose (Farina et al., 2003). The case of gray seals breeding and giving birth on Sable Island, Canada is of particular interest because of their high densities and the importance of seal-fertilized grasses to feral horses, the only land mammals present (Mansfield, 1967). The subpopulation of gray seals that breeds on Sable Island had grown to around 390,000 by 2014, distributed patchily across the 32-km^2 island (Figure 6.2). Marine-derived nitrogen is concentrated where

(b)

(a)

Figure 6.1 Brown bears and black bears are important agents for transporting salmon carcasses from streambeds into nearby forests, where marine-derived nutrients can be accessed by other scavengers and by plant roots. A: Spawned-out red salmon carcasses in a streambed in Katmai National Park, Alaska. Nutrients from carcasses that remain in such locations are flushed below ground and removed from the system. B: Carcasses that are carried into adjacent forest, in this case by a black bear, contribute phosphorus and nitrogen to soils, where plant roots access them. Photos: © Amy Gulick.

Figure 6.2 Gray seals and feral horses on Sable Island, Nova Scotia, Canada. This, the largest breeding colony of gray seals in the world, transports marine-derived nutrients to land, fertilizing vegetation locally. These hot spots of marine-derived nutrients are important to the horse population. Photo: Paul Gierszewski/Wikimedia Commons.

seal pups are born and adults loaf. When horse densities are not so high as to cause forage competition, they graze selectively where seal activity and marine-derived nitrogen are concentrated. The persistence of the Sable Island horse population may depend on seal-fertilized vegetation; by contrast, nesting seabirds play a relatively minor role in nutrient transport in this system (McLoughlin et al., 2016).

Otters perform similar but apparently smaller roles in transporting nutrients from aquatic to terrestrial communities. In coastal and inland freshwater environments, the North American river otter creates nitrogen hotspots, reflected in plant growth, by depositing urine and feces repeatedly in the same locations to form latrines. These latrines are important in intraspecific communication and have important functions in maintaining local vegetative heterogeneity (Ben-David et al., 1998; Crait and Ben-David, 2007). Although carnivorans provide largely redundant services in cycling organic materials, they accelerate these processes and influence the spatial distribution of vegetation. They do not provide mass transport of nutrients across most community boundaries but have measurable—and in some cases strong—local effects on vegetative heterogeneity and primary production near salmon spawning streams. Some of these effects have feedback to salmon, for example, the effects of

maintaining spawning habitat on salmon populations and benthic food chains. They also influence nutrient transport from water to land by other vertebrates (Section 9.1).

6.2 Direct effects on soil

Aside from cycling nutrients via ingestion, urination, defecation, and decomposition, some terrestrial carnivorans affect soils more directly in ways that alter landscape pattern. Specifically, carnivorans influence where prey animals die. Even after scavenging and disassembly by vertebrates, carcasses create nutrient hotspots that influence landscape pattern (Wilson and Wolkovich, 2010). Using over fifty years' worth of data, ecologists have showed that moose carcasses are deposited in some areas of Isle Royale at twelve times the density of other areas, affecting woody plant growth (Bump et al., 2009a, b). The effects on soil of digging and living below ground can be dramatic as well. European badgers and American badgers both excavate soil, altering chemical profiles in near-surface layers (Figure 6.3). European badgers contribute to increased acidity, potassium, calcium, and magnesium in near-surface layers (Figure 6.3A). American badgers create small-scale heterogeneity in C:N ratio, acidity, and metal concentration between excavated mounds and burrows (Figure 6.3B)

(a)

(b)

Figure 6.3 Badgers of various species alter soils directly. A: European badgers dig and maintain long-standing burrow complexes ("setts"). B: American badgers do not establish permanent burrow systems but excavate rodent burrows and occupy them for up to several weeks in winter. Both badger species mix near-surface soil horizons and create local topography and nutrient hot spots.
Photos: A: CreativeNature_nl/iStock. B: Max Allen/Shutterstock.

(Eldridge and Whitford, 2009; Kurek *et al.*, 2014). Burrowing activities may also affect seed germination. Siberian weasels, European badgers, and hog badgers all leave small indentations in the soil where they forage for soil animals (Box 6.1). Within the range of Liaodong oak, these pockets catch fallen acorns and facilitate germination by hiding acorns from seed predators and partially burying them. In this way, mechanical alteration of soil by small carnivorans is a key consideration in oak reproduction in parts of East Asia (Gao and Sun, 2005).

Box 6.1 Wolves and beavers

One of the most plausibly strong community-level effects of a single carnivoran species is that of wolves preying on North American beavers in the boreal zone. Gable and colleagues (2020) reported that wolves killed dispersing and newly colonizing beavers, and that whether a colonizing beaver survived wolf predation through the winter following dispersal determined whether dams would be maintained. Beaver damming of streams strongly affects water tables, stream flows, woody vegetation, nutrient cycling, and fish migration. No other northern hemisphere mammal warrants the designation "keystone" to the degree that the beaver does. That wolf predation is such a strong factor in the survival of dispersing and colonizing beavers in boreal North America suggests that wolves could have strong effects on streams and riparian zones via predation on a true keystone species. At the same time, wolves and beavers are sympatric across vast tracts of the boreal zone; no evidence suggests that wolves can exclude beavers from a suitable colony site for many years. Whereas the effects of carnivoran predation on herbivores is regarded as protective of vegetation in most discussions of trophic cascades (Section 9.2), wolf predation on colonizing beavers could delay or reduce the effects of beavers on vegetation. Ordinarily, scientists regard herbivory as damaging to vegetation, whereas in this example beaver herbivory leads to dam maintenance, which is considered an important driver of riparian community structure.

6.3 Disease ecology

Carnivorans play important roles in pathogen life cycles and affect the incidence and virulence of diseases in prey and non-prey hosts. They serve as definitive and intermediate hosts, prey on parasite hosts, and transmit disease via multiple pathways, the latter by biting another mammal in the case of rabies, or by shedding virus-laden droplets from respiratory passages in the case of canine distemper. Some species shed the eggs of hydatid tapeworms in their feces, where small mammals, the intermediate hosts, ingest them, completing the life cycle. Some species hold encysted nematodes in their muscles, where predators and scavengers ingest them, becoming infected and infectious. Each of these infective pathways has its own relationship to carnivoran predation and other behaviors.

Because of the complex potential interactions in pathogen–host systems, this field of study has benefitted strongly from modeling, which helps to identify the drivers of disease incidence and virulence in mammals, including humans. This complexity is suggested by the fact that, limiting our focus to the metazoan Subclass Coccidia, around 264 species are known from carnivoran hosts, including formally named species, incompletely identified species, and putative species requiring additional evaluation (Duszynski et al., 2018). Each of these taxa may cause disease symptoms under some circumstances, have its own transmission pathway, and show unique cross-infectivity. Here, I limit my discussion to pathogens that cause strong documented symptoms affecting the behavior, vigor, or vital rates of carnivorans or their prey.

6.3.1 Factors affecting disease prevalence in carnivorans

Low population densities and infrequent social behaviors should affect transmission of some diseases, and therefore their prevalence. Such relationships have been shown in various intraspecific comparisons involving mammals, although density is difficult to parse from sociality (Altizer et al., 2003). Packer and colleagues (1999) showed for lions in Tanzania that the proportion of lions infected with various feline viruses was positively related to population density—simple density-dependent prevalence. However, Watve and Sukumar (1995) tested for this relationship for the intestinal parasites of a community of carnivorans and herbivores in southern India and found contrary patterns. Herbivores (e.g. hares, chitals, and gaurs) occurred at higher densities and were more social than carnivorans, had lower parasite loads, and showed lower parasite diversities than carnivorans (e.g. leopards, sloth bears, and dholes). The relative importance of sociality and density, which mainly affect prevalence, immune competency, and parasite loads, was not clear.

Do phylogenetically related or ecologically similar carnivoran species share more parasites than others? If they did, it might affect patterns of predation or competition. Huang and colleagues (2014) addressed this issue by examining occurrence data for various parasite lineages across sixty-four carnivoran species. Parasites were more commonly shared across phylogenetically related host species pairs, although this pattern depended on the parasite lineage considered. Sharing of helminth and viral parasites depended strongly on phylogenetic similarity, whereas that of arthropod and bacterial parasites did to a lesser degree. The relationships for carnivorans were weaker than those found in earlier studies in Primates, possibly because of the far greater geographic range and number of niches occupied by carnivorans. So, possible parasite sharing across phylogenetically similar carnivoran species is an unresolved issue. By contrast, body size has been shown to affect infection rates and parasite loads (Malmberg *et al.*, 2021). Bioaccumulation of pathogen exposure over the lifetime of an animal occurs as does bioaccumulation of toxicants, and large carnivorans are particularly prone to accumulating pathogens by nature of their long lives and diverse prey species. This makes them prone to serve as sources of infection to other sympatric carnivorans.

6.3.2 Predation and disease prevalence in prey

The possible role of carnivorans as "sanitary engineers" is a recurring topic in predation ecology. Is carnivoran predation likely to interrupt pathogen life cycles, or have pathogens evolved to predatory behaviors, so that predation facilitates transmission of a parasite lineage? These questions have been addressed for various invertebrate and aquatic systems (e.g. Duffy *et al.*, 2011), but field studies of carnivorans and their prey are few. Clearly, the answers depend partly on whether a life stage of the parasite survives in the predator long enough to be transmitted—the cross-infectivity of the parasite. Rabies is characterized by multiple strains, and some strains have high cross-infectivity, whereas others do not (Brunker and Mollentze, 2018). Although not transmitted by predation per se, rabies is spread by biting and injecting virus-laden saliva. Nematodes of Genus *Trichinella* have more limited cross-infectivities. *T. britovi* infects diverse wild European carnivorans, as well as domestic dogs and cats, whereas *T. spiralis* infects fewer carnivoran species, but is more common than

T. britovi in wild boars and domestic pigs. *T. nativa* infects almost exclusively wild carnivorans in the Subarctic and Arctic; it is not found in sylvatic or domestic suids (Pozio *et al.* 2009). So, a mustelid that kills and consumes an infected mouse has little or no chance of contracting or transmitting *T. spiralis*, and therefore could reduce incidence via predation. By contrast, any of several European canids can well carry and transmit *T. britovi* obtained from infected rodents.

A disease with even narrower cross-infectivity is chronic wasting disease (CWD), an invariably fatal brain inflammation caused by misfolded proteins that convey their shape to normal proteins in the brain. In the wild, this disease is limited to cervids in Order Artiodactyla, although primates and genetically modified mice now have been infected in lab settings. Because its onset is gradual and causes impaired movement and reduced vigor (Figure 6.4), predation is a plausible influence on its incidence. Indeed, pumas appear to prey selectively on prion-infected mule deer in Colorado, US; deer of both sexes killed by pumas were more likely to be infected with CWD than same-sex deer killed by hunters in the same area (Krumm *et al.*, 2009). Modeling by Wild and colleagues (2011) predicted that expansion of the geographic range of wolves in the region would lead to greater reductions in the incidence of CWD than in deer abundance. That range expansion was well underway by 2022, aided by natural dispersal from Yellowstone to the north. Both wolves and CWD are expanding their geographic ranges in the western US, and the former has the potential to limit the prevalence of the latter.

Several studies have shown similar results for diseases with intermediate cross-infectivity. Watve and Sukumar (1995) showed such a pattern in their southern Indian study; predatory pressure (estimated from diet studies of carnivorans and population densities of potential prey) was the strongest predictor of parasite loads. The same was shown for a community of bank voles and wood mice preyed on by red foxes, two species of martens, and European polecats (Hofmeester *et al.*, 2017). The voles serve as intermediate hosts for ticks of Genus *Ixodes*, which serve as reservoirs for a range of diseases. Tick burdens on both species of rodents

Figure 6.4 Advanced chronic wasting disease in a mule deer in Wyoming, US. Infection of the brain with mis-folded proteins (prions) causes the host's own proteins to assume the mis-folded form, resulting in gradual deterioration of vigor and coordination. Pumas selectively remove symptomatic animals, plausibly reducing rates of incidence.
Photo: © Justin Binfet.

decreased with increasing activity levels of red foxes and stone martens, whereas European pine marten and polecat activity levels showed weaker correlations with tick burdens. Tick abundance is limited by abundance of small mammalian hosts. Two plausible factors—density-dependent tick transmission or carnivoran effects on rodent behavior—could explain the effect. The first explanation posits that predation reduces prey densities, whereas the second assumes that predators reduce prey activity and movement, reducing opportunities for ectoparasite sharing among hosts.

6.3.3 Trophic transmission and prey behaviors

Because so many carnivorans are carnivorous, we should expect to see high incidence of diseases transmitted via ingestion of infected body parts—"trophic transmission" (Lafferty, 1999). How trophic transmission affects the behaviors of prey species has been considered for several invertebrate hosts, but scarcely examined for carnivorans eating vertebrate prey. Various contingencies could affect how parasites affect transmission from prey to consumer, including multi-species infections and whether the consumer is an intermediate or definitive host. The ways in which parasites manipulate invertebrate intermediate hosts include some dramatic examples (Poulin, 2010). For example, raccoon roundworm (*Baylisascaris procyonis*) can

cause poor coordination and whirling behavior in rodent intermediate hosts (Kazacos, 2016). However, parasites may have less dramatic effects—loss of vigor or slower reaction times. The field of parasite manipulation of host behavior is well developed for invertebrate communities but has limited empirical support in systems in which carnivorans are predators (but see Section 6.3.2). The relative body size of carnivorans and their prey could mediate this interaction. Predators of prey much larger than themselves should be more inclined to kill prey displaying reduced vigor, because of the greater ease and safety of subduing them. By contrast, predators of small prey, for example, foxes and weasels feeding on mice or insects, may rely less on signals about the vigor of their prey before capturing them, because their success may not depend on avoiding injury from prey.

Pinnipeds present special patterns of parasite infection that reflect the dichotomous marine and terrestrial phases of their lives. Among pinniped tapeworms, the most common and widespread family is Diphyllobothriidae, which occurs worldwide. These worms are passed through the food chain, intermediate stages being found in shrimp, which are eaten by small and then larger fish. By contrast, the nematode faunas of pinnipeds resemble the faunas of land carnivorans; they are transmitted on land. For example, the infective stages of *Uncinaria*

(Ancylostomatidae) develop in the soil near otariid rookeries and enter the skin of the flippers. The larvae migrate to various host tissues, including milk glands, thereby infecting the pups, where adult worms develop in the gut (Raga *et al.*, 2018). High aggregations of animals at rookeries and haul-outs predispose to this kind of transmission.

6.4 Scavenging and access to carrion

Virtually all predatory mammals are scavengers to some extent; however, the importance of scavenging varies across carnivoran families, and depends on the quality of carrion and other factors (Allen *et al.*, 2015; Beasley *et al.*, 2016). Based on Prugh and Sivy (2020) and other sources, an approximate ranking of carnivoran family-level reliance on carrion is lowest for Felidae (2% of diet) < Mephitidae < Herpestidae ≈ Hyaenidaae < Mustelidae < Canidae. Ursids and Viverrids have been studied too little to infer family-level trends and have access to carrion that is too geographically variable to allow generalization. Coyotes and red foxes in some areas ingest most vertebrate foods as carrion (Hewson, 1984; Huegel and Rongstad, 1985). The most specialized scavengers may be striped hyenas and brown hyenas and the wolverine. Each of these species relies heavily on carrion killed by other species but kills some prey itself. The wolverine also relies on winterkill (Inman *et al.*, 2012; Jones *et al.*, 2016). These species show various scavenging adaptations, and all travel large distances alone, increasing their chance of discovering carcasses. They have stout premolars for cracking bone, acute nasal olfactory reception, and modifications to the immune system in hyenas to cope with bacteria and their toxins found in warm-climate carrion. Much of the carrion eaten by wolverines is preserved, in some cases for months, by cold temperatures.

The number and importance of trophic pathways involving vertebrate scavengers have been little appreciated until recently. Traditional food web models overestimated predatory pathways, and underestimated scavenging links, in some cases by orders of magnitude. In fact, ecologists now recognize scavenging as a comparable or greater fate of vertebrate prey biomass than ingestion by the killer (Wilson and Wolkovich, 2011). Carnivorans play

various roles in these guilds—facilitating as well as benefitting from carrion availability. The reciprocal benefits of scavenging are asymmetrical across carnivoran body sizes, in that small scavengers benefit from large carnivoran kills; the reverse is less-often true (Figure 6.5). Where scavengers compete at a carcass, access depends largely on body size, circadian activity, and habitat type (Selva *et al.*, 2005). As an example of the first of these, pumas suppress scavenging of carcasses by mid-sized carnivorans and birds, thereby facilitating scavenging by still smaller carnivorans. Brown bears and polar bears, the largest-bodied far-northern scavengers, dominate competitive interactions at large mammal carcasses, whether killed by them or by smaller carnivorans (Green *et al.*, 1997; Ballard *et al.*, 2003). At the population level, scavenging is not only an alternative to predation, but also a positive feedback that stabilizes the scavenger-predator population via multi-channel feeding. This can increase predatory pressure on the carrion species (Polis and Strong, 1996). For example, the introduced red fox population on an Australian island, supported by marine-derived carrion, reduced the abundances of avian and other species that were fox prey as well as carrion sources (Brown *et al.*, 2015). A further confounding effect is to increase predation risk to small-bodied vertebrates near carrion, whether they are attracted to the carrion or merely happen to live nearby (Forsyth *et al.*, 2014; Steinbeiser *et al.*, 2018). Thus, carrion creates rich foraging sites for diverse scavengers, carnivoran and otherwise, but increases the risk of conflict or predation for all but the largest-bodied species.

Why are large carnivorans important to the scavenging community at all? One might point out that all animals die eventually, and overall carrion availability might not depend on large predators. Wilmers and colleagues (2003) studied the system of wolves, elk, and scavengers in the Yellowstone area in the US. Wolves had the effect of reducing seasonal and annual variation in carrion availability. Before wolf reintroduction in 1995, carrion availability had been highest during late winter and during years of heavy snowfall. When Wilmers and Getz (2004) modeled this system for the period before and after wolf reintroduction, they found that wolves reduced the total amount of elk carrion by reducing

Figure 6.5 A brown bear steals the carcass of a wild boar that has been killed by wolves. More commonly, smaller-bodied scavengers benefit from predation by larger-bodied carnivorans. In some cases, entire packs of wolves are unable to reclaim carcasses from bears that weigh a fraction of their collective weight.
Photo: P. Ivanyi/Shutterstock.

elk populations. However, this decrease was partially offset by increased elk production, resulting from higher predation on elk with low reproductive value (calves and weak, old cows). Large predators may reduce the patchiness of carrion deposition across the landscape; lacking predators in the Yellowstone system, carrion tends to be more common in areas with overused late winter forage.

Competition for carrion at large mammal carcasses can influence the behavior of the predatory species; an example is the evolved relationship of common ravens and wolves preying on ungulates. Why do wolf packs hunt moose together, when only two to three pack members participate in prey capture and killing? Investing in such a high-risk activity would seem to be a poor choice for the killing wolves, considering that they can consume only a small part of a moose before it is eaten by scavengers, becomes degraded, or freezes. An explanation first considers that packs comprise close relatives, so that prey killed by one or two wolves, but eaten by kin, enhances the killers' inclusive fitness. Common ravens, a nearly omnipresent competitor at wolf kills, are so adept at tracking wolf movements and kills that they can quickly consume carcasses not attended by wolves (Vucetich *et al.*, 2004). Kaczensky and colleagues (2005) found that ravens were able to appropriate up to 75% of the edible biomass of moose carcasses killed by smaller wolf packs. They removed almost none of the carcasses

of moose killed by large wolf packs, because of wolf presence at the carcass. Therefore, group hunting permits wolf packs to utilize a larger proportion of a moose carcass than could the wolves that did the killing, increasing the group's fitness in the face of strong exploitative competition by ravens.

Another carnivoran behavior that reduces exploitative competition at carcasses is the complete burial of ungulates by American badgers. Many carnivorans cover parts of carcasses with soil or litter between feeding bouts, but the American badger takes this behavior further. A single 10-kg badger can inter completely the carcass of a domestic cow much larger than itself and access it underground, so that competitors are excluded. The badger enjoys exclusive access to a large carrion source for days or weeks (Figure 4.3) (Frehner *et al.*, 2017).

6.5 Direct effects on plant life cycles

6.5.1 Pollination

Carnivorans traditionally have not been considered important pollinators of flowers; however, several species have been reported to have traits that might qualify them as effective in this role. Evidence for an animal being a pollinator typically includes eating nectar or pollen without destroying the flower and carrying pollen on its body from one flower to another (Carthew and Goldingay, 1997). Plant species suspected of being pollinated by carnivorans tend to produce dull-colored inflorescences near the ground, and abundant sugar-rich nectars. Cape genets and Cape gray mongooses have been observed feeding on flowers of *Protea* spp. and have been recorded carrying pollen on their snouts (Steenhuisen *et al.*, 2015) and kinkajous and olingos have been described sequentially visiting flowers of *Quararibea* spp. in Peru (Janson *et al.*, 1981). Other herpestids, viverrids, and procyonids are known to eat flowers or nectar and may contribute to pollination.

6.5.2 Seed dispersal

Many vascular plants have coevolved seed dispersal mutualisms with birds and mammals. Animal-aided seed dispersal benefits plants by reducing inbreeding between close relatives. It also reduces density-dependent processes—competition among closely related plants and outbreaks of plant diseases near the parent. Over two-thirds of woody plant species in tropical forests have coevolved with dispersal mutualists (Bagchi *et al.*, 2018). Mammal-aided seed dispersal is so important in the tropics that the overhunting of mammalian seed dispersers in Amazonian forests is recognized as altering the dispersal of seeds of trees important to carbon sequestration. Therefore, seed dispersal by mammals, including carnivorans, plays a role in atmospheric carbon dynamics (Bello *et al.*, 2015; Peres *et al.*, 2016).

Many mammalian species can transport seeds on skin or fur. Carnivoran dispersal of ingested seeds takes two forms. In endozoochory, carnivorans ingest, transport, and defecate seeds, typically encased in fleshy fruits. Passing through the gut of an animal removes juice and pulp from seeds, which enhances germination, at the risk of the seed being crushed by cheek teeth (Tsuji, 2020). Which is more likely to occur depends on seed and dental morphology. If seeds are small relative to the grinding teeth (e.g. those of *Vaccinium*, *Opuntia*, or *Actinidia*), many omnivorous frugivores will disperse them in viable form. If the seeds are intermediate in size (e.g. paw paw), mid-sized carnivorans might not swallow them or may damage them

during mastication. Large-seeded fruits (e.g. avocado) are eaten by the coyote and other mid-sized carnivorans without swallowing any seeds, so any dispersal would depend on carrying the fruit to a feeding site. Tsuji and colleagues (2020) evaluated the seed dispersal potential of Japanese martens and several sympatric frugivores—carnivoran, primate, and avian—and inferred that martens were the most effective disperser of liana seeds. Even in the case of the large-seeded avocado, germination may be enhanced by removal of pulp from around the seed (Jordano, 2000; Nogeire *et al.*, 2013).

A second form of enteric seed transport has recently been emphasized: diploendozoochory occurs when a predator or scavenger consumes, transports, and defecates seeds contained in the gastrointestinal tracts of animals that consumed fruit. For this process to enhance seed dispersal, seeds secondarily ingested by carnivorans would need to have higher fitness than those that would have passed through the gut of the original consumer or would have remained in the carcass of the consumer, without secondary ingestion (Hämäläinen *et al.*, 2017). Evidence showing that carnivorans perform this function is sparse to date, but intriguing.

In North America, more mammalian than avian species are documented frugivores, in spite of bird species greatly outnumbering mammals (Willson, 1993). Carnivorans may even surpass birds in dispersing seeds that are ingested, because mammalian guts are more effective in stripping seeds of fruit pulp, and in scarifying seeds to facilitate germination (Enders and Vander Wall, 2012). Near North American salmon-spawning streams, black bears and brown bears attain such high seasonal densities that they surpass birds as seed dispersal agents for devil's club (Harrer and Levi, 2018). Due to their larger home ranges and meat-adapted dentition, carnivorans likewise may be more effective seed dispersers than sympatric mammalian herbivores. Carnivoran cheek teeth, being sectorial or bunodont, do not finely grind plant material, including small seeds. As a result, fruit seeds may be more likely to survive mastication by a carnivoran than an herbivore (Hickey *et al.*, 1999). Fredriani and Delibes (2009) showed this for vertebrate frugivores feeding on the Iberian pear: birds

Figure 6.6 Some bear species consume large quantities of berries and seeds while hyperphagic and defecate them in piles. Such concentrations should result in high relatedness of germinated seeds from a single fecal pile, as well as density-dependent competition and disease incidence among seedlings. Secondary dispersal of seeds from these fecal piles by rodents improves survival of dispersed propagules.
Photo: AmeliaM/CanStock.

ate only pulp near the fruit surface, rodents fed on pulp but left seeds beneath the tree, and wild boars ate seeds but damaged nearly half of them. European badgers ingested seeds along with pulp and defecated them intact, making them relatively important agents, considering their numerical scarcity.

Some ursids seem especially likely to disperse seeds because they consume large amounts of fleshy fruit in autumn (Figure 6.6). Garcia-Rodriguez and colleagues (2021) found that across the geographic range of the brown bear, its diet is about one-fourth fleshy fruits, produced by over 100 plant species. However, closer examination shows that bears deposit large numbers of seeds in a single fecal pile, so that the same issues of neighbor-relatedness and density dependence arise as if seeds germinated near the parent plant. Bear scats also tend to dry slowly, encouraging fungal growth that kills seeds. Bear-mediated dispersal is assisted by other animals, however. Viable seeds that have passed through a bear's gut are transported and buried by scatter-hoarders, particularly mice. These mice cache seeds singly or in small groups, only a few millimeters beneath the soil surface—optimal sites for germination. Data show that fruit seeds in bear scats are more likely to germinate if secondarily dispersed by small mammals (Enders and Vander Wall, 2012). These examples draw mostly from northern temperate studies, but the most common

and perhaps highly evolved mutualisms between carnivorans and endozoochorous plants occur in the tropics. There, the numbers of fruit-producing plant species is large, and their relationships with carnivoran dispersers, including some near-obligate frugivores, are poorly understood.

6.5.3 Seed exploitation

In contrast to seed dispersers, some carnivorans exploit hard mast, without serving a strong dispersal function. In the case of stone pines (Section 4.2.1) dispersal requires one of the holarctic nutcrackers to extract the large seeds from cones and cache them distant from the parent tree (Gernandt *et al.*, 2005). Pine squirrels are the primary exploiters of this relationship in North America, which involves whitebark pine in the western temperate mountains. Squirrels cut cones and cache them in piles of discarded cone bracts close to the parent tree; they later extract the seeds from the cones, leaving caches of dozens or hundreds of seeds. Bears and martens find and consume these seed caches; between pine squirrels and various carnivorans, pine nut survival is reduced (Schmidt and Holtmeier, 1994; Mattson and Reinhart, 1995). When cone crops are large, whitebark pine seeds are the major autumn food of brown bears where the two species co-occur in North America. Indeed, around two-thirds of brown bears in the Yellowstone area derive most of their assimilated protein in autumn from whitebark pine nuts (Felicetti et al., 2004). Sables of East Asian forests likewise consume the cached seeds of Siberian dwarf pine and likely other stone pine species, and black bears eat seeds of limber pine, the cones of which open at maturity (Buskirk *et al.*, 1996; McCutchen, 1996). Ursids similarly exploit acorns of oaks and nuts of beeches, hazels, and chestnuts across large swaths of northern hemisphere forests (Frackowiak and Gula, 1992; Koike, 2010; Arimoto *et al.*, 2011). Essentially all large conifer seeds, acorns, and other hard mast ingested by carnivorans are lost from the disperser pool.

Key points

- Carnivorans affect soils, plants, and scavengers directly; the strength of these effects varies with species and context. They transport nutrients out of streambeds, make carrion accessible to various vertebrate scavengers, and disperse seeds of fruit-producing plants.
- The community influences of carnivorans are newly appreciated and are more important at the community level than those initiated by birds or mammalian herbivores, at least in some studies.
- Bears transporting the carcasses of spawning or spawned Pacific salmon are the best documented examples of carnivoran nutrient transport across ecological boundaries. Movement of salmon biomass out of stream courses encourages growth of woody vegetation, including of dominant conifer species. This process supports salmon spawning habitat in the long term.
- Pinnipeds also transport marine fish remains to land, but to a much lesser extent, because pinnipeds defecate and urinate in water, and have limited mobility on land.
- Badgers and other burrowing carnivorans affect soils, mixing layers and creating local topography and avenues for water percolation.
- Carnivorans play diverse roles in disease ecology, serving as hosts, reservoirs, and transmitters of pathogens, and limiters of host abundance.
- Whether carnivorans increase or decrease disease incidence depends largely on cross-infectivity and whether the carnivoran holds disease organisms in infectious stages for long enough to transmit them.
- Carnivoran families vary in their tendency to scavenge, and body size tends to determine which species facilitate or benefit from carrion availability.
- Carnivorans play limited documented roles in pollination of flowering plants, but strong roles in seed dispersal. On the other hand, they tend to consume hard mast without serving a dispersal function.

References

Allen, M. L. *et al.* (2015) "The comparative effects of large carnivores on the acquisition of carrion by scavengers," *American Naturalist*, **185**, pp. 822–33.

Altizer, S. *et al.* (2003) "Social organization and parasite risk in mammals: integrating theory and empirical studies," *Annual Review of Ecology, Evolution, and Systematics*, **34**, pp. 517–47.

Arimoto, I. *et al.* (2011) "Autumn food habits and home-range elevations of Japanese black bears in relation to hard mast production in the beech family in Toyama Prefecture," *Mammal Study*, **36**, pp. 199–208.

Bagchi, R. *et al.* (2018) "Defaunation increases the spatial clustering of lowland western Amazonian tree communities," *Journal of Ecology,* **106**, pp. 1470–82.

Ballard, W.B., Carbyn, L.N. and Smith, D.W. (2003) "Wolf interactions with non-prey," in Mech, L.D. and Boitani, L. (eds.) *Wolves: behavior, ecology, and conservation.* Chicago: University of Chicago Press, pp. 259–71.

Beasley, J.C., Olson, Z.H. and DeVault, T.L. (2016) "Ecological role of vertebrate scavengers," in Benbow, M.E., Tomberlin, J.K. and Tarone, A.M. (eds.) *Carrion ecology, evolution, and their applications.* Boca Raton: CRC Press, pp. 107–27.

Bello, C. *et al.* (2015) "Defaunation affects carbon storage in tropical forests," *Scientific Advances*, **1**, p. e1501105

Ben-David, M. Hanley, T.A., and Schell, D.M. 1998. Fertilization of terrestrial vegetation by spawning Pacific salmon: the role of flooding and predator activity. Oikos 83:47–55.

Brown, M.B. *et al.* (2015) "Invasive carnivores alter ecological function and enhance complementarity in scavenger assemblages on ocean beaches," *Ecology*, **96**, pp. 2715–25.

Brunker, K. and Mollentze, N. (2018) "Rabies virus," *Trends in Microbiology*, **26**, pp. 886.

Bump, J.K., Peterson, R.O. and Vucetich, J.A. (2009a) "Wolves modulate soil nutrient heterogeneity and foliar nitrogen by configuring the distribution of ungulate carcasses," *Ecology*, **90**, pp. 3159–67.

Bump, J.K. *et al.* (2009b) "Ungulate carcasses perforate ecological filters and create biogeochemical hotspots in forest herbaceous layers allowing trees a competitive advantage," *Ecosystems*, **12**, pp. 996–1007.

Buskirk, S.W. *et al.* (1996) "Diets of, and prey selection by, sables (*Martes zibellina*) in northern China," *Journal of Mammalogy*, **77**, 725–30.

Carthew, S.M. and Goldingay, R.L. (1997) "Non-flying mammals as pollinators," *Trends in Ecology and Evolution*, **12**, pp. 104–8.

Chesson, P. and J. J. Kuang. (2008) "The interaction between predation and competition," *Nature*, **456**, pp. 235–8.

Crait, J.R., and Ben-David, M. 2007. Effects of river otter activity on terrestrial plants in trophically altered Yellowstone Lake. *Ecology* **88**: 1040–52.

Deacy, W.W. *et al.* (2018) "Phenological tracking associated with increased salmon consumption by brown bears," *Scientific Reports*, **8**, p. 11008.

Drake, D.C. and Naiman, R.J. (2007) "Reconstruction of Pacific salmon abundance from riparian tree-ring growth," *Ecological Applications*, **17**, pp. 1523–42.

Duffy, M.A. *et al.* (2011) "Unhealthy herds: indirect effects of predators enhance two drivers of disease spread," *Functional Ecology*, **25**, pp. 945–53.

Duszynski, D.A., Kvičerová, J. and Seville, R.S. (2018). *The biology and identification of the Coccidia (Apicomplexa) of the carnivores of the world*. London: Elsevier-Academic Press.

Eldridge, D.J. and Whitford, W.G. (2009) "Badger (*Taxidea taxus*) disturbances increase soil heterogeneity in a degraded shrub-steppe ecosystem," *Journal of Arid Environments*, **73**, pp. 66–73.

Enders, M.S. and Vander Wall, S.B. (2012) "Black bears *Ursus americanus* are effective seed dispersers, with a little help from their friends," *Oikos*, **121**, pp. 589–96.

Farina, J.M. *et al.* (2003) "Nutrient exchanges between marine and terrestrial ecosystems: the case of the Galapagos sea lion Zalophus wollebaecki," *Journal of Animal Ecology*, **72**, pp. 873–87.

Felicetti, L.A., Schwartz, C.C., Rye, R.O. Gunther, K.A., Crock, J.G., Haroldson, M.A., Waits, L., and Robbins, C.T. 2004. Use of naturally occurring mercury to determine the importance of cutthroat trout to Yellowstone grizzly bears. *Canadian Journal of Zoology* **82**: 493–501.

Forsyth, D.M. *et al.* (2014) "How does a carnivore guild utilise a substantial but unpredictable anthropogenic food source? Scavenging on hunter-shot ungulate carcasses by wild dogs/dingoes, red foxes and feral cats in south-eastern Australia revealed by camera traps," *PLoS ONE*, **9**, p. e97937.

Frackowiak, W. and Gula, R. (1992) "The autumn and spring diet of brown bear *Ursus arctos* in the Bieszczady Mountains of Poland," *Acta Theriologica*, **37**, pp. 339–44.

Fredriani, J.M. and Delibes, M. (2009) "Seed dispersal in the Iberian pear, *Pyrus bourgaeana*: A role for infrequent mutualists," *Ecoscience*, **16**, pp. 311–21.

Frehner, E.H. *et al.* (2017) "Subterranean caching of domestic cow (*Bos taurus*) carcasses by American badgers (*Taxidea taxus*) in the Great Basin Desert, Utah," *Western North American Naturalist*, **77**, pp. 124–9.

Gable, T.D. *et al.* (2020) "Outsized effect of predation: wolves alter wetland creation and recolonization by killing ecosystem engineers," *Science Advances*, **6**, p. eabc5439.

Gao, X. and Sun, S. (2005) "Effects of the small forest carnivores on the recruitment and survival of Liaodong oak (*Quercus wutaishanica*) seedlings," *Forest Ecology and Management*, **206**, pp. 283–92.

Garcia-Rodriguez, A. *et al.* (2021) "The role of the brown bear *Ursus arctos* as a legitimate megafaunal seed disperser," *Scientific Reports*, **11**, p. 1282.

Gernandt, D.S. *et al.* (2005) "Phylogeny and classification of Pinus," *Taxon*, **54**, pp. 29–42.

Green, G.I., Mattson, D.J. and Peek, J.M. (1997) "Spring feeding on ungulate carcasses by grizzly bears in Yellowstone National Park," *Journal of Wildlife Management*, **61**, pp. 1040–55.

Hämäläinen, A. *et al.* (2017) "The ecological significance of secondary seed dispersal by carnivores," *Ecosphere*, **8**, p. e01685.

Harrer, L.E.F. and Levi, T. (2018) "The primacy of bears as seed dispersers in salmon-bearing ecosystems," *Ecosphere*, **9**, p. e02076.

Helfield, J.M. and Naiman, R.J. (2001) "Effects of salmon-derived nitrogen on riparian forest growth and implications for stream productivity," *Ecology*, **82**, pp. 2403–9.

Hewson, R. (1984) "Scavenging and predation upon sheep and lambs in west Scotland," *Journal of Applied Ecology*, **21**, pp. 843–68.

Hickey, J.R. *et al.* (1999) "An evaluation of a mammalian predator, *Martes americana*, as a disperser of seeds," *Oikos*, **87**, pp. 499–508.

Hocking, M.D. and Reynolds, J.D. (2012) Nitrogen uptake by plants subsidized by Pacific salmon carcasses: a hierarchical experiment. *Canadian Journal of Forest Research*, **42**, pp. 908–17.

Hofmeester, T.R. *et al.* (2017) "Cascading effects of predator activity on tick-borne disease risk," *Proceedings of the Royal Society B: Biological Sciences*, **284**, p. 20170453.

Holtgrieve, G.W., Schindler, D.E. and Jewett, P.K. (2009) "Large predators and biogeochemical hotspots: brown bear (*Ursus arctos*) predation on salmon alters nitrogen cycling in riparian soils," *Ecological Research*, **24**, pp. 1125–35.

Huang, S. *et al.* (2014) "Phylogenetically related and ecologically similar carnivores harbour similar parasite assemblages," *Journal of Animal Ecology*, **83**, pp. 671–80.

Huegel, C.N. and Rongstad, O.J. (1985) "Winter foraging patterns and consumption rates of northern Wisconsin coyotes," *American Midland Naturalist*, **113**, pp. 203–7.

Inman, R.M. *et al.* (2012) "The wolverine's niche: linking reproductive chronology, caching, competition, and climate," *Journal of Mammalogy*, **93**, pp. 634–44.

Janson, C.H., Terborgh, J. and Emmons, L.H. (1981) "Non-flying mammals as pollinating agents in the Amazonian forest," *Biotropica*, **13** (2, Suppl.), pp. 1–6.

Jones, S.C., Strauss, E.D. and Holekamp, K.E. (2016) "Ecology of African carrion," in Benbow, M.E., Tomberlin, J.K. and Tarone, A.M. (eds.) *Carrion ecology, evolution, and their applications*. Boca Raton: CRC Press, pp. 461–91.

Jordano, P. (2000) "Fruits and frugivory," in Fenner, M. (ed.) *Seeds: the ecology of regeneration in plant communities*. 2nd edn. Wallingford: CABI Publishing, pp. 125–66.

Kaczensky, P., Hayes, R.D. and Promberger, C. (2005) "Effect of raven *Corvus corax* scavenging on the kill rates of wolf *Canis lupus* packs," *Wildlife Biology*, **11**, pp. 101–8.

Kazacos, K.R. (2016) "Baylisascaris Larva migrans," US Geological Survey Circular 1412. Reston: US Geological Survey.

Koike, S. (2010) "Long-term trends in food habits of Asiatic black bears in the Misaka Mountains on the Pacific coast of central Japan," *Mammalian Biology*, **75**, pp. 17–28.

Krumm, C.E. *et al.* (2009) "Mountain lions prey selectively on prion-infected mule deer," *Biology Letters*, **6**, pp. 209–11.

Kurek, P., Kapusta, P. and Holeksa, J. (2014) "Burrowing by badgers (*Meles meles*) and foxes (*Vulpes vulpes*) changes soil conditions and vegetation in a European temperate forest," *Ecological Research*, **29**, pp. 1–11.

Lafferty, K.D. (1999) "The evolution of trophic transmission," *Parasitology Today*, **15**, pp. 111–5.

Malmberg, J.L., White, L.A. and VedeWoude, S. (2021) "Bioaccumulation of pathogen exposure in top predators," *Trends in Ecology and Evolution*, **36**, pp. 411–20.

Mansfield, A.W. (1967) "The mammals of Sable Island," *Canadian Field-Naturalist*, **81**, pp. 40–9.

Mattson, D.J. and Reinhart, D.P. (1995) "Influences of cutthroat trout (*Oncorhynchus clarki*) on behaviour and reproduction of Yellowstone grizzly bears (*Ursus arctos*), 1975–1989," *Canadian Journal of Zoology*, **73**, pp. 2072–9.

McCutchen, H.E. (1996) "Limber pine and bears," *Great Basin Naturalist*, **56**, pp. 90–2.

McInturf, A.G. *et al.* (2019) "Vectors with autonomy: what distinguishes animal-mediated nutrient transport from abiotic vectors?", *Biological Reviews*, **94**, pp. 1761–73.

McLoughlin, P.D. *et al.* (2016) "Density-dependent resource selection by a terrestrial herbivore in response to sea-to-land nutrient transfer by seals," *Ecology*, **97**, pp. 1929–37.

Nogeire, T.M. *et al.* (2013) "Carnivore use of avocado orchards across an agricultural-wildland gradient," *PLoS ONE*, **8**, p. e68025.

Packer, C. *et al.* (1999) "Viruses of the Serengeti: patterns of infection and mortality in African lions," *Journal of Animal Ecology*, **68**, pp. 1161–78.

Peres, C.A. *et al.* (2016) "Dispersal limitation induces long-term biomass collapse in overhunted Amazonian forests," *Proceedings of the National Academy of Sciences of the United States of American*, **113**, pp. 892–7.

Polis, G.A. and Strong, D.R. (1996) "Food web complexity and community dynamics," *American Naturalist*, **147**, pp. 813–46.

Poulin, R. (2010) "Parasite manipulation of host behavior: an update and frequently asked questions," *Advances in the Study of Behavior*, **41**, pp. 151–86.

Pozio, E. *et al.* (2009) "Hosts and habitats of *Trichinella spiralis* and *Trichinella britovi* in Europe," *International Journal for Parasitology*, **39**, pp. 71–9.

Prugh, L.R. and Sivy, K. J. (2020) "Enemies with benefits: integrating positive and negative interactions among terrestrial carnivores," *Ecology Letters*, **23**, pp. 902–18.

Quinn, T.P. *et al.* (2009) "Transportation of Pacific salmon carcasses from streams to riparian forests by bears," *Canadian Journal of Zoology*, **87**, pp. 195–203.

Raga, J.A. *et al.* (2018). "Parasites," in Würsig, B., Thewissen, J.G.M. and Kovacs, K.M. (eds.) *Encyclopedia of marine mammals*. 3rd edn. London: Academic Press, pp. 821–30.

Schindler, D.E. *et al.* (2003) "Pacific salmon and the ecology of coastal ecosystems," *Frontiers in Ecology and the Environment*, **1**, pp. 31–7.

Schmidt, W.C. and Holtmeier, F.-K. (eds.) (1994) *Proceedings—international workshop on subalpine stone pines and their environment: the status of our knowledge*. General Technical Report, INT–309. Washington, DC: USDA Forest Service.

Schmitz, O.J., Hawlena, D. and Trussell, G.C. (2010) "Predator control of ecosystem nutrient dynamics," *Ecology Letters*, **13**, pp. 1199–209.

Selva, N. *et al.* (2005) "Factors affecting carcass use by a guild of scavengers in European temperate woodland," *Canadian Journal of Zoology*, **83**, pp. 1590–601.

Steenhuisen, S.-L. *et al.* (2015) "Carnivorous mammals feed on nectar of *Protea* species (Proteaceae) in South Africa and likely contribute to their pollination," *African Journal of Ecology*, **53**, pp. 602–5.

Steinbeiser, C.M. *et al.* (2018) "Scavenging and the ecology of fear: do animal carcasses create islands of risk on the landscape?" *Canadian Journal of Zoology*, **96**, pp. 229–36.

Tsuji, Y. *et al.* (2020) "Effects of Japanese marten (*Martes melampus*) gut passage on germination of *Actinidia arguta* (Actinidiaceae): implications for seed dispersal," *Acta Oecologica*, **105**, p. 103578.

Vucetich, J.A., Peterson, R.O. and Waite, T.A. (2004) "Raven scavenging favours group foraging in wolves," *Animal Behaviour*, **67**, pp. 1117–26.

Watve, M.G. and Sukumar, R. (1995) "Parasite abundance and diversity in mammals: correlates with host ecology," *Proceedings of the National Academy of Sciences of the United States of America*, **92**, pp. 8945–9.

Wild, M.A. *et al.* (2011) "The role of predation in disease control: a comparison of selective and nonselective removal on prion disease dynamics in deer," *Journal of Wildlife Diseases*, **47**, pp. 78–93.

Willson, M.F. (1993) "Mammals as seed-dispersal mutualists in North America," *Oikos*, **67**, pp. 159–76.

Wilmers, C.C. *et al.* (2003) "Trophic facilitation by introduced top predators: grey wolf subsidies to scavengers in Yellowstone National Park," *Journal of Animal Ecology*, **72**, pp. 909–16.

Wilmers, C.C. and Getz, W.M. (2004) "Simulating the effects of wolf-elk population dynamics on resource flow to scavengers," *Ecological Modelling*, **177**, pp. 193–208.

Wilson, E.E. and Wolkovich, E.M. (2011) "Scavenging: how carnivores and carrion structure communities," *Trends in Ecology and Evolution*, **26**, pp. 129–35.

CHAPTER 7

Interactions with non-prey animals

Carnivorans interact with community members via channels other than predation. Notably, they facilitate, coexist, and compete with other vertebrates, including other carnivorans. They increase the fitness of community members by multiple pathways and have obligatory mutualistic relationships with some. The nature of these interspecific interactions varies with context; two species that at times avoid each other might show no avoidance or even cooperate at other times. Some interactive behaviors arise as needed, while others are instinctive, having evolved over long periods, and show little variation.

7.1 Facilitation

Over much of western temperate North America, the coyote and American badger hunt cooperatively for ground squirrels, including prairie dogs. In the nineteenth century, naturalists noted this behavior, which is now photographed frequently. Travelling together (Figure 7.1), the more cursorial coyote forces squirrels down burrows where the fossorial badger can capture them, while the badger forces them to the surface, where the coyote awaits (Minta *et al.*, 1992). Otariids and seabirds, especially gulls, interact similarly. Sea birds locate concentrations of small fish, and circle above them, calling. California sea lions see and hear the gulls, swim toward them, and approach the fish schools from below, pushing them towards the surface. When birds and sea lions attack the fish from opposite directions, a chaotic feeding scene ensues (Figure 7.2) (Pierotti, 1988). The benefits to the respective participants are not easily quantified, but facilitation seems more likely than competition. Pierotti (1988) illustrated the possible fitness consequences of this interaction. Western gulls nesting in colonies away from sea lion

concentrations are less successful at foraging and produce fewer fledglings—despite similar or larger clutch sizes—than colonies near sea lion concentrations. While this mutualism has been reported for other otariid species, it is not known for any phocid seals, which eat different fish species than do gulls.

Another commensal or mutualistic relationship occurs between herpestids of two species and Cape ground squirrels, all of which are similar in size and jointly occupy burrow systems in southern Africa (Figure 7.3). The three species fear the same predators, produce alarm calls, and share vigilance duties to some extent. Ground squirrels relax their vigilance in the presence of yellow mongooses (Makenbach *et al.*, 2013) and the presence of meerkats reduces ground squirrel vigilance as well (Waterman and Roth, 2007). In these cases, carnivorans and a squirrel species of size suitable to be their prey have evolved commensalisms or mutualisms based on common threats and information sharing.

Perhaps the strongest interspecific behavioral mutualism reported for any terrestrial vertebrate occurs between dwarf mongooses and three species of hornbills (*Tockus* spp.) and involves reciprocal interspecific altruism. These four species are ground-foraging predators of invertebrates, small reptiles, and small mammals in Africa. The hornbill species wait for mongooses to emerge in groups from termite-mound resting sites (Figure 7.4), arousing them if the hornbills are impatient. Hornbills and mongooses then forage together, relaxing vigilance in each other's presence. Hornbills benefit by following mongooses and eating insects and small vertebrates that mongooses have flushed from low vegetation (Rasa, 1987; Goodale *et al.*, 2010). Mongooses benefit from hornbills perching in trees and warning of nearby predators.

Carnivoran Ecology. Steven W. Buskirk, Oxford University Press. © Steven W. Buskirk (2023). DOI: 10.1093/oso/9780192863249.003.0007

Figure 7.1 The coyote and American badger, shown here in Yellowstone National Park, US, are mutual facilitators, commonly hunting together. Badgers enter burrows and force ground squirrels to the surface, where coyotes await. Coyotes force squirrels below ground, where badgers can excavate them. Minor conflicts arise over the possession of carcasses.
Photo: © James Hager.

Figure 7.2 California sea lions and gulls hunting northern anchovies in Monterey Bay, US. Gulls circle and call above fish schools, attracting sea lions, which force the fish to the surface, where both sea lions and gulls can more easily capture them.
Photo: H. Minakuchi/Minden.

Carnivorans participate in commensalisms as well. Ethiopian wolves forage amid herds of gelada monkeys. The wolves mingle with geladas while seeking rodents, eliciting almost no alarm reaction from the monkeys, and seldom trying to capture them (Figure 7.5). By contrast, the geladas flee to nearby rock outcroppings at the approach of feral dogs, which are monkey predators. Wolves habituate geladas to their presence by appearing nonthreatening and enjoy foraging success rates two to three times as high as those while foraging away from monkeys (Venkataraman *et al.*, 2015). The mechanism underlying this increased foraging success is not clear and the monkeys seem to incur no harm or benefit from the wolves' proximity.

Carnivorans serve an intriguing but limited function for common waxbills of sub-Saharan Africa. The waxbills build nests of grass leaves and other plant materials and incorporate carnivoran feces on and around the nests. The fecal droppings may provide camouflage or olfactory deterrence to predators; the mechanism of their effect is not certain. Nests with these deterrents have lower rates of predation than those lacking them (Schuetz, 2005).

Figure 7.3 Yellow mongooses and South African ground squirrels share vigilance duties at their common burrow system in Kgalagadi Transfrontier Park, southern Africa.
Photo: B. Marcon/Biosphoto.

Figure 7.4 Dwarf mongooses and a red-billed hornbill forage together at a termite mound in Tsavo East National Park, Kenya. Mongooses and hornbills represent one of the strongest examples of interspecific behavioral mutualism among terrestrial vertebrates.
Photo: Jabruson/Minden.

Here, the waxbills have evolved the behavior of collecting and placing carnivoran feces to protect their eggs and nestlings at no plausible cost or benefit to the carnivoran.

One interspecies association has evolved convergently and appears on at least two continents. Small musteloids with deterrent anal sprays accompany foxes of various species on their foraging trips. In North America and Central America spotted skunks commonly are recorded by camera traps travelling with common gray foxes, whereas in South America Molina's hog-nosed skunks behave similarly with chilla foxes (Figure 7.6). Photographic anecdotes are corroborated by fine-scale telemetry data from Yucatan, Mexico (Mejenes-López et al., 2021). The nature of these associations is not clear, but they are remarkable because the typical rate of travel of foxes and skunks would seem to require effort on the parts of one or both participants to remain together. It is not clear which species is altering its movement rate or other behavior to accommodate the other.

Figure 7.5 An Ethiopian wolf mingles with geladas while foraging for rodents in the Ethiopian highlands. The monkeys largely ignore solitary wolves, which spend hours only a few meters away. While near monkeys, wolves capture rodents at two to three times the rate as when distant from them, whereas monkeys appear to experience no associated costs or benefits. The mechanism underlying the wolves' enhanced predation success is not clear. Photo: © Jeffrey Kerby.

7.2 Competition and coexistence

One of the strongest forces acting on carnivorans is interspecific competition. Competition between avian and carnivoran predators surely occurs but intercarnivoran competition is more common and better documented. That carnivoran species can be so limiting to each other was little recognized before 1966, when Rosenzweig (1966) wrote his pioneering account of niche partitioning in weasels. Earlier, ecologists tended to regard prey abundance as the limiting factor for carnivoran populations. In such forms as exploitation, interference, and intraguild predation, competition shapes behaviors, distributions, and abundances of even the largest-bodied species.

7.2.1 Exploitation competition

Exploitation refers to consuming resources that are limiting to another species and evidence exists regarding this behavior for a number of carnivoran species pairs. In southern Kenya, spotted hyenas and lions compete strongly, both species killing, scavenging, and usurping prey from the other. Hyenas in Amboseli National Park have more body fat and higher reproductive output than those in Masai Mara National Park because of a sixfold difference in the relative abundance of lions and hyenas—in Amboseli, twelve hyenas per lion, in Mara two

hyenas per lion. The higher relative abundance allows Amboseli hyenas to drive lions away from carcasses more successfully (Watts and Holekamp, 2008). One recurring exploitation mode involves the usurping of canid reproductive dens based on body size. Circumpolar arctic foxes have longstanding ties to their reproductive dens, in some cases occupying the same ones for centuries. The limited supply of dens seems to reflect the distribution of frozen soils, which limits new den excavation during spring. For more than forty years, and with a warming climate, red foxes across the Holarctic have expanded their distributions northward into tundra habitats and displaced arctic foxes from dens. In response, arctic foxes now tend to limit themselves to dens far from red foxes, so that red foxes are seen as a major threat to arctic foxes throughout their range, and den exploitation seems to be a factor (Frafjord, 2003; Stickney et al., 2014).

7.2.2 Interference competition

Crombie (1947) first described this type of interaction, where one guild member harasses, attacks, or kills another. Among mammals, the phenomenon is especially common among carnivorans because the behaviors and morphologies key to predaceous feeding are the same ones useful for interfering with guild members (Donadio and Buskirk, 2006). Interference also is common among rodents

Figure 7.6 Competition-coexistence of carnivoran species is mediated by factors other than body size, dentition, and partitioning. Multiple pairs of canids and musteloids, the latter having defensive anal sprays, travel together, suggesting commensal or mutualistic benefits. A: A Molina's hog-nosed skunk follows closely behind a South American gray fox in La Pampa, Argentina, suggesting lack of fear of the larger carnivoran. B: Western spotted skunks are frequently recorded accompanying common gray foxes in New Mexico, US. In both cases, anal glands are a trusted deterrent to attack by the larger carnivoran.
Photos: A: © D. Kloster, J. Zanón. B: Philmont Scout Ranch.

and primates (Ferretti and Mori, 2019). Reinforcing potential violence, various behaviors and sensory modes (e.g. scent marking, vocalizing, and other territorial signals) establish and maintain areas from which competitors can be excluded. These signals advertise the potential for violent and lethal enforcement of community structures, the most extreme form of which is interspecific killing (Figure 7.7).

Haswell and colleagues (2018) showed the effectiveness of signaling for wolves in Croatia. Wolf urine applied across the landscape caused red foxes to reduce foraging times in food patches and to leave 34% more food behind in feeding trays, relative to areas without wolf urine. In South Africa, bat-eared foxes are frequent victims of interspecific killing and left 38% more food behind on new-moon nights than on full-moon nights, showing the fitness trade-off between foraging vs. avoiding competitors under different ambient light regimes. Bat-eared foxes also left 16% less food behind when human observers were present than when not, illustrating that human presence was perceived as protective from larger carnivorans (Welch *et al.*, 2017).

Interspecific killing resembles intraguild predation, the latter being the term for similar interactions

Figure 7.7 A juvenile swift fox plays with an injured long-tailed weasel a parent has brought to the den. Interspecific killing is the most extreme form of interference competition among carnivorans, where it is more common than in other mammalian orders. It is most likely between carnivorans that differ in body size by a factor of about 4. Photo: J. Zipp/Science Source.

in invertebrates and fish (see Lourenço *et al.*, 2014 for a discussion of semantics). One difference is that carnivorans commonly leave their victims uneaten. As an example among many, Sunde and colleagues (1999) showed that the proportion of red fox carcasses left uneaten by Eurasian lynxes that had killed them (37%) was much higher than that for carcasses of hares (0%) or roe deer (2%) killed by the same lynx. Why should carnivorans not consume the carnivoran competitors that they kill? One possible reason is to reduce parasitic infection, a hypothesis that assumes infective parasite stages are transmitted via ingestion (trophic transmission, Section 6.3.3), and that the risk of infection is greater between carnivoran species than between carnivorans and their typical prey. Moleón and colleagues (2017) tested these ideas in field experiments, comparing mammalian scavenger responses to herbivore vs. carnivore carrion. They observed similar numbers of scavenging species visiting herbivore vs. carnivoran carcasses, but greatly reduced numbers of species willing to feed at carnivoran carcasses. Only 12–33% of visiting species ate carnivoran carcasses, compared with 100% of those visiting herbivore carcasses. Intraspecific scavenging was not observed at any carnivoran carcass. Carnivorans were similarly unlikely to scavenge carnivoran carcasses in a midwestern North American community featuring red foxes, coyotes, and northern raccoons (Olson *et al.*, 2016). Clearly, if

predation assumes ingestion, "intraguild predation" does not aptly describe many instances of carnivorans killing each other. In the case of some small mustelids, a further reason to leave them uneaten is their unpleasant odor (Waggershauser *et al.*, 2021).

However, there are exceptions to this pattern. The red fox co-occurs with domestic and feral cats over large areas of the Northern Hemisphere, but seldom has been reported to eat cats. In those rare cases, scavenging of road kills, rather than predation, tends to be suspected. In central Italy, however, the red fox eats domestic cats primarily (Sogliani and Mori, 2019), and predation is suspected over scavenging. That is because road-killed cat carcasses should not vary as much in seasonal availability as was observed in fox diets. In this case, domestic cats seem to compete with foxes for human foods, with foxes killing cats for their flesh and for increased access to human foods, so that the boundaries between predation, exploitation, and interference are blurred.

Published examples of interference among carnivorans have expanded rapidly in recent years, improving our understandings of this process. Web of Science (accessed September 2021) indexed four articles dealing with "competition+carnivore" for the period 1961–1980, which grew to 103 during 1981–2000, and spiked to 1000 for the period 2001–2021. This rise is due partly to technological change, but partly to a growing recognition of the ecological significance of this phenomenon. Published accounts use various kinds of evidence, including necropsy findings, field evidence, and occupancy in relation to syntopy.

7.2.3 Influences on interference

The most consistently influential factor affecting co-occurrence is relative body size (Arias-Alzate *et al.*, 2022). The general pattern in carnivoran communities is consistent with game theory (Smith, 1982). A carnivoran species that is much larger than another tends to ignore the smaller species, the resource needs of which are too different to warrant the effort of attacking. Also, the smaller species enjoys some protection because of its speed, agility, and use of habitat structure for shelter. A 500-kg bear trying to capture a 50-g weasel in most habitats would waste time and energy. Interspecific killing is most likely to involve pairs of carnivorans differing in body weight by a factor of 2.5–8 (Donadio and Buskirk, 2006). Prugh and Sivy (2020) found that a factor of 4 was the ratio maximizing the chance for lethal interference. The same relative body size was the threshold for local exclusion of smaller-bodied species in a community of eight carnivorans ranging in size from Egyptian mongoose to Iberian lynx in southwestern Europe (*Monterroso et al.*, 2020).

Exclusion can occur at multiple spatial scales. Very locally, cheetahs avoid areas of high densities of Thomson's gazelle because of the increased risk of encountering lions or spotted hyenas hunting there (Durant, 1998). More extensively, American black bears historically avoided treeless regions of North America because of potential brown bear attacks away from trees (Herrero, 1972). This behavioral tendency remains even a century after the extinction of brown bears from many areas and has left its trace in the population genetic structure of black bears in the Intermountain West (Larson *et al.*, 2018). At a nearly continental scale, the coyote expanded its geographic range northeastward by over 3000 km following the extirpation of the wolf in the early twentieth century (Gompper, 2002). However, a noteworthy exception to the clear advantage of large body size in these interactions involves

the fisher and Canada lynx. The fisher (mean body weight about 4.5 kg) inflicted most animal-caused deaths of lynx (9–11 kg) in Maine (McLellan *et al.*, 2018). Adding to the anomaly, most killed lynx were partially eaten by fishers.

The relatively large range of body sizes over which carnivorans interfere strongly with each other reflects their pre-adaptation to interference competition. The ratio range (2.5–8) for maximum interference between carnivoran species pairs, with some avoidance of interference at ratios ≈ 1 differs markedly from the predictions of Hutchinson and MacArthur (1959), who expected that similar species could co-occur at body-size ratios > 3 and did not account for behavioral avoidance at ratios near 1. Exploitation, rather than interference, may have accounted for most of the competitive exclusion that they considered.

If large carnivorans limit carnivorans the next size smaller, then large carnivorans should facilitate even smaller species by limiting their dominant competitors. Canids exhibit this pattern well—wolves strongly limit mid-sized canids, which in turn limit red foxes, so that wolf presence facilitates red fox abundance in North America and Eurasia, either through limiting intermediate-sized competitors or by providing access to carrion (Rossa *et al.*, 2021). Not surprisingly, group living is a force multiplier in interference interactions (Arias-Alzate *et al.*, 2022). Elbroch and Kusler (2018) examined the dominance relationships between the puma, the most widespread New World carnivoran, and other guild members. Pumas were subordinate to all larger carnivorans and dominant to all smaller carnivorans. Wolves, similar in size to pumas, dominated pumas by hunting and traveling in packs. Natural history accounts reflect this as well; wolves travelling together can kill young bear cubs by harassing and distracting the mother, but a lone wolf is unlikely to do so (Ballard *et al.*, 2003, Trinkel and Kastberger, 2005). Being the largest-bodied carnivoran in a community permits greater prey size selection. Large body size commonly is perceived to require specializing on large prey, but, in fact, it permits use of smaller prey without interference when prey conditions require it (Ferretti *et al.*, 2020).

Evidence for how size-structured interactions affect carnivorans—and how carnivorans affect other community members—comes from various kinds of evidence. Studies using before–after or with–without (a larger carnivoran) comparisons are common. Data on known-cause mortality are compelling, as are studies of occupancy, often measured in recent years via camera trapping, in relation to syntopy. Spatio-temporal partitioning is a common method for inferring either exclusion or coexistence (Arias-Alzate et al., 2022). Many, but not all, studies show both spatial and temporal partitioning; Vissia and van Langevelde (2022) studied a community of ten Botswanan carnivorans and found strong spatial, but weak temporal, partitioning of most species pairs. Some studies (e.g. Tian et al., 2022) show an inverse relationship between spatial and temporal partitioning. Black bears in Nevada, US were shown to have differing effects on competitive interactions between smaller-bodied carnivorans depending on whether the bears were active or in winter dens. High black bear activity during autumn reduced competitive pressure exerted by coyotes and bobcats on the smallest carnivoran, the common gray fox (Moll et al., 2021). In this case, the seasonality of bear activity provides the with–without treatment effect that is the basis of the inference.

7.2.4 Mesopredator release

"Mesopredator release" describes some of these interactions and has been extended to various systems in which the putative dominant predator is a bird (e.g. Cassano et al., 2016), a shark (Ferretti et al., 2010), or a crayfish (Ficetola et al., 2012). The term was coined by Soulé and colleagues (1988) and has become shorthand for the increased abundance of a smaller predator when a larger one—to which it is competitively inferior—is absent or leaves. Some use it to include all consequent effects of the absence of carnivoran competition. Unfortunately, a term for the reverse process, in which the smaller predator becomes restricted by the presence or arrival of a larger one, has not gained traction. "Larger" and "smaller" are best understood here to refer to the range of body sizes that predispose to strong competition; an ermine is not a mesopredator to a wolf, nor a dwarf mongoose to a leopard. Some use "mesocarnivore" in ways that do not denote body size at all; they use it to refer to an intermediate level

of carnivory (Malmberg et al., 2021). A few studies (e.g. Brashares et al., 2013, Ripple et al., 2016) suggest links between mesopredator release and the concept of trophic cascades, although the two processes are unrelated. The former refers to changes in the dominant predator regime, the latter to predator-caused effects on prey that in turn affect plants. Colman and colleagues (2014) describe one cascade resulting from a carnivoran competitive interaction, but examples tend to be rare.

How common is mesopredator release? Jachowski and colleagues (2020) reviewed thirty-eight papers reporting mesopredator release involving North American carnivorous mammals and concluded that about one-half lacked strong evidence. Examples of weak support included contradictory findings regarding such response variables as distribution, abundance, and behavior of the smaller species. They acknowledged that studies of short-term shifts in behavior were suggestive of larger-scale and longer-term effects but cautioned against using "mesopredator release" in the absence of effects at the population level. Castle and colleagues (2021) studied the effects of dingo eradication on introduced red foxes and feral cats in Australia and reported a striking absence of mesopredator release in a large-scale study. They found no effects of dingo suppression—either temporary and repeated or sustained—on abundance of the smaller carnivorans, which were well within the body size range to elicit strong interference effects.

The primacy of body size as a structuring force extends to intraspecific interactions as well. Tarugara and colleagues (2021) showed that leopards attracted to bait stations in Zimbabwe responded to the presence of spotted hyenas, lions, and male (but not female) leopards. They spent the longest times feeding if no competitors were present but spent the shortest times feeding if male leopards and spotted hyenas were nearby. Lions alone and spotted hyenas alone had intermediate effects on feeding bout duration. They also considered the distance from a carcass at which leopards rested between feeding bouts, which also depended on the presence of competitors. Nearby lions caused leopards to rest four times as far from bait stations as if male leopards were present. Spotted hyenas had a still weaker effect on resting distance from baits.

7.2.5 Demographic effects of interference

Interspecific killing is seldom witnessed, and the killing species often is inferred. Some traumatic injuries of carnivorans cannot even be attributed to a mammalian vs. avian killer. Further, demographic effects of these events are poorly reflected by the frequency of carnivoran remains in the stomachs or feces of other carnivorans. Waggershauser and colleagues (2021) reviewed carnivoran and raptor diet studies and found that a single carnivoran species seldom rose above 5% of the species occurrences in another carnivoran's diet. Nevertheless, the authors concluded, and more direct investigations have shown, that interspecific killing is a major or the primary mortality source for some carnivoran populations. In the Serengeti, Tanzania, cheetah cubs suffered 95% mortality before reaching independence, 73% of which was caused by larger carnivorans (Laurenson, 1994, 1995). Ralls and White (1995) showed that larger canids (coyote, red fox, and domestic dog) accounted for eighteen of twenty-three known-cause deaths of kit foxes in California, and Trinkel and Kastberger (2005) found that five out of seven spotted hyena deaths in a single year were lion-inflicted. Farias and colleagues (2005) found that eleven out of twelve common gray fox deaths in southern California were attributable to coyotes or bobcats, and red foxes caused 50% of European pine marten deaths in Scandinavia (Lindström *et al.*, 1995). Reviewing fifty-eight intraguild killing estimates involving thirty-six carnivoran species pairs, Prugh and Sivy (2020) found that one-third of deaths (means of species-pair means) were caused by a larger carnivoran. The compounded effects of direct killing, denial of access to resources, and other forms of intimidation make interference competition the most consistently plausible driver of carnivoran presence and abundance. Reinforcing this, Alston and colleagues (2019) reviewed how extinction and subsequent restoration of large carnivorans affected various community interactions. They found that the most reciprocal effects between extinction and restoration of carnivorans—effects lost with extinction but restored with reintroduction—were suppressive effects on smaller carnivorans.

7.2.6 Coexistence or interference?

Between carnivoran species that are similar in body size—differing by less than twofold—lethal interactions are uncommon. This is expected, because carnivorans tend not to initiate heterospecific attacks unless success is assured. Absent some clear size difference, co-occurring carnivoran species tend to coexist via several mechanisms: partitioning of habitats or of times active or selecting non-shared prey items. Evidence exists of four Andean carnivorans (Andean cat, pampas cat, culpeo, and puma) that partitioned time active to avoid conflict. Only the Andean cat synchronized time active with that of its primary prey, the mountain vizcacha (Lucherini *et al.*, 2009). Various studies have shown similar effects, as well as interesting confounding variation (Table 7.1). Some studies demonstrate that the largest-bodied carnivoran species synchronizes its activity to its preferred prey, while smaller carnivorans (within the size range to elicit interference behaviors) restrict their active times to avoid larger guild members.

Aside from relative body size, several factors determine whether sympatric carnivoran species pairs tend toward conflict or coexistence:

1. **The number of large carnivores in the system**. Mid-sized carnivorans experience synergistic additive effects of exposure to multiple large carnivorans. Occurring with three large carnivoran species causes a more than threefold increase in risk of interspecific killing relative to one large carnivoran (Prugh and Sivy, 2020).

2. **Other morphological traits**. In general, carnivorans direct interference behaviors, including killing, toward close phylogenetic relatives (Donadio and Buskirk, 2006). Particularly within the Canidae, body shape (using relative leg length as a proxy) is similar across most species (Figure 7.8); morphological differences are mostly in size, which varies less than across the Mustelidae. Sympatric canid species tend to forage in similar ways for similar prey in similar habitats, increasing the likelihood of competition. This has been documented especially well in North America, where the wolf is competitively dominant over the coyote, the coyote dominant

Table 7.1 Studies reporting spatio-temporal partitioning among sympatric carnivoran species

Region	Species present (descending body size)	Patterns	Reference
Southern Africa	Wildcat, Cape gray mongoose, common genet, striped polecat, yellow mongoose	Partitioning by time of day (nocturnal, diurnal), spatial avoidance within those groups	de Satgé et al. (2017)
Botswana	Lion, spotted hyena, cheetah, African wild dog	Strong temporal overlap, including nocturnal hunting, but smaller-bodied wild dogs and cheetahs were sensitive to moon phase. Lions and hyenas were not.	Cozzi et al. (2012)
Madagascar	Fossa, falanouc, Malagasy civet, ring-tailed mongoose, broad-striped Malagasy mongoose	Two smallest species overlap minimally in activity time but have similar diets. Two medium-sized species overlap temporally, but diets differ. Fossa was crepuscular, but otherwise idiosyncratic in activity time, but highly sensitive to presence of domestic dogs.	Gerber et al. (2012)
South American pampas	Crab-eating fox, pampas fox	Two similar-sized species with similar diets partitioned habitat types and times active. Crab-eating foxes associated more with gallery forest and shrubland, pampas foxes with grassland.	Di Bitetti et al. (2009)
Taiwan	Masked palm civet, crab-eating mongoose, small Indian civet, Chinese ferret badger	Partitioning by time of day, elevation, slope, and canopy cover.	Chen et al. (2009)
Midwestern North America	Coyote, bobcat, common gray fox, northern raccoon, red fox, striped skunk	Most partitioning was along habitat—especially anthropogenic—attributes. Bobcats avoided human development, whereas red foxes, gray foxes, and striped skunks were attracted to them. Coyotes and raccoons were ubiquitous. Gray foxes avoided sites preferred by coyotes, suggesting tree-climbing to escape attacks.	Lesmeister et al. (2015)
Southern Africa	Black-backed jackal, bat-eared fox, cape fox	Cape foxes avoided core home range areas of jackals. Bat-eared foxes hunt in groups and specialize on insects.	Kamler et al. (2012, 2013)

over the red fox, and the red fox over smaller foxes (Berger and Gese, 2007; Levi and Wilmers, 2012; Newsome and Ripple, 2015). Prugh and Sivy (2020) showed that the risk of interspecific killing from large felids was the same for mid-sized felids vs. non-felids. By contrast, the risk of being killed by large canids was over five times greater for smaller canids than for smaller non-canids.

3. **Robustness of canine teeth** (Section 2.1). Species with weak canine dentition, such as the bat-eared fox and northern raccoon, are more likely to be victims than initiators of attacks (Figure 7.9).

Wolves kill northern raccoons (which prey little on smaller mammals) easily and the eradication of the wolf across the mostly treeless Great Plains of North America by 1900 allowed the raccoon to expand its geographic range broadly (reviewed by Buskirk, 2016, p. 322).

4. **Climbing and burrowing adaptations**. These traits facilitate escape. Boreal forest martens can escape from most competitors, including canids and male fishers, by climbing slender tree branches and jumping from tree to tree. Many felids, procyonids, and the common gray fox climb trees with lower bole branches to escape.

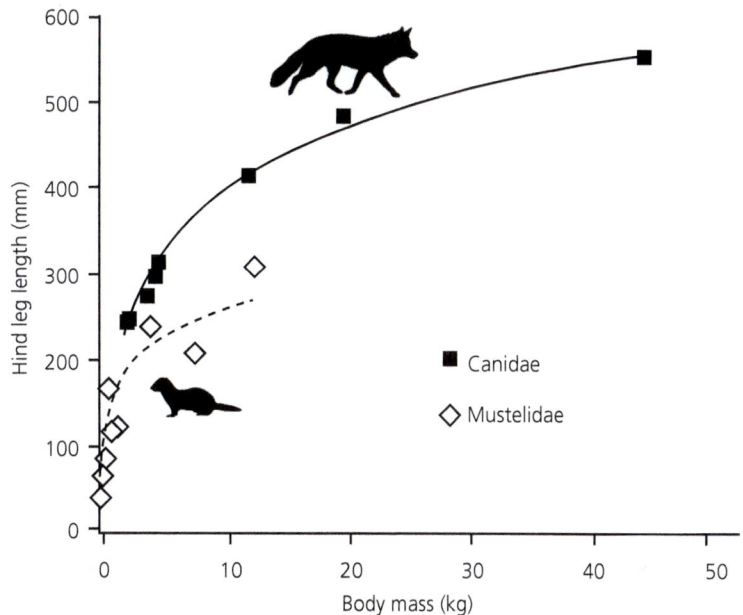

Figure 7.8 Hind leg length in relation to body mass of eight species of canids and nine species of mustelids from North America. This relationship is a proxy for morphological diversity and is consistent with differences between North American canids (r = 0.92) and mustelids (r = 0.88).
Data from Harris and Steudel, 1997.

Figure 7.9 A brown hyena carries the head of a freshly killed bat-eared fox in the Kalahari Desert, South Africa. Bat-eared foxes and other small foxes are commonly victims of interspecific attacks.
Photo: J. Swanepoel/Alamy.

Moehrenschlager and colleagues (2007) found that the kit fox in Mexico had lower mortality rates than the similar swift fox in Canada due to coyote and golden eagle predation on the latter. They attributed this to the high availability of black-tailed prairie dog burrows in their Mexican study area, which facilitated kit foxes escaping coyotes and avian predators.

5. **Chemical defenses**. Mephitids and some mustelids spray foul-smelling and toxic compounds from their anal glands. The volatile compounds in skunk sprays are not merely annoying, but can cause blood pathologies, blindness, and death in carnivorans (Fierro *et al.*, 2013). This defensive weapon allows coexistence with larger carnivoran species (Figure 7.6).

Musteloid species that employ chemical defense (Box 7.1) have wider niche spaces—more diverse diets and broader circadian activity—than species that lack chemical defenses (Arbuckle et al. 2013). Similarly, mephitids are more likely to co-occur with other mid-sized carnivorans than are species lacking noxious sprays (Hunter and Caro 2008). Noxious sprays are so effective that skunks are the only carnivorans more likely to be killed by raptors, which lack keen olfaction, than by other carnivorans (e.g. Lesmeister et al. 2010).

6. **Diet and habitat heterogeneity**. Heterogeneity of natural (as opposed to human-altered) habitats has been shown to increase the potential for carnivoran coexistence. Manlick and colleagues (2020) showed that American martens and fishers, sympatric forest mustelids across the southern boreal zone, co-occur more where various natural habitat attributes—for example, topography and vegetation type—were heterogeneous. The same was not true for human-altered habitats. Although guild-level analyses are lacking, co-occurring carnivorans in landscapes that lack heterogeneity and physical structure seem to predispose to more interference and less co-occurrence than if habitats are structurally complex and diverse (Figure 7.11) (Box 7.2).

7. **The spatial scale of the analysis**. Larger-scale studies appear more likely to detect negative correlations between occurrence of large- and mid-sized carnivorans. Smaller-scale studies tend to detect elevated activity of mid-sized carnivorans in response to the presence of the carrion that large carnivorans make available. However, activity near carrion tends to increase the risk of conflict and interspecific killing (Prugh and Sivy, 2020).

8. **The temporal scale of the analysis**. Arias-Alzate and colleagues (2022) showed that the functional niche constraints on carnivoran co-occurrence at temperate latitudes in the Americas weakened sharply during the last glacial maximum. They attributed this to continental ice sheets that caused range contractions and local extinctions. Coexistence of carnivorans was destabilized by abiotic factors.

The various predictors of coexistence vs. competition show interactive effects. For example, Di Bitetti and colleagues (2010) found that in a guild of neotropical felids, the most morphologically similar species had the most contrasting activity patterns. Where body form is dissimilar, the range of body sizes that predisposes to competition is reduced (Figure 7.13). Where relative body size and required foraging times predispose to conflict, tree-climbing or chemical defenses can permit coexistence.

Importantly, exploitation and interference can co-occur; a single competitive relationship can have multiple mechanisms. American black bears and brown bears have strongly overlapping diets in the Rocky Mountains, and black bears can reach higher densities by being more efficient exploiters of small berries. Larger-bodied brown bears are more efficient at using meat and roots, but black bears can limit brown bear abundance via exploitation. Brown bears are dominant in confrontations with black bears, affecting use of space (Mattson *et al.*, 1992, 2005). Likewise, a single interspecific interaction can have multiple risk–reward tradeoffs. Ruprecht and colleagues (2021) demonstrated this in coyotes co-occurring with pumas in Washington, US. Coyotes scavenged puma-killed ungulate carcasses so frequently that the diets of the two carnivorans were similar, even though coyotes seldom killed ungulates. However, pumas killed coyotes often; 23% of the coyote population died of puma predation yearly. For coyotes, sympatric pumas represented major sources of reward and risk, requiring complex decision making to maximize fitness.

7.2.7 Abiotic factors and interference

Spatio-temporal partitioning can reflect differential species responses to abiotic factors. Snow provides several examples and is important because it covers 33% of Northern Hemisphere lands from November to April (Lemke *et al.*, 2007) and affects locomotion and access to the snow–ground interface, where many boreal small carnivorans find their prey. Temperature adaptations aside, carnivorans vary strongly in how they live in and move through snow. Adaptations include long limbs and large feet (e.g. Canada lynx) and behavioral avoidance of deep snow. The coyote is an example of

Box 7.1 Aposematic coloration

Along with noxious chemical sprays, some carnivoran lineages have evolved pelage coloration schemes that advertise this defense. Markings are found across dorsa, flanks, and tails of diverse musteloids with anal gland sprays, but also are found on species that are stout and aggressive for their size (Figure 7.10; Stankowich *et al.*, 2011). Species with mostly white dorsa tend to be nocturnal, enhancing the display function in darkness. Distinct horizontal stripes across the trunk are common in species that can spray from anal glands accurately. Species with strong facial markings tend to be burrow-dwellers living in open country, which can deter a burrow invader.

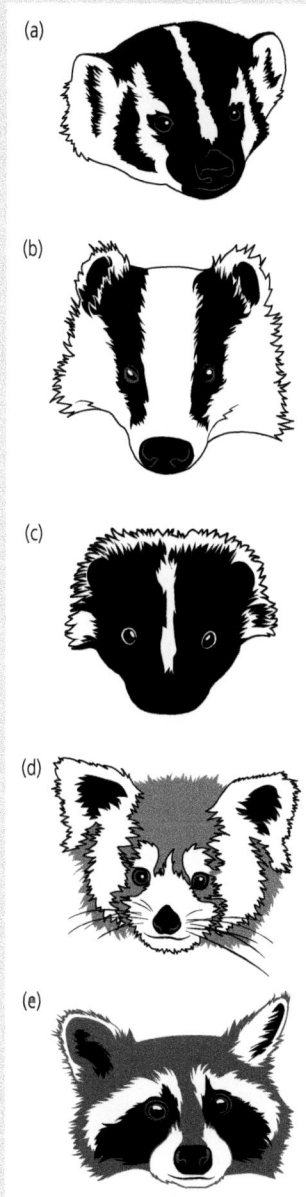

Figure 7.10 Facial markings of carnivorans reflect various postulated adaptive mechanisms. The American badger (A) and European badger (B) are fossorial and tend to defend themselves from burrow entrances, where their faces send strong aposematic signals. The striped skunk (C) sprays noxious chemicals from its anal glands; its facial markings are extensions of bicolored markings on its dorsum and tail. The red panda (D) and North American raccoons (E) seem not to fit the aposematic hypothesis.

Figure 7.11 Tree boles, including horizontal ones, are important for the coexistence of leopards with other large carnivorans. A female leopard uses a fallen tree to remain out of reach of a spotted hyena.
Photo: © Donna Passero.

a species that avoids soft snow (e.g. Pozzanghera *et al.*, 2016). Two small weasel species occupy the subnivean space most of the winter, traveling along rodent tunnels and resting in insulated microsites (Pruitt, 1960). The fisher is widely regarded as a competitor dominant to North American martens, but the two species respond differently to deep, soft snow. In both New England and the western North American parts of the range, fishers occupy areas lower in elevation, with shallower snow, or more crusted snow than do martens (Pauli *et al.*, 2022). In the absence of elevational zonation of snow characteristics (for example, in Michigan and Wisconsin, US, and Québec, Canada), martens and fishers show little or no partitioning of habitats (Manlick *et al.*, 2017; Croose *et al.*, 2019; Suffice *et al.*, 2020). It is not clear whether the absence of a spatial isolating mechanism results in greater conflict between these two species.

7.2.8 Human influence on interference

Humans and their actions strongly mediate carnivoran coexistence and interference. Commonly, larger-bodied carnivorans are more averse to contact with humans, perhaps because they are more likely to be harassed or persecuted. As a result, smaller carnivorans can avoid larger competitors by associating with roads, developments, or human activity. For example, in the Kenai Peninsula of Alaska, wolves were extirpated around 1915, and coyotes, absent from Alaska in the nineteenth century, arrived in the Kenai from north of the Chugach Mountains in the 1920s. By the 1970s, wolves had recolonized the peninsula and began competing with coyotes. Wolves killing coyotes became common, but the two species coexisted by eating different prey and because wolves avoided heavily travelled roads. The relative abundance of coyotes was fourteen times greater near roads with traffic than near roads closed to traffic (Thurber *et al.*, 1992). In this case, coexistence depended partly on wolves ceding areas near high human use to coyotes. A similar pattern occurs among smaller canids—a larger species is persecuted near human developments, so that a smaller canid takes advantage of human activities—and possibly human-provided foods—to find a competition-free environment. In North America, humans have tended to tolerate red foxes more than coyotes near towns and ranches (Gosselink *et al.*, 2003; Goad *et al.*, 2014). However, when suburban-urban red foxes undergo epizootic die-offs, even smaller-bodied swift foxes occupy urban areas they have vacated. Carnivoran occupancy of rural and suburban areas represents clear trade-offs between harassment by humans vs. by larger carnivorans.

Some exceptions to this pattern appeared in the last few decades. The increase in the commonness

Box 7.2 Raising of the dead

Among tree-climbing carnivorans, the leopard is especially noteworthy for its use of habitat structure to avoid competitors. As the world's most widely distributed wild felid, the leopard co-occurs with various carnivoran species (several of them group hunters) that can usurp its kills or attack it outright: lions, spotted hyenas, African wild dogs, tigers, wolves, and bears. In response, leopards have evolved tree-climbing behaviors, combined with caching prey above the reach of competitors (Figure 7.12). Caching in trees is more common in males than females and more common where competitors are abundant than where rare (Stein *et al.*, 2015). Hoisting carcasses also hides them from the sight of vultures. Lacking trees of suitable size for escape and hoisting prey, leopards are vulnerable to exploitative and interference competition.

Figure 7.12 A female leopard hoists an antelope that it has killed into a tree in Botswana, where the carcass and leopard are protected from competitors. The availability of trees for hoisting is an important predictor of the presence of leopards, where they are sympatric with larger carnivorans.
Photo: M. and C. Denis-Huot/Minden.

of coyotes in various US and Canadian cities is associated with their greatly expanded geographic range as well as relaxed human persecution (Gehrt, 2007). Where coyotes find sufficient habitat and food, they can become a permanent presence and the largest carnivoran in a major city,

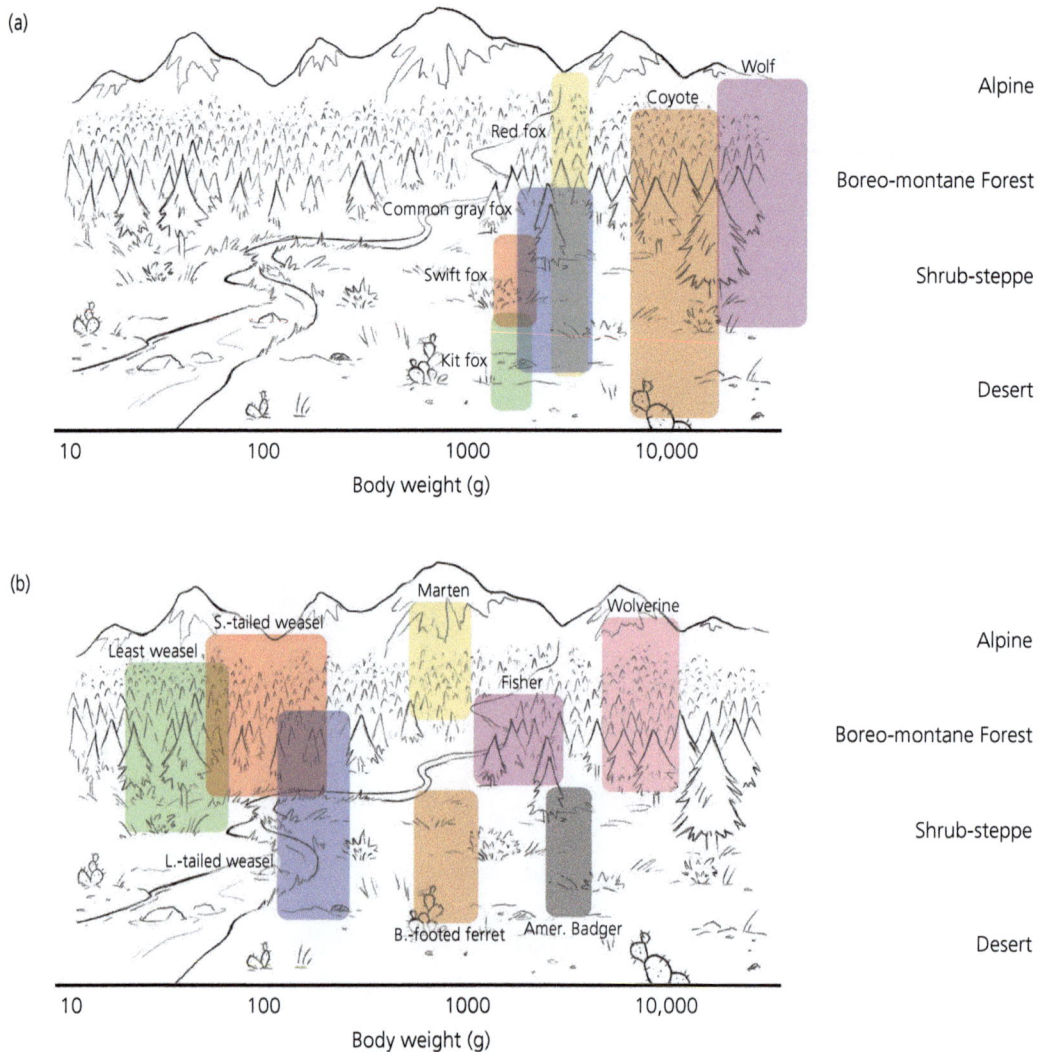

Figure 7.13 Habitat partitioning in relation to body size for (A) canids and (B) mustelids in western temperate North America. American minks and northern river otters occur in aquatic habitats that traverse a wide range of ecological zones, differing primarily in that minks occur closer to shorelines, while river otters occur farther from shore and in larger bodies of water. Both semi-aquatic species show minimal competition with other carnivorans. Kit foxes and swift foxes are sibling species that hybridize; niche partitioning is across a gradient of aridity. Gray foxes overlap with three *Vulpes* spp., coexistence being facilitated by the tree-climbing ability of gray foxes. Canids tend to show greater morphological similarity across species and greater niche breadth within species than is true for mustelids. Mustelids have a wider range of body sizes (across species and between sexes), narrower ecological tolerances within species, and higher levels of syntopy with other mustelids than is the case for canids. In the boreal zone, as many as eight species of mustelids occur in local sympatry.

which can reduce or exclude red foxes (Moll *et al.*, 2018). Coyotes can live for generations where extensive systems of urban parks and preserves are densely vegetated and interconnected (Figure 7.14) (Tigas *et al.*, 2002). Although human tolerance of urban coyotes is a novel development historically, it shows how body size interacts with landscape pattern to determine which carnivoran species becomes dominant (Gehrt *et al.*, 2013).

Figure 7.14 Home ranges of three resident coyotes in Cook County (Chicago), Illinois, US during 2004, showing the spatial association of coyotes with an extensive system of forest reserves. The westernmost home range incorporates mostly residential neighborhoods but shows strong selection for forested areas. The eastern home ranges are almost completely contained in the 15-km^2 Ned Brown Forest Preserve. Home range boundaries often follow major roads or other barriers, although extra-boundary forays are apparent.
Modified from Gehrt *et al.*, 2009, Figure 6.

7.3 Domestic dogs as competitors with wild carnivorans

Domestic dogs occupy special positions in carnivoran communities, in part because of the wide range of circumstances under which they live. At one extreme lie owned dogs, the movements and nutrition of which are controlled by humans. These dogs have limited interactions with wildlife, except when they venture into natural areas with humans, or if they affect disease reservoir size because of being unvaccinated (this discussion is based largely on Vanak and Gompper, 2009a). Near the other extreme are feral dogs, including dogs introgressed with wild *Canis*, which are wild-born, free-ranging, and perhaps even independent of human foods. The wildest dogs are the Australian dingo and related Australasian types, which are independent of humans, having lost any trace of their former domestication. Dogs seem not to be major exploitative competitors with wild carnivorans, as was shown for Ethiopian wolves (Atickem *et al.*, 2009), although studies that could detect such competition have been rare. Not surprisingly, how much domestic dogs and

cats compete with wild forms depends on their human-supplied foods (Silva-Rodriquez and Sieving, 2011). Dogs can inflict locally high mortality on wild prey. Particularly where they are the largest-bodied carnivoran present (e. g. dingoes) and occur at high densities, they are dominant to native marsupial predators as well as smaller-bodied carnivorans.

Interference competition between dogs and wild carnivorans is well documented, although its population-level consequences are uncertain. Red foxes in Australia avoid bait stations that have been visited by larger dingoes, suggesting fear of attacks (Mitchell and Banks, 2005), and red foxes in India likewise avoid food sources if dogs are present (Vanak and Gompper, 2009b). Bobcats in Colorado avoided areas where dogs were allowed to accompany hikers. Reports of interspecific killing between dogs and wild carnivorans are fairly common, but many of these are complicated by the use of dogs to hunt wild carnivorans, or by dogs protecting their owners (Figure 7.15). Owned dogs roaming in protected areas of Brazil showed little avoidance of sympatric pumas, and wild carnivorans the size of dogs or smaller showed little spatiotemporal

Figure 7.15 A domestic dog chases a coyote in an urban park in San Francisco, US. Dogs occasionally kill coyotes and other small canids, and vice versa, but the population-level consequences of these events are not clear.
Photo: © Janet Kessler.

avoidance of dogs (de Cassia Bianchi *et al.*, 2020). The effects on tayras and coatis were particularly weak, perhaps explained by the forested habitat and strong arboreal tendencies of those species. A last possible form of competition is apparent competition, in which a shared competitor favors one carnivoran over another. For example, carrion from dingo-killed carcasses benefits feral cats in Australia during droughts, increasing cat competition with native marsupials (Glen and Dickman, 2005). Similarly, disease-mediated apparent competition involves a shared parasite that affects dogs or wild carnivorans more severely. Rabies and canine distemper are enzootic in unvaccinated dog populations worldwide, and dog reservoirs of those pathogens have contributed to major die-offs of dozens of wild carnivoran species across the Canidae, Felidae, Hyaenidae, Phocidae, Mustelidae, Viverridae, and Procyonidae (Funk *et al.*, 2001; Cleaveland *et al.*, 2007). Some of these outbreaks have had major conservation significance, driving a few species to near extinction.

7.4 Carnivorans: apex, meso-, and other

The terminology invoked by ecologists to describe ecological roles of carnivorans can be disorienting. "Apex predators," "top predators," or "top-order predators" are loosely considered those that either are the largest-bodied or occupy the highest trophic levels in their communities. Because both criteria are used, the ranking of a focal carnivoran species among guild members can be difficult to assess repeatably. For example, the top predator considered by Crooks and Soulé (1999) in coastal Southern California was the coyote. But the coyote is only a mesopredator in the presence of Yellowstone wolves, which are two to three times their body mass—a ratio that predisposes to competition. Likewise, brown bears are considered apex predators in many North American and Eurasian studies, because of their large body sizes and carnivory (Tallian *et al.*, 2017). However, where they are sympatric with tigers in the Russian Far East, their "top" status is downgraded to "meso" because bears are important foods in the diet of some tigers, but not vice versa. Some male tigers even specialize on bears as food, killing mostly females and cubs (Seryodkin *et al.*, 2018). Clearly, "apex" and "top" describe context as much as a focal carnivoran itself, with few conventions observed in scientific use. A further complication is that large carnivorans need not function as predators to have a strong effect on smaller carnivorans. Engebretsen and colleagues (2021) found that large prey killed by pumas in Nevada, US were often usurped by larger-bodied

American black bears, although it was not clear whether bears displaced pumas from carcasses. That female pumas accompanied by young had briefer feeding bouts than those without young could be due to quicker consumption of freshly killed prey, or maternal anticipation of being surprised by a bear while kittens are in tow. The effect of recolonizing black bears on feeding bout duration and prey selection by pumas shows that the apex carnivoran predator can be subordinate to the apex scavenger. The scavenger outcompetes the predator through some combination of exploitation and interference; body size trumps predaceous nature in determining apex status. Lastly, previously established populations of subordinate carnivorans may withstand competition from a newly arriving dominant species better than vice versa. This has been inferred for several species pairs; cheetahs can tolerate the return of lions without major spatiotemporal shifts (Parker *et al.*, 2020). By contrast, reintroduced American martens in Wisconsin persist poorly where there are pre-existing populations of the fisher, a strong competitor (Manlick *et al.*, 2017).

Key points

- Aside from making carrion available to diverse vertebrate scavengers, carnivorans facilitate non-prey animals in diverse relationships, some of them mutualistic and some obligatory. Commensal carnivoran interactions with non-carnivorans are also known.
- Exploitative competition has been shown between carnivoran species, but interference competition is more commonly reported. It is inferred from studies of diets and movements, from necropsies, and from habitat-level competitive exclusion.
- Interspecific killing between carnivorans typically does not result in ingestion; carnivorans tend to avoid eating carnivoran flesh, although exceptions are common.
- Body size is the strongest predictor of whether carnivoran species coexist without conflict, co-occur with conflict, or avoid encounters via spatio-temporal partitioning. The ratio of body sizes that maximizes the likelihood of lethal attacks is about 4.

- Group-living species can dominate carnivorans larger than their individual sizes. The collective body mass is more predictive of competitive dominance than individual mass.
- Interspecific killing commonly is an important or the single most important cause of death among species prone to being victims. The compounded effects of killing, denial of access to resources, and exploitation make competition a conspicuous leading factor in carnivoran species abundances.
- Contextual factors that determine whether carnivoran species conflict or coexist include habitat diversity, habitat structural complexity, and the number of large-bodied carnivorans in the community.
- Species with robust canine teeth or noxious chemical sprays have reduced vulnerability to being excluded or killed. Access to trees or burrows for escape from other carnivorans is important for many small-bodied species.
- Humans influence competitive interactions strongly, by excluding larger-bodied carnivorans and providing food subsidies that favor some species, especially domestic dogs.
- Exploitation competition between domestic dogs and wild carnivorans is seldom reported, but interference competition is common.
- The vocabulary for carnivoran ecological roles—"top," "apex," "meso-," and "keystone"—is imprecise and context-dependent.

References

Alston, J.M. *et al.* (2019) "Reciprocity in restoration ecology: When might large carnivore reintroduction restore ecosystems?" *Biological Conservation*, 234, pp. 82–9.

Arbuckle, K., Brockhurst, M. and Speed, M.P. (2013) "Does chemical defence increase niche space? A phylogenetic comparative analysis of the Musteloidea," *Evolutionary Ecology*, 27, pp. 863–81.

Arias Alzate, A. *et al.* (2022) "Functional niche constraints on carnivore assemblages (Mammalia: Carnivora) in the Americas: what facilitates coexistence through space and time?" *Journal of Biogeography*, 49, pp. 497–510.

Atickem, A., Bekele, A. and Williams, S.D. (2009) "Competition between domestic dogs and Ethiopian wolf (*Canis simensis*) in the Bale Mountains National Park, Ethiopia," *African Journal of Ecology*, 48, pp. 401–7.

Ballard, W.B., Carbyn, L.N. and Smith, D.W. (2003) "Wolf interactions with non-prey," in Mech, L.D. and Boitani, L. (eds.) *Wolves: behavior, ecology, and conservation*. Chicago: University of Chicago Press, pp. 259–71.

Berger, K.M. and Gese, E.M. (2007) "Does interference competition with wolves limit the distribution and abundance of coyotes?" *Journal of Animal Ecology*, 76, pp. 1075–85.

Brashares, J.S. *et al.* (2013) "Ecological and conservation implications of mesopredator release," in Terborgh, J. and Estes, J.A. (eds.) *Trophic cascades: predators, prey and the changing dynamics of nature*. Washington, DC: Island Press, pp. 221–40.

Buskirk, S.W. (2016) *Wild mammals of Wyoming and Yellowstone National Park*. Oakland: University of California Press.

Cassano, C.R. *et al.* (2016) "Bat and bird exclusion but not shade cover influence arthropod abundance and cocoa leaf consumption in agroforestry landscape in northeast Brazil," *Agriculture, Ecosystems and Environment*, 232, pp. 247–53.

Castle, G. *et al.* (2021) "Terrestrial mesopredators did not increase after top-predator removal in a large-scale experimental test of mesopredator release theory," *Scientific Reports*, 11, p. 18205.

Chen, M.-T. *et al.* (2009) "Activity patterns and habitat use of sympatric small carnivores in southern Taiwan," *Mammalia*, 73, pp. 20–6.

Cleaveland, S. *et al.* (2007) "The conservation relevance of epidemiological research into carnivore viral diseases in the Serengeti," *Conservation Biology*, 21, pp. 612–22.

Colman, N. J. *et al.* (2014) "Lethal control of an apex predator has unintended cascading effects on forest mammal assemblages," *Proceedings of the Royal Society B: Biological Sciences*, 281, p. 20133094.

Cozzi, G. *et al.* (2012) "Fear of the dark or dinner by moonlight? Reduced temporal partitioning among Africa's large carnivores," *Ecology*, 93, pp. 2590–9.

Crombie, A.C. (1947) "Interspecific competition," *Journal of Animal Ecology*, 16, pp. 44–73.

Crooks, K.R. and Soulé, M.E. (1999) "Mesopredator release and avifaunal extinctions in a fragmented system," *Nature*, 400, pp. 563–6.

Croose, E. *et al.* (2019) "American marten and fisher do not segregate in space and time during winter in a mixed-forest system," *Ecology and Evolution*, 9, pp. 4906–16.

de Cassia Bianchi, R. *et al.* (2020) "Dog activity in protected areas: behavioral effects on mesocarnivores and the impacts of a top predator," *European Journal of Wildlife Research*, 66, p. 36.

de Satgé, J., Teichman, K. and Cristescu, B. (2017) "Competition and coexistence in a small carnivore guild," *Oecologia*, 184, pp. 873–84.

Di Bitetti, M.S. *et al.* (2010) "Niche partitioning and species coexistence in a Neotropical felid assemblage," *Acta Oecologica*, 36, pp. 403–12.

Di Bitetti, M.S. *et al.* (2009) "Time partitioning favors the coexistence of sympatric crab-eating foxes (*Cerdocyon thous*) and pampas foxes (*Lycalopex gymnocercus*)," *Journal of Mammalogy*, 90, pp. 479–90.

Donadio, E. and Buskirk, S.W. (2006) "Diet, morphology, and interspecific killing in Carnivora," *American Naturalist*, 136, pp. 524–36.

Durant, S.M. (1998) "Competition refuges and coexistence: an example from Serengeti carnivores," *Journal of Animal Ecology*, 67, pp. 370–86.

Elbroch, L.M. and Kusler, A. (2018) "Are pumas subordinate carnivores, and does it matter?" *PeerJ*, 6, p. e4293.

Engebretsen, K.N. *et al.* (2021) "Recolonizing carnivores: is cougar predation behaviorally mediated by bears?" *Ecology and Evolution*, 11, pp. 5331–43.

Farias, V. *et al.* (2005) "Survival and cause-specific mortality of gray foxes (*Urocyon cinereoargenteus*) in southern California," *Journal of Zoology*, 266, pp. 249–54.

Ferretti, F. *et al.* (2020) "Only the largest terrestrial carnivores increase their dietary breadth with increasing prey richness," *Mammal Review*, 50, pp. 291–303.

Ferretti, F. and Mori, E. (2019) "Displacement interference between wild ungulate species: does it occur?" *Ethology, Ecology & Evolution*, 32, pp. 2–15.

Ferretti, F. *et al.* (2010) "Patterns and ecosystem consequences of shark declines in the ocean," *Ecology Letters*, 13, pp. 1055–71.

Ficetola, G.F. *et al.* (2012) "Complex impact of an invasive crayfish on freshwater food webs," *Biodiversity Conservation*, 21, pp. 2641–51.

Fierro, B.R. *et al.* (2013) "Skunk musk causes methemoglobin and Heinz body formation in vitro," *Veterinary Clinical Pathology*, 42, pp. 291–300.

Frafjord, K. (2003) "Ecology and use of arctic fox *Alopex lagopus* dens in Norway: tradition overtaken by interspecific competition?" *Biological Conservation*, 111, pp. 445–53.

Funk, S.M. *et al.* (2001) "The importance of disease in carnivore conservation," in: Gittleman, J.L. *et al.* (eds.) *Carnivore conservation*. Cambridge: Cambridge University Press, pp. 11–34.

Gehrt, S.D. (2007) "Ecology of coyotes in urban landscapes," in Nolte, D.L., Arjo, W.M. and Stalman, D.H. (eds.) *Proceedings of the 12th Wildlife Damage Management Conference*. Lincoln: University of Nebraska, pp. 303–11.

Gehrt, S.D., Anchor, C. and White, L.A. (2009) "Home range and landscape use of coyotes in a metropolitan landscape: conflict or coexistence?" *Journal of Mammalogy*, 90, pp. 1045–57.

Gehrt, S.D. *et al.* (2013) "Population ecology of free-roaming cats and interference competition by coyotes in urban parks," *PLoS ONE*, 8, p. e75718.

Gerber, B.D., Karpanty, S.M. and Randrianantenaina, J. (2012) "Activity patterns of carnivores in the rain forests of Madagascar: implications for species coexistence," *Journal of Mammalogy*, 93, pp. 667–76.

Glen, A.S. and Dickman, C.R. (2005) "Complex interactions among mammalian carnivores in Australia, and their implications for wildlife management," *Biological Reviews*, 80, pp. 387–401.

Goad, E.H. *et al.* (2014) "Habitat use by mammals varies along an exurban development gradient in northern Colorado," *Biological Conservation*, 176, pp. 172–82.

Gompper, M.E. (2002) "Top carnivores in the suburbs? Ecological and conservation issues raised by colonization of northeastern North America by coyotes," *BioScience*, 52, pp. 185–90.

Goodale, E. *et al.* (2010) "Interspecific information transfer influences animal community structure," *Trends in Ecology and Evolution*, 25, pp. 354–61.

Gosselink, T.E. *et al.* (2003) "Temporal habitat partitioning and spatial use of coyotes and red foxes in east-central Illinois," *Journal of Wildlife Management*, 67, pp. 90–103.

Haswell, P.M. *et al.* (2018) "Fear, foraging and olfaction: how mesopredators avoid costly interactions with apex predators," *Oecologia*, 187, pp. 573–83.

Herrero, S. (1972) "Aspects of evolution and adaptation in American black bears (*Ursus americanus* Pallas) and brown and grizzly bears (*U. arctos* Linné.) of North America," in Herrero, S. (ed) *Bears, their biology and management*. New Series 23. Morges: International Union for Conservation of Nature and Natural Resources Publications, pp. 221–31.

Hunter, J. and Caro T. (2008) "Interspecific competition and predation in American carnivore families," *Ethology Ecology and Evolution*, 20, pp. 295–324.

Hutchinson, G.E. and MacArthur, R.H. (1959) "A theoretical ecological model of size distributions among species of animals," *American Naturalist*, 93, pp. 117–25.

Jachowski, D.S. *et al.* (2020) "Identifying mesopredator release in multi-predator systems: a review of evidence from North America," *Mammal Review*, 50, pp. 367–81.

Kamler, J.F. *et al.* (2012) "Resource partitioning among cape foxes, bat-eared foxes, and black-backed jackals in South Africa," *Journal of Wildlife Management*, 76, pp. 1241–53.

Kamler, J.F., Stenkewitz, U. and Macdonald, D.W. (2013) "Lethal and sublethal effects of black-backed jackals on cape foxes and bat-eared foxes," *Journal of Mammalogy*, 94, pp. 295–306.

Larson, R.C. *et al.* (2018) "The genetic structure of American black bear populations in the southern Rocky Mountains," *Intermountain Journal of Sciences*, 24, pp. 56–66.

Laurenson, M.K. (1994) "High juvenile mortality in cheetahs (*Acinonyx jubatus*) and its consequences for maternal care," *Journal of Zoology*, 234, pp. 387–408.

Laurenson, M.K. (1995) "Implications of high offspring mortality for cheetah population dynamics," in Sinclair, A.R.E. and Arcese, P. (eds.) *Serengeti II: dynamics, management and conservation of an ecosystem*. Chicago: University of Chicago Press, pp. 385–99.

Lemke, P. *et al.* (2007) "Observations: changes in snow, ice and frozen ground," in Solomon, S. *et al.* (eds.) *Climate change 2007: the physical science basis. Contribution of working group I to the fourth assessment report of the Intergovernmental Panel on Climate Change*. Cambridge: Cambridge University Press, pp. 337–83.

Lesmeister, D.B. *et al.* (2010) "Eastern spotted skunk (*Spilogale putorius*) survival and cause-specific mortality in the Ouachita Mountains, Arkansas," *American Midland Naturalist*, 164, pp. 52–60.

Lesmeister, D.B. *et al.* (2015) "Spatial and temporal structure of a mesocarnivore guild in midwestern North America," *Wildlife Monographs*, 191, pp. 1–61.

Levi, T. and Wilmers, C.C. (2012) "Wolves–coyotes–foxes: a cascade among carnivores," *Ecology*, 93, pp. 921–9.

Lindström, E.R. *et al.* (1995) "Pine marten–red fox interactions: a case of intraguild predation?" *Annales Zoologici Fennici*, 32, pp. 123–30.

Lourenço, R. *et al.* (2014) "Lethal interactions among vertebrate top predators: a review of concepts, assumptions and terminology," *Biological Reviews*, 89, pp. 270–83.

Lucherini, M. *et al.* (2009) "Activity pattern segregation of carnivores in the high Andes," *Journal of Mammalogy*, 90, pp. 1404–9.

Makenbach, S.A., Waterman, J.M. and Roth, J.D. (2013) "Predator detection and dilution as benefits of associations between yellow mongooses and Cape ground squirrels," *Behavioral Ecology and Sociobiology*, 67, pp. 1187–94.

Malmberg, J.L., White, L.A. and VandeWoude, S. (2021) "Bioaccumulation of pathogen exposure in top predators," *Trends in Ecology and Evolution*, 36, pp. 411–20.

Manlick, P.J. *et al.* (2020) "Can landscape heterogeneity promote carnivore coexistence in human-dominated landscapes?" *Landscape Ecology*, 35, pp. 2013–27.

Manlick, P.J. *et al.* (2017) "Niche compression intensifies competition between reintroduced American martens

(*Martes americana*) and fishers (*Pekania pennanti*)," *Journal of Mammalogy*, 98, pp. 690–702.

Mattson, D.J., Herrero, S. and Merrill, T. (2005) "Are black bears a factor in the restoration of North American grizzly bear populations?" *Ursus*, 16, pp. 11–30.

Mattson, D.J., Knight, R.R. and Blanchard, B.M. (1992) "Cannibalism and predation on black bears by grizzly bears in the Yellowstone ecosystem, 1975–1990," *Journal of Mammalogy*, 73, pp. 422–5.

McLellan, S.R. *et al.* (2018) "Fisher predation on Canada lynx in the northeastern United States," *Journal of Wildlife Management*, 82, pp. 1775–83.

Mejenes-López, S.d.M.A. *et al.* (2021) "First record of the coexistence of two mesocarnivores in the Yucatán Peninsula, México," *Therya Notes*, 2, pp. 79–84.

Minta, S.C., Minta, K.A. and Lott, D.F. (1992) "Hunting associations between badgers (*Taxidea taxus*) and coyotes (*Canis latrans*)," *Journal of Mammalogy*, 73, pp. 814–20.

Mitchell, B.D. and Banks, P.B. (2005) "Do wild dogs exclude foxes? Evidence for competition from dietary and spatial overlaps," *Austral Ecology*, 30, pp. 581–91.

Moehrenschlager, A., List, R. and Macdonald, D.W. (2007) "Escaping intraguild predation: Mexican kit foxes survive while coyotes and golden eagles kill Canadian swift foxes," *Journal of Mammalogy*, 88, pp. 1029–39.

Moleón, M. *et al.* (2017) "Carnivore carcasses are avoided by carnivores," *Journal of Animal Ecology*, 86, pp. 1179–91.

Moll, R.J. *et al.* (2018) "Humans and urban development mediate the sympatry of competing carnivores," *Urban Ecosystems*, 21, pp. 765–78.

Moll, R.J. *et al.* (2021) "An apex carnivore's life history mediates a predator cascade," *Oecologia*, 196, pp. 223–34.

Monterroso, P. *et al.* (2020) "Ecological traits and the spatial structure of competitive coexistence among carnivores," *Ecology*, 101, p. e03059.

Newsome, T.M. and Ripple, W.J. (2015) "A continental scale trophic cascade from wolves through coyotes to foxes," *Journal of Animal Ecology*, 84, pp. 49–59.

Olson, Z.H., Beasley, J.C. and Rhodes, O.E., Jr. (2016) "Carcass type affects local scavenger guilds more than habitat connectivity," *PLoS ONE*, 11, p. e0147798.

Parker, D.M., van de Vyver, D.B. and Bissett, C. (2020) "The influence of an apex predator introduction on an already established subordinate predator," *Journal of Zoology*, 313, pp. 224–35.

Pauli, J.N. *et al.* (2022) "Competitive overlap between martens *Martes americana* and *Martes caurina* and fishers *Pekania pennanti*: a rangewide perspective and synthesis," *Mammal Review*, 52, pp. 392–409.

Pierotti, R. (1988) "Interactions between gulls and otariid pinnipeds: competition, commensalism, and cooperation," in Burger, J. (ed.) *Seabirds and other marine vertebrates: competition, predation, and other interactions*. New York: Columbia University Press, pp. 213–39.

Pozzanghera, K.J., Lindberg, M.S. and Prugh, L.R. (2016) "Variable effects of snow conditions across boreal mesocarnivore species," *Canadian Journal of Zoology*, 94, pp. 697–705.

Prugh, L.R. and Sivy, K.J. (2020) "Enemies with benefits: integrating positive and negative interactions among terrestrial carnivores," *Ecology Letters*, 23, pp. 902–18.

Pruitt, W.O. (1960) "Animals in the snow," *Scientific American*, 202(1), pp. 60–9.

Ralls, K. and White, P.J. (1995) "Predation on San Joaquin kit foxes by larger canids," *Journal of Mammalogy*, 76, pp. 723–9.

Rasa, O.A.E. (1987) "The dwarf mongoose: a study of behavior and social structure in relation to ecology in a small, social carnivore," *Advances in the Study of Behavior*, 17, pp. 121–63.

Ripple, W.J. *et al.* (2016) "What is a trophic cascade?" *Trends in Ecology and Evolution*, 31, pp. 842–9.

Rosenzweig, M.L. (1966) "Community structure in sympatric carnivora," *Journal of Mammalogy*, 47, pp. 602–12.

Rossa, M., Lovari, S. and Ferretti, F. (2021) "Spatiotemporal patterns of wolf, mesocarnivores and prey in a Mediterranean area," *Behavioral Ecology and Sociobiology*, 75, p. 32.

Ruprecht, J. *et al.* (2021) "Variable strategies to solve risk-reward tradeoffs in carnivore communities," *Proceedings of the National Academy of Sciences of the United States of America*, 118, p. e2101614118.

Schuetz, J.G. (2005) "Common waxbills use carnivore scat to reduce the risk of nest predation," *Behavioral Ecology*, 16, pp. 133–7.

Seryodkin, I.V. *et al.* (2018) "Interspecific relationships between the Amur tiger (*Panthera tigris altaica*) and the brown (*Ursus arctos*) and Asiatic black bears (*Ursus thibetanus*)," *Biology Bulletin*, 45, pp. 853–64.

Silva-Rodriguez, E.A. and Sieving, D.E. (2011) "Influence of care of domestic carnivores on their predation on vertebrates," *Conservation Biology*, 25, pp. 808–15

Smith, J.M. (1982) *Evolution and the theory of games*. Cambridge University Press, Cambridge.

Sogliani, D. and E. Mori. (2019) "'The fox and the cat': sometimes they do not agree," *Mammalian Biology*, 95, pp. 150–4.

Soulé, M.E. *et al.* (1988) "Reconstructed dynamics of rapid extinctions of chaparral-requiring birds in urban habitat islands," *Conservation Biology*, 2, pp. 75–92.

Stankowich, T., Caro, T. and Cox, M. (2011) "Bold coloration and the evolution of aposematism in terrestrial carnivores," *Evolution*, 65, pp. 3090–9.

Stein, A.B., Bourquin, S.L. and McNutt. J.W. (2015) "Avoiding intraguild competition: leopard feeding ecology and prey caching in northern Botswana," *African Journal of Wildlife Research*, 45, pp. 247–57.

Stickney, A.A., Obritschkewitsch, T. and Burgess, R.M. (2014) "Shifts in fox den occupancy in the greater Prudhoe Bay area, Alaska," *Arctic*, 67, pp. 196–202.

Suffice, P. *et al.* (2020) "Habitat, climate, and fisher and marten distributions," *Journal of Wildlife Management*, 84, pp. 277–92.

Sunde, P., Overskaug, K. and Kvam, T. (1999) "Intraguild predation of lynxes on foxes: evidence of interference competition?" *Ecography*, 22, pp. 521–3.

Tallian, A. *et al.* (2017) "Competition between apex predators? Brown bears decrease wolf kill rate on two continents," *Proceedings of the Royal Society B: Biological Sciences*, 284, pp. 20162368.

Tarugara, A. *et al.* (2021) "The effect of competing carnivores on the feeding behaviour of leopards (*Panthera pardus*) in an African savanna," *Ecology and Evolution*, 11, pp. 7743–53.

Thurber, J.M. *et al.* (1992) "Coyote coexistence with wolves on the Kenai Peninsula, Alaska," *Canadian Journal of Zoology*, 70, pp. 2494–8.

Tian, J. *et al.* (2022) "Spatial and temporal differentiation are not distinct but are covariant for facilitating coexistence of small and medium-sized carnivores in Southwestern China," *Global Ecology and Conservation*, 34, p. e02017.

Tigas, L.A., Van Vuren, D.H. and Sauvajot, R.M. (2002). "Behavioral responses of bobcats and coyotes to habitat fragmentation and corridors in an urban environment," *Biological Conservation*, 108, pp. 299–306.

Trinkel, M. and Kastberger, G. (2005) "Competitive interactions between spotted hyenas and lions in the Etosha National Park, Namibia," *African Journal of Ecology*, 43, pp. 220–4.

Vanak, A.T. and Gompper, M.E. (2009a) "Dogs *Canis familiaris* as carnivores: their role and function in intraguild competition," *Mammal Review*, 39, pp. 265–83.

Vanak, A.T. and Gompper, M.E. (2009b) "Dietary niche separation between sympatric free-ranging domestic dogs and Indian foxes in central India," *Journal of Mammalogy*, 90, pp. 1058–65.

Venkataraman, V.V. *et al.* (2015) "Solitary Ethiopian wolves increase predation success on rodents when among grazing gelada monkey herds," *Journal of Mammalogy*, 96, pp. 129–37.

Vissia, S. and van Langevelde, F. (2022) "The effect of body size on co-occurrence patterns within an African carnivore guild," *Wildlife Biology*, 2022, p. e01004.

Waggershauser, C.N. *et al.* (2021) "Lethal interactions among forest-grouse predators are numerous, motivated by hunger and carcasses, and their impacts determined by the demographic value of the victims," *Ecology and Evolution*, 11, pp. 7164–86.

Waterman, J.M. and Roth, J.D. (2007) "Interspecific associations of Cape ground squirrels with two mongoose species: benefit or cost?" *Behavioral Ecology and Sociobiology*, 61, pp. 1675–83.

Watts, H.E. and Holekamp, K.E. (2008) "Interspecific competition influences reproduction in spotted hyenas," *Journal of Zoology*, 276, pp. 402–10.

Welch, R.J. *et al.* (2017) "Hunter or hunted? Perceptions of risk and reward in a small mesopredator," *Journal of Mammalogy*, 98, pp. 1531–7.

CHAPTER 8

Interactions with prey

Carnivorans participate in trophic webs that include plants, herbivores, and other carnivorous taxa, all potentially influencing each other. Some of these interactions, for example, the effects of predation on prey populations, have been long studied and debated. Others are newly recognized, involve behavioral or physiological mechanisms in prey species, and emphasize more indirect effects than killing prey. Some of these interactions cause or result from predator–prey coevolution. Here, I consider the effects exerted by carnivorans on their prey, typically herbivores, but which do not meet the criteria for trophic cascades (Chapter 9). This is one of the most rapidly advancing fronts in carnivoran ecology, involving many recently described examples of carnivorans influencing prey morphology, behavior, physiology, and demography. One problem in this subject area has been semantic inconsistency across studies (Fauth et al., 1996), so that what is a cascade to one ecologist is simple trophic interaction, competition, or facilitated nutrient transport to another. A further limitation is that community ecology tends to be reductionist and to describe local phenomena so that patterns that are general across geographic areas and across taxa are difficult to identify (Simberloff, 2004). Regardless of how we label them, these community interactions can be strong and illustrate the ecological importance of predation by carnivorans.

8.1 Who eats whom?

Most broadly, the answer to this question involves body size and metabolic rate, but includes many morphological, physiological, and behavioral factors, in some cases looping back in unexpected ways. Predators, including carnivorans, generally are larger than their prey, but that trend varies with the body size of the carnivoran. Carnivorans with body masses less than 20 kg tend to prey on invertebrates and vertebrates smaller than themselves, for example, foxes prey on vertebrates the size of mice or rabbits, and lions prey on buffalo. The yellow-throated marten presents an exception (Twining and Mills, 2021). This, the largest-bodied (2–5 kg) of the otherwise solitary martens, travels in groups of up to three and cooperatively kills small cervids and primates larger than itself (Figure 8.1). Carnivorans larger than 20 kg mostly eat vertebrates around their own size or larger (Carbone et al., 2007). This trend translates into differences in foraging styles and sociality across body sizes. Preying on animals much larger than oneself requires high rates of energy expenditure, cooperative behaviors, and increased risk of injury. Large predaceous carnivorans tend to hunt in groups, despite the added costs of supporting non-hunting group members. The largest extant land mammals (e.g. adult elephants, rhinos, and hippos) are too large for even the largest group-hunting carnivorans to kill (le Roux et al., 2018); these "megaherbivores" are nearly immune from predation. The same was true during the late Pleistocene, when the large carnivoran community was more speciose than today's (Widga et al., 2017). Giant ground sloths and mammoths had no effective predators before encountering humans.

Second, across all metazoan predatory interactions, prey tend to have the same or a lower metabolic "type" than their predators, where types comprise (in ascending trophic order) invertebrate, vertebrate ectotherm, and vertebrate endotherm. Thus, mammals commonly kill and eat invertebrates and vertebrate ectotherms, but the opposite is rare (Cohen et al., 1993). Exceptions include reptilian

Carnivoran Ecology. Steven W. Buskirk, Oxford University Press. © Steven W. Buskirk (2023). DOI: 10.1093/oso/9780192863249.003.0008

Figure 8.1 Two yellow-throated martens jointly subdue and kill a Rhesus macaque in Corbett National Park, India. Martens and other small carnivorans are generally regarded as solitary hunters, but this cooperative behavior in the largest-bodied marten is increasingly documented.
Photo: © Chris Mills.

predators of mammals and nestlings: snakes, monitor lizards, and crocodilians, but these are relatively unimportant ecologically because over most of the land area of Earth and especially in the oceans, they occur at such low biomass densities (Bar-On *et al.*, 2018). Carnivorans, as vertebrate endotherms that have widely varied body sizes, collectively occupy a vast trophic niche, eating metazoans from zooplankton to primates, plus a wide range of plant species and parts.

Kill rates are potentially important attributes of carnivoran predation, the values being potentially useful in discussions of functional responses, prey limitation, energy requirements, and competition with other predators, including humans. While such rates are easily visualized (how many prey animals are killed per carnivoran per unit time), they are difficult to measure for several reasons. For small carnivorans, predation events are nearly invisible to humans, and for large carnivorans, they are challenging to record as well. An observer following a wolf pack in an airplane might obtain excellent data during the day, but nighttime events would be hidden, as would predation in dense forest or followed by heavy snowfall. Detailed kill rates have been obtained using GPS telemetry on pumas (e.g. Andreasen *et al.*, 2021), because as each kill requires one or more days for the puma to consume the carcass, a pause is detectible in puma movement patterns. However, a pack of wolves can kill and

completely consume an ungulate the size of a deer in an hour or less, which would not be detectable from such data. Consumption of small prey items likewise is difficult to infer from any means except direct observation. As a result, kill rates for all but a few carnivoran species are unavailable.

8.2 Do carnivorans limit prey abundance?

That predators affect prey numbers over short time periods seems obvious enough, but what evidence shows that carnivoran predation sets an upper limit on prey abundance across years? Some mid-twentieth-century ecologists believed that predators killed only old or weak animals that were about to die anyway (contrast Flader, 1974 with Errington, 1946). Errington proposed that American minks mostly killed a "doomed surplus" of muskrats, which were replaced with each birth pulse. If he were correct, the within-year effect of mink predation on muskrat numbers might be strong, yet a multi-year effect could be nil. In recent decades, the notion of such an ill-fated population segment has fallen out of favor; it is not clear that a doomed surplus exists for any population. Even if it does, predation removes animals before their fates otherwise, so that population size and food competition are temporarily reduced until the next birth pulse, with consequent ecological effects (Sinclair and Pech, 1996).

Modern estimates of predators limiting prey suggest strong effects—Hill and colleagues (2019) found that across 305 terrestrial vertebrate species, predation caused 55% of all deaths. More specific to carnivorans, Linnell and colleagues (1995) showed that north temperate neonatal ungulates lost 47% of their numbers to all causes in environments where predators occurred (mean of sixty-eight studies) but suffered only 19% mortality where predators were absent. Overall, predators—overwhelmingly carnivorans—accounted for 67% of neonatal deaths. By contrast, in the Serengeti ecosystem carnivoran predation accounted for only 36% of the biomass of ungulate carrion (Houston, 1979). Likewise, also referring to ungulate prey, two early 1990s reviews reached disparate conclusions. Boutin (1992) found that wolf predation did not limit North American moose abundance as much as that by bears and human hunting. Examining the same system, Van Ballenberghe and Ballard (1994) concluded that the way in which wolf predation limited prey abundance varied with context. Skogland (1991) considered ungulates more broadly and proposed several factors that mediate how carnivorans affect prey. These include spatial factors—migrating prey, heterogeneous habitat, prey refuges, and gaps between adjacent carnivoran territories—all of which reduce potential limiting effects. Other factors include the body size of the focal prey, pulse births, and a lack of alternative prey. Skogland found poor agreement between field data and the predictions of simple predator–prey models.

The most dramatic evidence for limitation comes from carnivoran arrivals where they were previously absent. Domestic cats released on islands or escaped from ships, small Asian mongooses introduced for rat or snake control, feral domestic dogs (including dingoes) and other newly arrived carnivorans have driven scores of avian and mammalian species to rarity or extinction (Frenot et al., 2005; Doherty et al., 2016). This effect is stronger on islands, especially small ones that lack habitat diversity. A similar outcome occurs when carnivoran species are introduced to where eutherians were formerly absent or rare. For example, carnivorans tend to drastically limit marsupials in Australia and adjacent islands, driving some populations to extinction. The invasion of southwestern Australia by red foxes and feral domestic cats caused such an effect, reducing or eradicating several native small marsupials (Wayne et al., 2017). Introduced red foxes likewise affect some Australian reptile populations strongly. Lace monitors (4–7 kg) were 1.8 times more abundant in areas with fox control than without it, although smaller reptiles (< 150 g) were unaffected by foxes (Hu et al., 2019). Salo and colleagues (2007) concluded from their meta-analysis that alien predators had stronger limiting effects on prey populations than native predators, and that effects were stronger on mainlands than on islands. However, this result was confounded by the inclusion of Australian mainland communities, which show the effects of eutherian-marsupial competition. In this case physiological differences confound ecological comparisons. Eutherians are physiologically superior to marsupials, and any consideration of predation on marsupials involves novelty of the interaction as well as relative physiological efficiency.

Excluding Australia and considering mainlands with diverse habitats and native eutherian predators and prey, limitation of prey by a single carnivoran species can be difficult to discern. The most reliable studies have tended to be addition or removal experiments, but these are costly if done solely for the purpose of research. The colonization of northern Europe by the invasive American mink has provided one test. This species arrived in northern Poland in around 1984, and within three years caused measurable decreases in waterbird numbers. The effect was strongest fifteen years after the arrival of invasive minks and averaged a 53% decline in abundance across thirteen waterbird species (Brzezinski et al., 2012). Between fifteen and twenty-five years post-arrival, American minks declined in abundance, and waterbird numbers recovered slightly. This effect is remarkable, considering that the European mink, an ecological analogue with some convergently evolved traits, produced no comparable effects. It raises questions of how the two mink species—occupying similar Grinnellian niches—could have evolved such different Eltonian ones. Natural events present additional opportunities that approach addition–removal experiments in clarity. A major die-off of red foxes in Fennoscandia revealed the responses of prey populations to a single predator species. Red fox abundance was limiting to two species of hares, two species of grouse, and roe deer fawns (Lindström et al., 1994).

8.2.1 Bottom-up vs. top-down effects on herbivores

Whether and under what circumstances carnivorans limit prey abundances often is framed in terms of bottom-up vs. top-down effects; are herbivores limited by forage, predation, or (in some cases) disease? For carnivoran predators, much evidence is based on ungulates: large canids, large cats, and bears preying on artiodactyls and perissodactyls. Many of these ungulates are subject to natural as well as human predation, and the effects of "nature" are often confounded. Krebs (2002) considered how ecologists have studied prey population dynamics and found two salient approaches: the density paradigm vs. the multi-factor mechanistic paradigm. The density approach considers bottom-up forces and searches for density-dependent effects, for example, by regressing some demographic variable (e.g. growth rate) on density. The slope or intercept conveys whether forage is limiting. The mechanistic approach identifies candidate external factors suspected of causing dynamics—predation, disease, hunting, or extreme weather events—and searches for support for each factor.

One Alaskan study examined the effect of translocation of brown bears out of a study area on the survival of moose calves inside it. It had been shown earlier that neonatal moose calves in that region suffered high losses to brown bear predation (Ballard et al., 1981). The follow-on study considered whether calf survival could be increased by translocating 60% of the brown bear population out of the study area. The treatment was accomplished by relocating forty-seven brown bears, but treatment efficiency was dampened by bears trickling back into the study area almost immediately, so that around 60% of the translocated bears had returned to the study area within a few months. Still, the translocations reduced calf mortality (birth

to November) from 55% before the translocation to 12% after it (Ballard and Miller, 1990). Part of this effect is simply that, by temporarily moving some bears away, predation risk is postponed until maturing calves are more predation resistant. This study is an example of Krebs's (2002) second approach—the researchers specifically suspected predation as a limiting factor, but did not estimate population-level effects of the improved calf survival or consider density dependence related to the removal.

A similar outcome was shown in Idaho, US, where six years of removal of pumas increased survival of mule deer fawns and does, but did not increase deer numbers. Similarly, removal of 24–75% of coyotes had no effect on either fawn–doe ratios or deer numbers. The absence of population-level effects may have been due to confounding bottom-up (weather) effects (Hurley et al., 2011). Krebs (2002) argues that the potential is high for these mixed effects; animals suffering from poor nutrition (bottom-up) become more vulnerable to predation (top-down), so that an uncertain mix of factors limits prey numbers. Such a relationship was shown for the Yellowstone wolf–elk system, in which density of adult elk affected pregnancies among yearlings, producing effects suggestive of forage competition. Adult pregnancy increased following years of high precipitation (bottom-up) but did not vary across a range of carnivoran densities (lack of top-down). The proportion of cows lactating in the fall (an index of how many still had calves) decreased with an increasing ratio of brown bears to elk, and calf recruitment decreased with increasing wolf–elk ratio, both patterns suggestive of top-down effects. However, calf recruitment decreased with increasing winter severity, consistent with bottom-up effects (Proffitt et al., 2014).

Additional studies provide contradictory messages about how ungulate densities are limited. Mule deer in California, US underwent a decline that was consistent with bottom-up forcing (competition for forage). However, the population also showed top-down effects from mountain lion predation during the subsequent recovery, which slowed population expansion (Pierce et al., 2012). Letnic and Ripple (2017) compared large herbivore responses to primary production where canid predators were common vs. rare. In low-predation settings, herbivores increased with primary production, but under high predation they did not, suggesting that carnivoran limitation of prey populations is mediated by interacting bottom-up and top-down forces. Similarly, Melis and colleagues (2009) studied roe deer and predator densities across Europe and found that large predators exerted weak effects in highly productive environments, but strong effects in less productive environments prone to harsh winters. Sinclair and Krebs (2002) generalized that bottom-up limitation should be considered the default control system for herbivore populations, but that predation and environmental disturbance can intervene. These examples serve to reinforce that bottom-up vs. top-down is not a dichotomous effect pathway. Limiting forces can switch between pathways, and strong herbivory–predation interactions are possible.

Hopcraft and colleagues (2009) examined how various resource gradients and the relative body sizes of ungulates and their carnivoran predators affected bottom-up vs. top-down forcing in African savannah ecosystems. They argued that small carnivorans were limited to killing small prey, whereas large carnivorans (e.g. lions), while they preferred prey of their own size or slightly larger, were also capable of killing smaller prey. Therefore, small ungulates (e.g. oribi, 18 kg) had many potential predators and showed top-down effects on population growth, whereas buffalo and giraffes had few predators and were less likely to be affected. Elephants, hippos, and rhinos are limited only by bottom-up factors. Bottom-up forcing may also differ between moist and dry environments in that dry climates tend to produce nutritious vegetation of low biomass, whereas wet climates tend to produce abundant vegetation of lower nutritive value. Much of the latter biomass is out of reach of or structurally defended against ungulates. The emergent effects of these interacting forces include the potential for small herbivores, feeding on nutritious forage, to persist in the face of top-down limitation in some areas of Africa. However, where soils and plants are nutrient-poor, small-bodied species cannot withstand predation losses and are rare. In other words, soil chemistry and predator body sizes combine to affect distributions

and abundances. So confounded and context-specific are our understandings of carnivoran–prey interactions that Krebs (2002) warned that population biologists risk becoming historians, rather than builders of successful predictive models.

In lieu of effective predictive models, ecologists commonly use indices to discern bottom-up from top-down effects (Bowyer *et al.*, 2005). One approach considers ratios of predator-to-prey biomass, the premise of which is that regressing the two values yields a relationship that can be used to predict predator and prey abundances in another area or at a future time. Problems arise, however, when multiple species of predators or prey interact, if numerical responses of predators to prey are lagged, or if competition among ungulates for forage is not considered. White-tailed deer support higher densities of wolves than does an equivalent biomass of moose because of their higher rates of population growth (Person *et al.*, 2001). The approach also requires estimates of predator kill rates (functional responses), which are regressed on the numerical abundance of prey or prey–predator ratios. Even in the case of simplified communities such as that on Isle Royale (see Section 8.6), these models do not account for the complexity of community interactions, especially the role of ungulate competition for forage.

Some ungulate ecologists use even simpler approaches, inferring limiting mechanisms for a population from physiological, reproductive, or demographic traits. This is possible because predation leaves telltale signs at the organismal or population level, as does forage limitation. Ungulate populations limited by predation tend to feature adult females with good energy reserves, especially fat depots. They have high pregnancy rates, especially in subadults, large litters, and large-bodied neonates. Survival of neonates commonly is low, but food is sufficient in quality and quantity. Populations limited by bottom-up forces tend toward females in poor body condition, with postponed first parturition, low pregnancy rates, small litters, and small neonates. In other words, bottom-up limitation tends to operate via nutrition affecting body condition, which in turn drives conception, gestation, parturition, and lactation. Top-down limitation operates via predators removing animals—particularly neonates and weakened adults—outright. Food-limited ungulate populations also tend to have low male:female ratios, because males are more vulnerable to seasonal physiological stresses than females. This effect is exaggerated under poor food conditions and can become so severe than not all females are impregnated due to scarcity of males.

8.3 Trophic diversity and limiting effects

A diverse diet stabilizes carnivoran populations in variable environments, increasing the likelihood of top-down limitation of prey. Particularly where alternate prey have low enough profitability that the focal predator cannot sustain itself solely by eating it (e.g. Fryxell and Lundberg, 1994), switching prey provides a buffer against shortages of the primary prey species. This effect has been shown for sea otters eating multiple invertebrate species (Ostfeld, 1982), red foxes preying on voles as well as ungulate fawns (Kjellander and Nordström, 2003), and wolves killing multiple ungulate prey species (Garrott et al., 2007). Ecologists now recognize some curious indirect effects of preying on multiple prey species. Apparent facilitation occurs when one prey species is abundant, causing predators to shift away from a less common, but more temporally reliable, prey species. In one example, lions killed mostly blue wildebeest and Burchell's zebras when those species were migrating through the Tarangire Ecosystem of northern Tanzania. However, when migratory ungulates were seasonally absent, calves of non-migratory giraffes were the primary prey (Lee et al., 2016). Essentially, migrant wildebeest and zebras provided dilution for giraffes when all three prey species co-occurred.

A more broadly cast example is provided by the black-footed ferret and American badger on North American grasslands and shrub-steppe. Formerly occupying most of this region, the ferret eats almost exclusively prairie dogs (*Cynomys* spp.); the two taxa have geographic ranges that correspond closely. The eradication of prairie dogs over most of this region led to the near extinction of the ferret by 1970 (Thorne et al., 1989). However, the same fate did not befall the American badger, which occupied the ferret's entire pre-settlement distribution. The badger preys on other ground squirrels, pocket gophers, and some leporids in addition to prairie dogs (Goodrich and Buskirk, 1998). Whereas the prey-specialized ferret has been reduced to <1% of its pre-settlement range, a diverse prey base allows the badger to persist where prairie dogs and ferrets have disappeared. Switching between animal and plant foods is likewise stabilizing for omnivorous carnivoran populations, as has been shown for brown bears switching between meat and fruit in Alberta, Canada (Nielsen et al., 2017).

8.4 How herbivores avoid predation

8.4.1 Antipredator structures

Species that are carnivoran prey have evolved diverse means to reduce the risk of being eaten, including structures and behaviors used in combination. The head ornamentations (horns, antlers, ossicones, and tusks) of most ungulates enhance breeding opportunities for males, but also defend against predators in one or both sexes (Mukherjee and Heithaus, 2013). How important are these antipredator weapons? Bull elk in the Yellowstone area that shed antlers early in winter are at higher risk of predation by wolves, although they have better body condition than bulls that retain antlers longer (Figure 8.2). Bull groups that included at least one animal that had shed its antlers were nearly ten times more likely to face attack by wolves than groups entirely of antlered bulls (Metz et al., 2018). The antipredator function of head ornamentation is especially conspicuous in female ungulates, in which it serves little mating function. Reviewing the traits of female bovid species as regards lacking horns, Stankowich and Caro (2009) found that large-bodied species that occur in open habitats are most likely to be horned, leading them to conclude that the primary function of horns in this guild is predator defense. Protecting neonates also is important; in caribou-reindeer, the only cervid taxon in which females grow antlers, nearly all cows giving birth in a year retain their antlers through the calving season, whereas nearly all non-pregnant cows shed

Figure 8.2 Many large herbivores use defensive weapons to fend off carnivoran attacks. A bull elk faces cautious wolves with its antlers in Yellowstone National Park, US. Elk that shed their antlers early in winter suffer higher predation losses than those that retain them longer. Photo: US National Park Service.

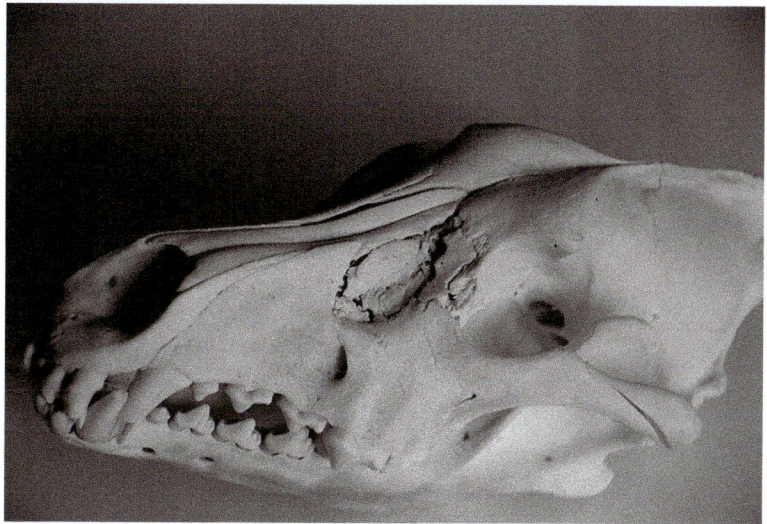

Figure 8.3 A partially healed depression fracture of the maxilla and frontal bones of a wolf from western Alaska, US. Wolves that prey primarily on moose have high rates of skull fractures, the result of moose kicking with their forelegs.

them early. This suggests some calf-defense function, although this could include defense against young bulls, rather than against predators (Whitten, 1995; Schaefer and Mahoney, 2001).

Hooves and teeth are also potent defensive weapons against carnivorans. Kicks with fore- or hind legs can injure or kill even African lions, which has led to co-evolved methods by which carnivorans approach and distract prey. Teeth can be used to bite (e.g. zebras), slash (e.g. peccaries, suids), or gore (e.g. elephants) various carnivorans, sometimes with lethal effect. It is difficult to quantify the population-level effects of these injuries; however, in Alberta, Canada, where wolves eat mostly moose, 14% of wolves had suffered skull fractures, presumably from moose hooves (Figure 8.3) (Losey *et al.*,

2014). Head injuries are less common where wolves mostly prey on caribou or deer, which do not flail their hooves at wolves.

These kinetic weapons against carnivorans have arisen convergently across mammalian orders. Australian kangaroos and wallabies evolved a body plan with robust hind limbs, but vestigial forelimbs. Their defensive tools—long, sharp claws—on fore- and hind limbs are wielded with powerful kicks (Figure 8.4) (Wright, 1993). Echidnas, hedgehogs, tenrecs, and porcupines carry protective spines, and a wide range of mammals and reptiles feature protective dermal armor against carnivoran attack. Some of these defenses are deadly deterrents; carnivorans ranging from martens to lions carry imbedded porcupine quills and some quill injuries

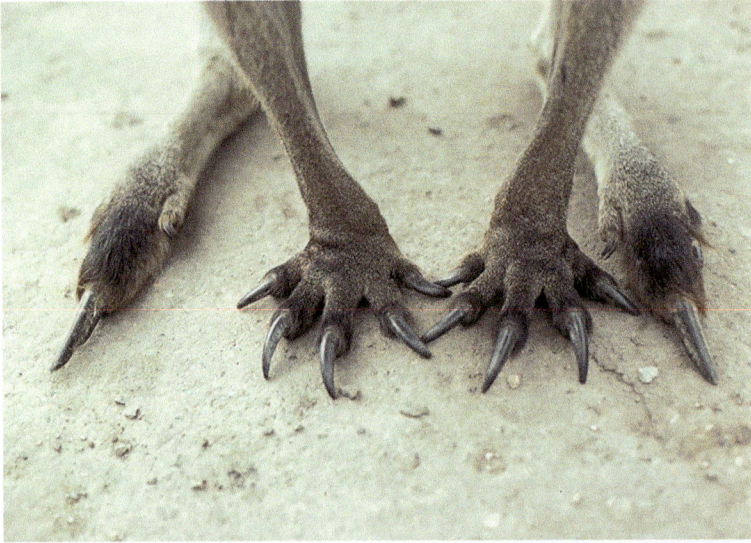

Figure 8.4 The fore and hind claws of an eastern gray kangaroo, which are wielded against attackers with a slashing motion. The defensive weapons of large herbivores are diverse.
Photo: K. Griffiths/Science Source.

Figure 8.5 A honey badger kills an African rock python in Kruger National Park, South Africa. Honey badgers in this area prey heavily on snakes, many of which are venomous.
Photo: © Susan McConnell.

are lethal (Quick, 1953; Mori *et al.*, 2014). This risk has led some carnivoran species, notably the fisher and mountain lion, to evolve behavioral counter-measures, including using the forelimb claws to manipulate North American porcupines onto their backs.

8.4.2 Antipredator chemicals

Potential prey and competitors of carnivorans deploy a range of chemical weapons—noxious sprays, venoms, and toxic body parts. Various insects produce toxins that repel mammals, for example, the protease inhibitors produced by the tracheal glands of some North American grasshoppers (Polanowski *et al.*, 1997). The skin glands of many amphibians produce toxins with inflammatory or paralytic effects—amines, pep-tides, steroids, and alkaloids (Daly, 1995). Various serpents have evolved venoms effective against ver-tebrate predators generally, and carnivorans are especially important predators of venomous snakes

(Figure 8.5). For example, snakes represented 67% of the diet of honey badgers in one southern African study, with venomous adders and cobras representing 13% of the ingested biomass (Begg *et al.*, 2003). The African crested rat, which was long known to elicit toxic reactions in domestic dogs that bit them, is now understood to carry toxins on its fur. After biting them, domestic dogs suffer symptoms ranging from minor loss of coordination to swift death, apparently of cardiac arrest. Kingdon and colleagues (2012) explored the toxin responsible and discovered that it was a cardenolide found in the roots and bark of a tree (*Acokanthera schimperi*) that the rats gnaw, mix with saliva, and slather onto specialized guard hairs. Inasmuch as the spiny rat is active mostly at night, carnivorans are the most likely predator toward which this evolutionary strategy is directed.

8.4.3 Induced predator defenses

Across metazoans, many species adjust their developmental trajectories, behaviors, and other traits in response to predation risk (Lima and Dill, 1990). Some invertebrates grow enlarged defensive structures (e.g. spines) in the presence of predators or produce more toxic deterrents if predation risk is high. Among the vertebrate prey of carnivorans, by contrast, developmental responses to predation risk are more likely to be behavioral. Adaptive behaviors require the ability of prey species to recognize potential predators, including rare or novel ones, but how do prey animals do so? Their deep evolutionary experience with predators of various taxa allows them to suppress antipredation behaviors when not needed, but to relearn them rapidly if any predator, familiar or novel, appears in their environment. This ability is reflected in the multipredator hypothesis (Atkins *et al.*, 2016), which recognizes that most prey species have encountered various predation threats over their evolutionary histories.

Risk-mediated prey behaviors are common across diverse taxa (Lima, 2002). Caro (2005) catalogued the behaviors of prey responding to predators, including carnivoran ones, during encounters. These behaviors begin with detection and include,

in rough chronological order, signaling to conspecifics, fleeing, seeking escape habitats, standing at bay, attacking (with mechanical and chemical weapons), and feigning death. Antipredator behaviors also include anticipatory ones—behaviors that reduce encounters with predators, based on recognized patterns. These include selecting or avoiding sites based on predation risk, altering movement rates, migrating, aggregating, disaggregating, and altering diel activity times. Universal rules governing antipredator behaviors are elusive. Prey species have been shown to perceive greater risk during daytime vs. at night, in open vs. closed habitats, under full moons vs. dark nights, and with vs. without humans present (Welch *et al.*, 2017). In Kenya, Günther's dik-diks responded to urine of African wild dogs by restricting movements within territories, avoiding scent marks, and avoiding overhead cover (Ford and Goheen, 2015a). North American moose respond similarly to wolves, avoiding areas with the highest likelihood of encountering wolves (Ditmer *et al.*, 2018). Roe deer respond variously to potential encounters with Eurasian lynx, depending on time of day. At night, they seek out low-risk areas, but during daylight do not, perhaps because of differing sensory acuities under circadian light variation (Gehr *et al.*, 2018).

The most consistent responses of prey species to the presence of predators are decreased activity, increased vigilance, avoidance of specific habitats, and avoidance of moonlit nights—sometimes interacting. These effects have been studied intensively in carnivoran–ungulate systems and fitness effects considered (Prugh *et al.*, 2019). Related to small carnivorans, Shapira and colleagues (2008) showed that native red foxes were more abundant and active near farms in an arid region of Israel because of food and water subsidies. Two species of gerbils were more abundant and active distant from foxes and farms. Gerbils showed the typical avoidance behaviors on full-moon nights only where red foxes were abundant; the concentration of predators produced site-specific avoidance behaviors.

8.4.3.1 Movements

Migrating is a common ungulate behavior in seasonal environments, and avoiding predators is one of two factors believed to favor it, the other being access to high-quality plants with geographically varying phenology. Migration can reduce predation risk in several ways. First, many carnivorans care for their dependent, semi-mobile young for weeks or months each year, restricting their ability to follow migrating prey. Further, carnivorans tend to defend stationary territories that do not saturate the landscape, within which migratory ungulates are a transient presence. Ungulates can pass through predator territories more quickly, then pause where predators are less common. Lastly, the lowest ungulate densities accessible to carnivorans during the year—those when ungulates are not migrating through—may limit carnivoran densities, so that the effective predation pressure over the entire migration route is lower than if predators had year-round access to the same prey (Fryxell, 1995; Fryxell and Sinclair, 1988).

Aggregation (herding) is another common antipredator defense among large herbivores in open habitats, because grouped animals benefit individually from collective vigilance, permitting more time for foraging. A further benefit is the reduction in risk resulting from being among many potential prey animals—the dilution effect. The assumptions underlying the presumptive benefits of aggregating suggest that benefits accrue unequally to participants (Lima, 1995). Such behaviors as cheating and selfish positioning can increase the inequality. Evidence shows that mixed-species groups enjoy additional protective benefits for at least two reasons. First, species may differ in sensory acuities by modality, so that a less-perceptive species can relax its vigilance in the presence of a more perceptive one (Stensland *et al.*, 2003). Second, the multiple species in a group may differ in desirability as prey, so that a more elusive or less preferred species may enjoy relative safety near a less elusive or more preferred one (Figure 8.6) (Fitzgibbon, 1990); this set of conditions can give rise to apparent competition (Section 8.8). Schmitt and colleagues (2014) were able to parse the effects of dilution vs. detection in Burchell's-zebra-only vs. zebra–bovid herds by considering vigilance in relation to herd composition. In zebra-only herds, vigilance declined with herd numbers, suggesting dilution, whereas in zebra–bovid herds, zebra vigilance was much lower overall, particularly when grouped with species preferred by lions (e.g. wildebeest). Vigilance did not vary strongly with herd size. For Burchell's zebras in mixed-species herds, detection is more important than dilution, particularly where the other species is wildebeest. Ng'weno and colleagues (2019b) found that African buffalo were less likely, but hartebeest more likely, to be killed by lions when found near high zebra densities. For neither hartebeest nor buffalo did conspecific density predict predation rate. Where lion activity was low, zebras and hartebeest were killed only in the densest vegetation, with the lowest visibility. Where lions were most active, zebra kills declined with the openness of habitat, whereas hartebeest kills increased in open habitats.

Figure 8.6 A mixed herd of Grant's gazelles (foreground) and Thomson's gazelles. Mixed-species groups of African bovids experience lower predation risk relative to single-species groups because of shared vigilance and differing sensory acuities. Photo: 1001slide/iStock.

The opposite pattern occurs as well—herbivores avoiding other guild members with the same predators to reduce risk. Caribou in northern British Columbia, Canada seek dispersed high-elevation sites for calving. This reflects high densities of moose at lower elevations, where wolves and brown bears concentrate their hunting during the post-calving period (Bergerud *et al.*, 1984). Animals dispersed at high elevations benefit further from mountain topography that forms visual barriers and wind eddies, limiting visual and olfactory detection by wolves and brown bears. Many possible interactions arise among carnivoran predation, habitat factors, and distribution and species composition of ungulate prey.

8.4.3.2 Spatio-temporal avoidance

"Landscape of fear" has gained currency as a descriptor of spatio-temporal variation in risk perceived by prey, particularly prey of carnivorans. While useful, this concept is a truism, in that fear—an animal's response to risk—must be heterogeneous to be adaptive. Spatially uniform risk requires no changes in prey response at all. This "fear landscape" experienced by prey is overlain on other landscapes: those of nutritional opportunity, mating opportunity, and aggressive interactions with other community members. Albery and colleagues (2020) have even proposed a "landscape of disgust"—heterogeneity in perceived risk of disease infection. All these landscapes interact in the cognition of prey animals, the two most commonly contrasted in the carnivoran literature being fear of predation and attraction to food. The most commonly described fear response by prey is spatio-temporal avoidance, commonly of discrete habitat types. Dense cover may reduce detection or present obstacles to pursuit, especially for small-bodied prey (e.g. snowshoe hares; Litvaitis *et al.*, 1985). By contrast, vicuñas hunted by pumas in the high Andes avoid areas with rock outcroppings and tall shrubs, which provide ambush cover for pumas (Smith *et al.*, 2019). In areas with steep topography, prey species from bovids to rodents seek to remain near this type of terrain where predators risk injury to hunt them (Mukherjee and Heithaus, 2013).

Abiotic factors mediate some of these habitat-based avoidance behaviors. In Serengeti National Park, Tanzania, lunar phase mediates spatio-temporal habitat selection among species killed by lions, but not by other species. Buffalo were the most formidable prey species studied and responded least to lunar phase, primarily via small spatial shifts. Warthogs were active strictly in daylight and sheltered underground at night. Wildebeest responded only to proportion of moon surface illuminated, without respect to ambient light levels, but only at certain seasons. By contrast, Thomson's gazelles and Burchell's zebras responded to ambient light levels in non-intuitive ways (Palmer *et al.*, 2017). In a North American study, snowshoe hares responded in intuitive ways to moon phase and snow cover. Under snowy conditions, hares reduced their activity levels during full moons, presumably reflecting their greater visibility on a white background under strong illumination. Under snow-free conditions, moon phase did not affect activity levels. These results were consistent with observed mortality rates, which were four times as high under full moons as under new moons when snow covered the ground (Griffin *et al.*, 2005).

These spatio-temporal shifts can become more permanent. White and colleagues (2012) found that elk in the northern Yellowstone area largely abandoned high-elevation areas with deep snow over the period of wolf reintroduction. They increased their use of low-elevation areas outside the national park, which the authors attributed to elk avoiding wolf attacks in deep snow. Effects such as these were widely predicted (e.g. Laundré *et al.*, 2001) for the Yellowstone system around the time of wolf reestablishment, but not all have been borne out by data. For example, elk were predicted to show small-scale transient movement responses to wolf presence, but multiple intensive studies did not confirm such a scenario. Middleton and colleagues (2013) examined small-scale elk movements and foraging behavior in response to wolf presence. Wolves, typically hunting in packs, approached within 1 km of a focal elk once per nine days, on average. Elk increased their movement rates and vigilance levels when wolves were within that distance but

Figure 8.7 Large herbivores as divergent as kangaroos and ungulates escape to water bodies when under attack by carnivorans. Here, a cow and calf moose use a pond to deter a pack of wolves in Denali National Park, Alaska. Photo: © Patrick Endres.

returned to normal movement rates within twenty-four hours. Elk did not alter their habitat use in response to close wolf encounters unless they were chased. Another analysis of the same system showed that while a few elk avoided open habitats during daylight hours, most showed no small-scale spatio-temporal response to perceived risk of wolf predation. Elk neither selected habitats associated with long-term patterns of wolf activity (an anticipatory response) nor changed habitats in response to immediate wolf threats (Cusack *et al.*, 2020).

8.4.3.3 Risk state and avoidance

Optimization theory predicts that behavioral responses of herbivores to predation risk should depend on their overall risk state; well-nourished animals should avoid risky foraging sites, and vice versa. In the Yellowstone system, moose were studied over late winter in relation to habitat types and encounters with wolves. Early during that period, moose avoided concentrating in riparian shrubs to access high-quality food, but as late winter progressed and body reserves drew down, they reentered riparian habitats to forage (Oates *et al.*, 2019). Other ungulate studies have corroborated that ungulates take contingency-dependent risks; however, herbivores exposed to homogeneous predation risk should choose places to forage

without regard to it (McArthur *et al.*, 2014). Further, having predators nearby differs from being under attack, the latter being when most herbivores make emergency decisions about habitats. Many ungulates flee to emergency habitats, for example, steep topography, burrows, trees, water bodies, and dense vegetation, when immediately threatened (Bleich, 1999). Terrestrial mammals as phylogenetically diverse as kangaroos, hippos, and moose have convergently evolved the behavior of escaping to bodies of water (Figure 8.7) (Mech, 1970, p. 208 for North American cervids, Wright, 1993 for macropodids). Javan rusa deer respond similarly to attack by Komodo dragons (Ariefiandy *et al.*, 2020). Staying near escape habitats involves complex trade-offs, however. Areas near predation refuges can be heavily grazed or browsed, but also can have higher concentrations of soil and plant nutrients, because of urine and feces deposited there (Hamel and Côté, 2007).

8.5 Physiological and demographic responses to risk

Chronic exposure to predation risk—along with other stressors—has long been thought to reduce the fitness of prey animals. Indeed, some of the proposed responses of prey animals to predation risk, if valid, suggest maladaptation—prey species become so stressed by the proximity of predators that they forego eating in favor of vigilance, lose body condition, terminate pregnancies, and in some cases die (Creel *et al.*, 2007 for an example). Physiological stress responses that could contribute to loss of fitness are of two types (Herman *et al.*, 2003). Reactive responses are direct physiological consequences of events, such as predator attacks or forgone feeding opportunities. Anticipatory responses result from animals perceiving the presence of predators and processing the potential threat through limbic-associated structures, without being chased, food-deprived, or otherwise physiologically stressed (Boonstra, 2013). Their putative physiological responses include hypertension, insulin resistance, and reduced immune competency. This anticipatory paradigm assumes that animals use higher-level processing to gauge future predation risk, and possibly react to the threat to an extent that reduces fitness. Do wild animals miscalculate their trade-offs between maintenance and predation avoidance? Some biologists are skeptical of such maladaptation and point out an overreliance on human biomedical models in our thinking about wild vertebrates (Boonstra, 2013). Among humans, psychological state and anticipation of future threats can indeed affect stress responses, with pathological consequences.

The population-level effects of combined reactive and anticipatory responses make up nonconsumptive effects (NCEs). In recent years, some authors have argued that carnivorans affect prey numbers more strongly via NCEs than by predation, and some evidence supports such a view (Peers *et al.*, 2018). Various mechanisms underlying NCEs are well documented, especially for invertebrates and ungulates. Most intuitively, prey animals forage less if they spend more time vigilant, and reduced foraging worsens body condition, leading to reduced disease resistance, terminated pregnancies, and starvation (Preisser *et al.*, 2005). Other mechanisms can be involved as well. In one study, a replica of a domestic cat placed near nesting European blackbirds reduced parental visits to nests by one third, which resulted in a 40% decrease in nestling growth. A further, unexpected effect was that nests where cat replicas were placed subsequently suffered high losses to various non-cat

predators. This may have occurred because predators witnessed the alarm reactions of parent birds, which advertised the presence of occupied nests (Bonnington *et al.*, 2013). Therefore, the presence of a simulated predator caused a reduction in parental care of young, as well as increased nest predation by other predators, the latter having used information carried by alarms from distressed parents. The relative strength of these mechanisms is an intriguing avenue for further research.

One long-recognized mechanism for NCEs is via the hypothalamic–pituitary–adrenal (HPA) axis, which responds to stressors—for example, perceived risk of predation—via elevated circulating cortisol. A stress event causes the hypothalamus to signal the pituitary to release adrenocorticotropic hormone, which stimulates the adrenal cortex to release cortisol. Cortisol mediates many metabolic, inflammatory, and immune processes, but at sustained high circulating levels leads to various pathologies (Suarez-Bregua *et al.*, 2018). While the HPA axis may play a role in the responses to carnivorans of many prey species, it has received close study in the snowshoe hare (Boonstra *et al.*, 1998). By the early twentieth century, hares were recognized to suffer illness and death associated with the crash phases of their wide population fluctuations in boreal North America (Green and Larson, 1938). Dramatic die-offs, attributed to "shock disease," were variously blamed on the stresses of overcrowding, starvation, or constant exposure to abundant predators. Several lines of evidence now have shown that proximity to carnivorans contributes to this syndrome; captive pregnant female hares exposed to predator cues tend to die before giving birth, and offspring of the females that do give birth die at high rates before weaning. In one Yukon, Canada experiment, exposure to domestic dogs under captive conditions reduced recruitment among captive hares by 87% (Sheriff *et al.*, 2009; MacLeod *et al.*, 2017). Some have noted, however, that the cue provided by the experimenters—the frequency or proximity of exposure to domestic dogs, and the amount and kind of habitat structure provided in the pens—differed markedly from what free-ranging hares experience. To examine this possible effect, a follow-on study monitored free-ranging hares exposed to encounters with an unleashed dog that chased them for a few seconds. Hares showed the expected stress responses, but no effects on body condition (Boudreau *et al.*, 2019). Clearly, encounters with carnivorans inflict stress responses in hares, which may dampen the predator-avoidance behaviors among hares, so that NCEs and consumptive effects could be linked. Some of these effects lead to lagged density dependence—a key component of prey population cycles.

8.6 Carnivorans and prey population cycles

How and to what extent carnivorans affect multi-annual population cycles of prey is one of the most enduring topics in carnivoran–prey ecology. Cyclic herbivore populations are known from various regions, particularly the far north, although cyclic patterns there have become dampened in recent decades. For example, populations of voles in Scandinavia and lemmings in Greenland have cycled with less amplitude in the twenty-first century than decades earlier (Hörnfeldt *et al.*, 2005; Gilg *et al.*, 2009). The essential elements of an herbivore population cycle are fluctuations in abundance with a regular period, the period being positively correlated with body size. The fluctuations have high but variable amplitudes, with lows and highs differing by an order of magnitude or more. The cycles exhibit synchrony across at least some sympatric species of prey and predator, and across geographic areas; they are not highly local or limited to single species. Lastly, the cycles are characterized by summer die-offs of the primary herbivore. This seems puzzling, because the most dramatic crash occurs at a time of year when temperatures are mild and food is abundant.

The synchrony of cycles across sympatric prey species can be puzzling because it can involve prey species of different trophic levels or body sizes (Ford and Goheen, 2015b). For example, vole populations in Britain and Scandinavia cycle with local interspecific synchrony. This is attributed to depletion of primary prey, typically a vole of Genus *Microtus* followed by high predation on alternate prey, for example, bank voles. Common shrews also

cycle with lagged synchrony with voles, presumably because of territorial interactions with voles (Korpimäki *et al.*, 2005). Because voles and shrews have different diets, bottom-up processes are not likely to affect them similarly. In Sweden, a major die-off caused by mange in red foxes showed how important fox predation was in transferring three- to four-year vole cycles to those of grouse and roe deer fawns (Lindström *et al.*, 1994).

Three kinds of vertebrate communities with cyclic population characteristics have been studied intensively: small mammals (voles and lemmings) and their predators in the circumpolar North; snowshoe hares and their predators in boreal North America; and ungulates and their predators in North America. How much we know about each system is inversely related to the body size of the focal herbivore. Small mammal population cycles are known from multiple studies, some experimental, and across two continents, but we have only the most basic understandings of any ungulate population cycle, which is based on one North American study. This knowledge gap is partially due to cycle periods—three to five years for voles, lemmings, and their predators (Korpimäki and Krebs, 1996), but several decades for moose preyed on by wolves (Post *et al.*, 2002). Small mammals are much more tractable subjects than ungulates.

What is the mechanistic basis of carnivoran effects on herbivore population cycles, and how strong are those effects? A basic, overarching rationale for prey population cycling is a delay in the system's density-dependent response to abundances of herbivores—lagged density dependence. The following proposed mechanisms attempt to isolate the factors that incorporate delayed density dependence and carnivoran predation. I do not consider proposed mechanisms that are intrinsic to herbivores (e.g. Inchausti and Ginzburg, 2009), or those that involve herbivore–plant interaction (Seldal *et al.*, 1994, but also see Klemola *et al.*, 2000). All of these mechanisms, predator-driven and otherwise, share the property of density-dependent responses being delayed after herbivore density changes (Korpimäki *et al.*, 2002; Gilg *et al.*, 2003).

Specialist predators are more likely than generalists to contribute to herbivore population cycles, because their numbers are more likely to track prey cycles. By nature of preying on multiple prey, generalists tend to stabilize their own populations and those of their prey over time. The most frequently mentioned specialist predators of voles are short-tailed weasels and especially least weasels, whereas generalist vole predators include foxes, owls, and jaegers (Korpimäki, 1993; Heikkilä et al., 1994). The major snowshoe hare specialist predator is the Canada lynx. Coyotes and great-horned owls are generalist predators that are thought to dampen multi-annual cycles of hares and other prey (Patterson et al., 1998). For moose, wolves are their only predators on Isle Royale, US, where cyclic moose populations have been described. Generalist predators are more common in more southern boreal forests, and this increasing predator generality at lower latitudes may dampen predation-mediated cycles there. Likewise, snow depth increases with latitude and may mediate predatory influences, because voles, lemmings, and snowshoe hares enjoy some protection from generalist—but not specialist—predators in the northernmost parts of their ranges (Hanski et al., 1991).

Predator numbers are more likely to track prey populations if the predator is resident, rather than seasonally migratory. Examples of migratory predators of cyclic rodent populations include Tengmalm's owl and juvenile raptors in Fennoscandia. Inasmuch as carnivoran predators of small mammals and snowshoe hares migrate short distances, if at all, they are more likely to respond and contribute to local prey population cycles than are avian predators (Korpimäki and Krebs, 1996).

Many studies have reported delayed numerical responses of carnivoran populations to fluctuations in prey numbers, cyclic or otherwise. Examples include red foxes preying on voles in Britain (O'Mahony et al., 1999), European pine martens preying on voles (one-year lag, Zalewski et al., 1995), and fishers preying on lagomorphs (two-year lag, Bowman et al., 2006). Gilg and colleagues (2006) compared the functional and numerical responses of arctic foxes, ermines, and two avian predators to lemming population fluctuations

in Greenland. They found that only ermines, the most specialized predator, showed clearly delayed numerical responses to lemming fluctuations. This prominence of a delayed implanter among those showing delayed responses suggests that the two delays might be causally linked. Specialized predators that exhibit delayed implantation seem likely to contribute to cycles in small mammals.

Predator-induced breeding suppression is another possible factor. Ylönen (1989) first reported that proximity of predators affects reproduction in prey species, observing that odors of weasels suppressed reproduction in captive bank voles. This later was found to be an artifact of the laboratory setting (Ylönen et al., 2006). However, similar phenomena were recognized in other systems, and it became clear that the reproductive failure occurred among female rather than male herbivores (Jochym and Halle, 2012). Norrdahl and Korpimäki (2000) proposed that the mechanism underlying these effects might not be endocrine, but via reduced mobility of small herbivores when predation risks were high. This could reduce foraging, affecting nutrition, maturation, and litter sizes. Sheriff and colleagues (2010, 2011) showed that female snowshoe hares had fecal cortisol levels that varied positively and synchronously with predator densities, and that high cortisol levels in females at parturition increased the responsiveness of the HPA axis among their offspring. Remarkably, female hares taken into captivity at high-stress phases of the cycle have litters, years later, that are smaller than those of females taken from the wild during low-stress phases (Sinclair et al., 2003). Further, the generational duration of the breeding suppression is longer if the predator-mediated stresses are greater (Sheriff et al., 2015).

In a now-classic multi-factor experiment conducted in Yukon, Canada, exclusion of carnivoran and avian predators from a fenced site increased snowshoe hare densities twofold during cyclic peaks, whereas chemical fertilization of plants increased densities threefold. Combining predator exclusion with fertilization increased hare densities elevenfold, so that food supply and predation had more than additive effects on hare abundance (Krebs et al., 1995). Predation causes almost all hare deaths during the peak and decline phases of the cycle, but food supply (quantity and quality) can influence predation mortality via body condition, infectious disease state, and stress levels (Krebs, 2001).

8.7 "Prudent" and "wasteful" predators

Two particular categories of predation deserve special mention, because both are mentioned in the literature, and because the adaptive value of both to the carnivoran predator is moot. Predation is considered "prudent" if a predator group kills in patterns that increase the sustainability of prey availability. The idea was proposed in the late 1960s, and became linked to the concept of kin selection, whereby animals behave in ways that maximize inclusive fitness, rather than individual fitness (Slobodkin, 1968; Mertz and Wade, 1976). The prudence might be in the form of reduced predation rates, or selection for prey with the lowest reproductive value. The potential for such a phenomenon is linked to several factors, including the ability of the predator to discern reproductive value of prey, the ability of kin to benefit from prudent behaviors, and—related to this second factor—the presence of competitors that could exploit any prudent behaviors. Tests of the concept were conducted mostly with invertebrates or microbes, but in no cases with carnivorans. The concept has faded to obscurity over the past thirty years, reflecting the general decline in kin-based explanations for behaviors of all but the most social species.

Surplus killing represents the nearly opposite phenomenon, and examples of it are widespread—geographically and taxonomically—in the carnivoran literature. Krebs (2001) refers to it as a common feature of lynx and coyote predation on snowshoe hares during the peak and decline phases of the snowshoe hare cycle. Two superficially similar surplus-killing behaviors should be distinguished. Carnivorans (e.g. arctic fox, least weasel) at times kill more prey than they (or their kin) can consume immediately, but cache carcasses in protected environments for future use. In Poland, least weasels tended to kill voles in summer and autumn at around the rate they consumed them. In late autumn and winter they killed far more than they consumed, and during severe winter weather they killed almost none, relying on carcasses cached in nests (Jędrzejewska and Jędrzejewski, 1989).

The superficially similar case is that of wild or free-ranging domestic carnivorans that kill far more prey animals than they consume (Kruuk, 1972); they may not consume any of multiple prey animals killed. At least two factors predispose to such surplus killing. First, the prey have not been exposed to the focal predator, or a similar one, for many generations—or ever—and have lost antipredator behaviors. Examples include introduced red foxes killing Australian marsupials (Short et al., 2002) and feral domestic cats killing seabirds on previously predator-free islands (Matias and Catry, 2008). The second factor is weakness or restricted mobility of prey. Neonatal caribou and Thomson's gazelles are intrinsically vulnerable and can suffer high local losses to predators (Kruuk, 1972; Miller et al., 1985). Domestic birds and hatchery fish stocks live in enclosures that limit escape (Andelt et al., 1980; Kloskowski, 2005), and animals weakened by weather conditions can be victims of mass killings. None of these factors, however, seems to explain anomalous surplus killing behaviors. For example, the author has investigated multiple cases of brown bears killing individual or groups of adult cow moose that weighed >350 kg each in Denali National Park, Alaska, but leaving their carcasses unopened for days. In each case, the bear killed and consumed the accompanying neonatal calf or calves (< 25 kg). Brown bears killing adult moose several times their size involves risk of injury and leaving > 90% of the prey biomass uneaten does not seem to justify such risk. Surplus killing is usually regarded as an aberrant and ecologically unimportant behavior, but it can exert population-level effects. For example, a single family of kit foxes killed ninety-one giant kangaroo rats in California, only to leave their carcasses uneaten around two den entrances (Endicott et al., 2014). The endangered, isolated population of giant kangaroo rats underwent a synchronous decline, a suspected result of the surplus killing.

8.8 Apparent competition

Apparent competition results when two prey species affect each other via a common predator, in a manner that superficially suggests competing prey species (Holt, 1977). Effects on the two prey species can be symmetrical but are more commonly asymmetrical—a more abundant prey population is relatively unaffected while a less abundant species experiences high predation losses (DeCesare *et al.*, 2010). In addition to differences in abundance, trait differences aggravate this asymmetry; the rarer species may be more palatable, easier to capture, or otherwise preferred over the more common one. The effect is visualized by the differences between Type II and Type III functional responses (Figure 8.8) (Box 8.1). When prey reach low densities, a predator may curtail hunting them, shifting to more common prey elsewhere, illustrated by the Type III response. If a predator does not curtail its hunting for the rare species at very low densities, the functional response follows a Type II curve. Type II responses push rare prey toward extinction. Examples are known from many systems: herbivore-plants, insectivore-insects, and a few with carnivoran predators (McLellan *et al.*, 2010). The uncommon, preferred vertebrate species may be rare enough to pose a conservation concern,

receiving close study. Examples include Przewalski's horses, which wolves in Mongolia prefer to more abundant livestock and red deer (Van Duyne *et al.*, 2009). In a North American system featuring the northern raccoon as the predator and American robins and wood thrushes as prey, thrushes suffered increased predation with increasing abundance of artificial nests of both species, whereas American robins showed no such effect (Schmidt and Whelan, 1998). Lions feature prominently in several African examples: a declining roan antelope population received increased lion predation following increased wildebeest and zebra numbers, which in turn reflected new human-created water sources (Harrington *et al.*, 1999). Similarly, large carnivorans, particularly lions, recolonized Laikipia, Kenya during 1995–2015, which contributed to declines in numbers of Jackson's hartebeest. Hartebeest are rarer there than Burchell's zebras, the primary prey of lions, but are more easily killed and preferred by lions (Ng'weno *et al.* 2017; 2019a,b). Hartebeest are sufficiently impacted by apparent competition that they exhibit lower rates of increase at lower abundances—predation-mediated Allee effects. Examples of these effects involving carnivoran predators are increasingly recognized (Courchamp *et al.* 2008).

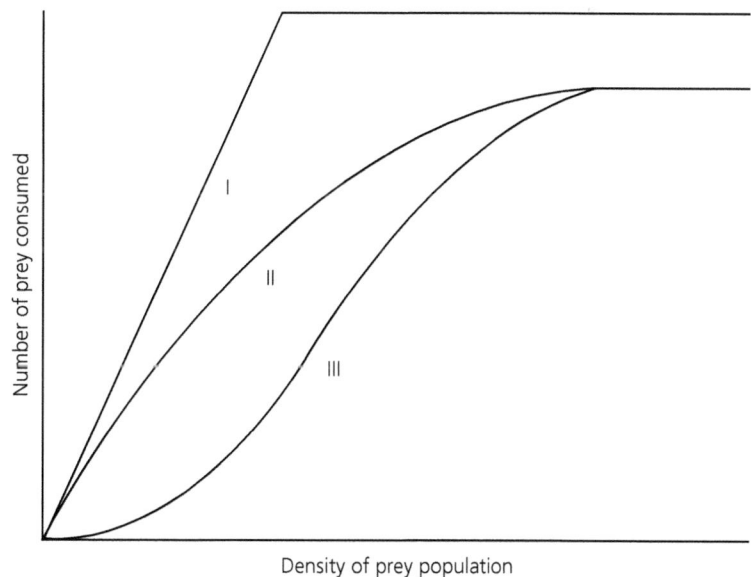

Figure 8.8 Three types of functional responses of consumers to variation in prey availability, following Holling (1959). Type I shows linear increase to a maximum, above which consumption remains constant. Type II shows decreasing consumption as availability increases. Type III resembles Type II, but with accelerating consumption at the lowest prey availabilities.

Box 8.1 The Macquarie Island parakeet

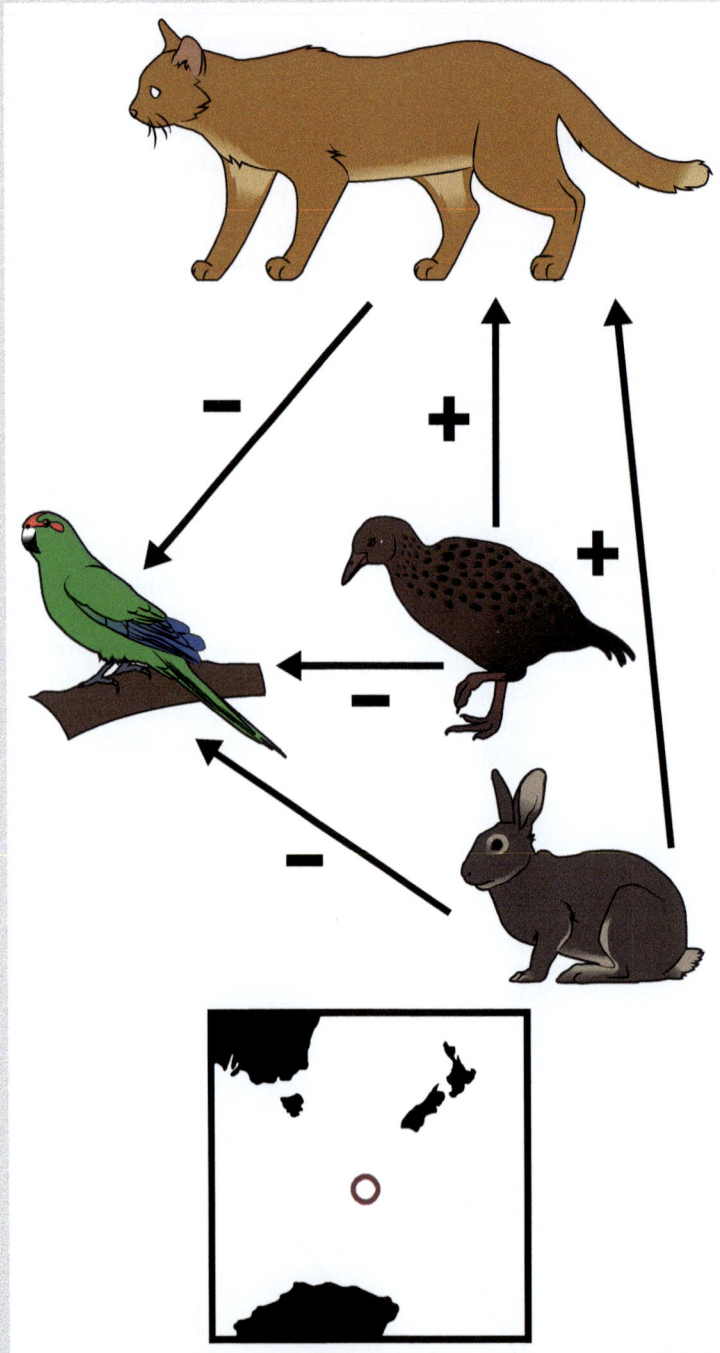

Figure 8.9 Apparent competition on Macquarie Island, southwest Pacific Ocean, which led to the extinction of the Macquarie Island parakeet in the nineteenth century. Domestic cats (and some domestic dogs) were introduced shortly after 1810 and preyed on the endemic parakeet, as well as on seasonally nesting penguins and petrels. The feralized carnivorans were food-limited in winter, which kept populations low year-round. When the European rabbit and the flightless weka were introduced around 1870, they provided abundant year-round foods, which allowed cat populations to increase. The parakeet became extinct within a decade.

Box 8.1 *Continued*

At least once during the historical period, apparent competition involving carnivorans has led to the extinction of a named species. When it was discovered by Europeans in 1810, Macquarie Island, midway between Tasmania and Antarctica, was home to the endemic, ground-nesting Macquarie Island parakeet. By 1820, both feral cats and dogs were established, but the parakeet persisted in the face of those exotic predators until the 1870s. The tipping point occurred around 1880, just after the introduction of the European rabbit and a flightless New Zealand bird, the weka (Figure 8.9). These new transplants thrived and provided alternate and year-round prey for the cats. Before their arrivals, cats had relied on seasonal penguins and petrels, but were food-limited in winter. By the mid-1880s, rabbits and wekas were common island-wide, and the Macquarie Island parakeet was last observed around 1890 (Taylor, 1979).

Key points

- Mammals commonly kill and eat invertebrates and vertebrate ectotherms, but the reverse is less common.
- Carnivorans weighing less than 20 kg tend to prey on invertebrates and vertebrates much smaller than themselves. Carnivorans larger than that threshold eat mostly vertebrates about their own size or larger. The latter predatory style requires higher energy expenditures, cooperative behaviors, and increased risk of injury of the carnivoran.
- Estimates of the overall proportion of vertebrates—including ungulates—killed by predators are high, commonly around half of all deaths.
- The most dramatic evidence of limitation of prey by carnivorans comes from introductions into novel environments. Feral domestic cats, small Asian mongooses, and stoats have driven many avian and mammalian species or populations to extinction. These effects are especially strong on islands.
- Carnivorans are superior competitors to metatherian ones because of physiological differences.
- Some introductions of carnivorans to mainland habitats with diverse eutherian communities have had strong effects on native birds and mammals. Among these, the American mink in Europe stands out. By contrast, managed reductions of single carnivoran species in diverse mainland communities can produce small or undetectable responses in prey abundances.
- Disease-caused die-offs of wild carnivorans have produced strong rebounds in prey numbers, in a few cases.
- The effects of carnivoran predation on prey are often framed in terms of "bottom-up" vs. "top-down" effects. Which factor is more important can depend on body sizes of carnivoran and prey, as well as on habitat productivity.
- Managers commonly rely on indices to discern bottom-up vs. top-down control in ungulate populations. Populations limited by predation tend to feature elevated pregnancy rates, particularly in subadults relative to those limited by nutrition. They also exhibit high birth rates and large-bodied neonates. Populations limited by forage tend toward poor body condition, postponed first reproduction, and small litters. Forage limitation tends to produce sex ratios among adult ungulates favoring females.
- Diverse prey tends to stabilize carnivoran populations through time, increasing the likelihood that they will limit prey. Diverse non-prey food sources, including soft mast, have similar effects.
- Animals use a wide range of structures, chemicals, and behaviors to avoid being eaten by carnivorans.
- Migration is an important carnivoran-avoidance behavior among ungulates in seasonal environments. Aggregation allows ungulates and other prey species use collective vigilance, especially if the herd comprises multiple species. Other ungulate species avoid mixed-species aggregations.

- Spatio-temporal avoidance—staying away from places where and when carnivorans are active—is a common prey strategy. This can involve selecting safe habitats or reducing time active.
- Carnivoran-avoidance behaviors depend on the risk state of the prey species. Avoiding carnivorans has costs, and prey species must trade off those costs against benefits of access to food.
- Chronic exposure to predation risk can have effects that reduce survival and reproduction of prey species. These non-consumptive effects in some instances are believed to affect populations more strongly than the direct effects of predation.
- Carnivorans contribute to prey population cycles in some instances. These effects are strongest in the far north, if the predator is a prey specialist, if the predator does not migrate, and if the predator employs delayed implantation. Small mustelids that specialize on small rodent prey are believed to exert such effects. Prey generalists such as the coyote are unlikely to do so.
- A key feature of prey population cycles is lagged density-dependence. Carnivorans can contribute to this lag in several ways.
- The interaction of bottom-up and top-down forces produces stronger population cycles in the snowshoe hare than either force operating independently.
- Apparent competition occurs when two prey species of a single predator show apparent interaction that reflects predator preferences for one species over the other. A less common, but more preferred, prey type can be more strongly affected, at times to the point of becoming a conservation concern.

References

Albery, G.F. *et al.* (2020) "Negative density-dependent parasitism in a group-living carnivore," *Proceedings of the Royal Society B: Biological Sciences*, 287, p. 20202655.

Andelt, W.F. *et al.* (1980) "Surplus killing by coyotes," *Journal of Mammalogy*, 61, pp. 316–77.

Andreasen, A.M. *et al.* (2021) "Prey specialization by cougars on feral horses in a desert environment," *Journal of Wildlife Management*, 85, pp. 1104–20.

Ariefiandy, A. *et al.* (2020) "Knee deep in trouble: rusa deer use an aquatic escape behavior to delay attack by Komodo dragons," *Australian Mammalogy*, 42, pp. 103–5.

Atkins, R. *et al.* (2016) "Deep evolutionary experience explains mammalian responses to predators," *Behavioral Ecology and Sociobiology*, 70, pp. 1755–63.

Ballard, W.B. and Miller, S.D. (1990) "Effects of reducing brown bear density on moose calf survival in southcentral Alaska," *Alces*, 26, pp. 9–13.

Ballard, W.B., Spraker, T.H. and Taylor, K.P. (1981) "Causes of neonatal moose calf mortality in southcentral Alaska," *Journal of Wildlife Management*, 45, pp. 335–42.

Bar-On, Y.M., Phillips, R. and Milo, R. (2018) "The biomass distribution on Earth," *Proceedings of the National Academy of Sciences of the United States of America*, 115, pp. 6506–11.

Begg, C.M. *et al.* (2003) "Sexual and seasonal variation in the diet and foraging behaviour of a sexually dimorphic carnivore, the honey badger (*Mellivora capensis*)," *Journal of Zoology*, 2260, pp. 301–16.

Bergerud, A.T., Butler, H.E. and Miller, D.R. (1984) "Antipredator tactics of calving caribou: dispersion in mountains," *Canadian Journal of Zoology*, 62, pp. 1566–75.

Bleich, V.C. (1999) "Mountain sheep and coyotes: patterns of predator evasion in a mountain ungulate," *Journal of Mammalogy*, 80, pp. 283–9.

Bonnington, C., Gaston, K.J. and Evans, K.L. (2013) "Fearing the feline: domestic cats reduce avian fecundity through trait-mediated indirect effects that increase nest predation by other species," *Journal of Applied Ecology*, 50, pp. 15–24.

Boonstra, R. (2013) "Reality as the leading cause of stress: rethinking the impact of chronic stress in nature," *Functional Ecology*, 27, pp. 11–23.

Boonstra, R. *et al.* (1998) "The impact of predator-induced stress on the snowshoe hare cycle," *Ecological Monographs*, 68, pp. 371–94.

Boudreau, M.R. *et al.* (2019) "Experimental increase in predation risk causes a cascading stress response in free-ranging snowshoe hares," *Oecologia*, 191, pp. 311–23.

Boutin, S. (1992) "Predation and moose population dynamics: a critique," *Journal of Wildlife Management*, 56, pp. 116–27.

Bowman, J., Donovan, D. and Rosatte, R.C. (2006) "Numerical responses of fishers to synchronous prey dynamics," *Journal of Mammalogy*, 87, pp. 480–4.

Bowyer, R.T., Person, D.K. and Pierce, B.M. (2005) "Detecting top-down versus bottom-up regulation of ungulates by large carnivores: implications for conservation of biodiversity," in Ray, J.C. *et al.* (eds.) *Large carnivores and the conservation of biodiversity*. Washington, DC: Island Press, pp. 342–61.

Brzezinski, M. *et al.* (2012) "The expansion wave of an invasive predator leaves declining waterbird populations behind," *Diversity and Distributions*, 26, pp. 138–50.

Carbone, C., Teacher, A. and Rowcliffe, J.M. (2007) "The costs of carnivory," *PLoS Biology*, 5, p. e22.

Caro, T. (2005) *Antipredator defenses in birds and mammals*. Chicago: University of Chicago Press.

Cohen, J.E. *et al.* (1993) "Body sizes of animal predators and animal prey in food webs," *Journal of Animal Ecology*, 62, pp. 67–78.

Courchamp, F., Berec, L. and Gascoigne, J. (2008) *Allee effects in ecology and conservation*. New York: Oxford University Press.

Creel, S. *et al.* (2007) "Predation risk affects reproductive physiology and demography of elk," *Science*, 315, p. 960.

Cusack, J.J. *et al.* (2020) "Weak spatiotemporal response of prey to predation risk in a freely interacting system," *Journal of Animal Ecology*, 89, pp. 120–31.

Daly, J.W. (1995) "The chemistry of poisons in amphibian skin," *Proceedings of the National Academy of Sciences of the United States of America*, 92, pp. 9–13.

DeCesare, N.J. *et al.* (2010) "Endangered, apparently: the role of apparent competition in endangered species conservation," *Animal Conservation*, 13, pp. 353–62.

Ditmer, M.A. *et al.* (2018) "Moose movement rates are altered by wolf presence in two ecosystems," *Ecology and Evolution*, 8, pp. 9017–33.

Doherty, T.S. *et al.* (2016) "Invasive predators and global biodiversity loss," *Proceedings of the National Academic of Sciences of the United States of America*, 113, pp. 11261–5.

Endicott, R.L., Prugh, L.R. and Brashares, J.S. (2014) "Surplus-killing by endangered San Joaquin kit foxes (*Vulpes macrotis mutica*) is linked to local population decline of endangered giant kangaroo rats (*Dipodomys ingens*)," *Southwestern Naturalist*, 59, pp. 110–15.

Errington, P.L. (1946) "Predation and vertebrate populations," *Quarterly Review of Biology*, 21, pp. 221–45.

Fauth, J.E. *et al.* (1996) "Simplifying the jargon of community ecology: a conceptual approach," *American Naturalist*, 147, pp. 282–6.

Fitzgibbon, C.D. (1990) "Mixed-species grouping in Thomson's and Grant's gazelles: the antipredator benefits," *Animal Behaviour*, 39, pp. 1116–26.

Flader, S.L. (1974) *Thinking like a mountain: Aldo Leopold and the evolution of an ecological attitude toward deer, wolves, and forests*. Madison: University of Wisconsin Press.

Ford, A.T. and Goheen, J.R. (2015a) "An experimental study on risk effects in a dwarf antelope, Madoqua guentheri," *Journal of Mammalogy*, 96, pp. 918–26.

Ford, A.T. and Goheen, J.R. (2015b) "Trophic cascades by large carnivores: a case for strong inference and mechanism," *Trends in Ecology and Evolution*, 30, pp. 725–35.

Frenot, Y. *et al.* (2005) "Biological invasions in the Antarctic: extent, impacts and implications," *Biological Reviews*, 80, pp. 45–72.

Fryxell, J.M. (1995) "Aggregation and migration by grazing ungulates in relation to resources and predators," in Sinclair, A.R.E and Arcese, P. (eds.) *Serengeti II: dynamics, management and conservation of an ecosystem*. Chicago: University of Chicago Press, pp. 257–73.

Fryxell, J.M. and Lundberg, P. (1994) "Diet choice and predator–prey dynamics," *Evolutionary Ecology*, 8, pp. 407–21.

Fryxell, J.M. and Sinclair, A.R.E. (1988) "Causes and consequences of migration by large herbivores," *Trends in Ecology and Evolution*, 3, pp. 237–41.

Garrott, R.A. *et al.* (2007) "Evaluating prey switching in wolf–ungulate systems," *Ecological Applications*, 17, pp. 1588–97.

Gehr, B. *et al.* (2018) "Evidence for nonconsumptive effects from a large predator in an ungulate prey?", *Behavioral Ecology*, 29, pp. 724–35.

Gilg, O., Hanski, I. and Sittler, B. (2003) "Cyclic dynamics in a simple vertebrate predator–prey community," *Science*, 392, pp. 866–8.

Gilg, O. *et al.* (2006) "Functional and numerical responses of four lemming predators in high arctic Greenland," *Oikos*, 113, pp. 193–216.

Gilg, O., Sittler, B. and Hanski, I. (2009) "Climate change and cyclic predator–prey population dynamics in the high Arctic," *Global Change Biology*, 15, pp. 2634–52.

Goodrich, J.M. and Buskirk, S.W. (1998) "Spacing and ecology of North American badgers (*Taxidea taxus*) in a prairie-dog (*Cynomys leucurus*) complex," *Journal of Mammalogy*, 79, pp. 171–9.

Green, R.G. and Larson, C.L. (1938) "A description of shock disease in the snowshoe hare," *American Journal of Epidemiology*, 28, pp. 190–212.

Griffin, P.C. *et al.* (2005). "Mortality by moonlight: predation risk and the snowshoe hare," *Behavioral Ecology*, 16, pp. 938–44.

Hamel, S. and Côté, S.D. (2007) "Habitat use patterns in relation to escape terrain: are alpine ungulate females trading off better foraging sites for safety?", *Canadian Journal of Zoology*, 85, pp. 933–43.

Hanski, I., Hansson, L. and Henttonen, H. (1991) "Specialist predators, generalist predators, and the microtine rodent cycle," *Journal of Animal Ecology*, 60, pp. 353–67.

Harrington, R. *et al.* (1999) "Establishing the causes of the roan antelope decline in the Kruger National Park, South Africa," *Biological Conservation*, 90, pp. 69–78.

Heikkilä, J., Below, A. and Hanski, I. (1994) "Synchronous dynamics of microtine rodent populations on islands in Lake Inari in northern Fennoscandia: evidence for regulation by mustelid predators," *Oikos*, 70, pp. 245–52.

Herman, J.P. *et al.* (2003) "Central mechanisms of stress integration: hierarchical circuitry controlling hypothalamo–pituitary–adrenocortical responsiveness," *Frontiers in Neuroendocrinology*, 24, pp. 151–80.

Hill, J.E., DeVault, T.L. and Belant, J.L. (2019) "Cause-specific mortality of the world's terrestrial vertebrates," *Global Ecology and Biogeography*, 28, pp. 680–9.

Holling, C.S. (1959) Some characteristics of simple types of predation and parasitism. *Canadian Entomologist*, **91**, pp. 385–98.

Holt, R.D. (1977) "Predation, apparent competition, and the structure of prey communities," *Theoretical Population Biology*, 12, pp. 197–229.

Hopcraft, J.G.C., Olff, H. and Sinclair, A.R.E. (2009) "Herbivores, resources and risks: alternating regulation along primary environmental gradients in savannas," *Trends in Ecology and Evolution*, 25, pp. 119–28.

Hörnfeldt, B., Hipkiss. T. and Eklund, U. (2005) "Fading out of vole and predator cycles?", *Proceedings of the Royal Society B: Biological Sciences*, 272, pp. 2045–9.

Houston, D.C. (1979) "The adaptations of scavengers," in Sinclair, A.R.E. and Griffiths, M.N. (eds.) *Serengeti: dynamics of an ecosystem*. Chicago: University of Chicago Press, pp. 263–86.

Hu, Y., Gillespie G. and Jessop, T.S. (2019) "Variable reptile responses to introduced predator control in southern Australia," *Wildlife Research*, 46, pp. 64–75.

Hurley, M.A. *et al.* (2011) "Demographic response of mule deer to experimental reduction of coyotes and mountain lions in southeastern Idaho," *Wildlife Monographs*, 178, pp. 1–33.

Inchausti, P. and Ginzburg, L.R. (2009) "Maternal effects mechanism of population cycling: a formidable competitor to the traditional predator–prey view," *Philosophical Transactions of the Royal Society B: Biological Sciences*, 364, pp. 1117–24.

Jędrzejewska, B. and Jędrzejewski, W. (1989) "Seasonal surplus killing as hunting strategy of the weasel *Mustela nivalis*—test of a hypothesis," *Acta Theriologica*, 34, pp. 347–59.

Jochym, M. and Halle, S. (2012) "To breed, or not to breed? Predation risk induces breeding suppression in common voles," *Oecologia*, 170, pp. 943–53.

Kingdon, J. *et al.* (2012) "A poisonous surprise under the coat of the African crested rat," *Proceedings of the Royal Society B: Biological Sciences*, 279, pp. 675–80.

Kjellander, P. and Nordström, J. (2003) "Cyclic voles, prey switching in red fox, and roe deer dynamics—a test of the alternative prey hypothesis," *Oikos*, 101, pp. 338–44.

Klemola, T., Norrdahl, K. and Korpimäki, E. (2000) "Do delayed effects of overgrazing explain population cycles in voles?", *Oikos*, 90, pp. 509–16.

Kloskowski, J. (2005) "Otter *Lutra lutra* damage at farmed fisheries in southeastern Poland, I: an interview survey," *Wildlife Biology*, 11, pp. 201–6.

Korpimäki, E. (1993) "Regulation of multiannual vole cycles by density-dependent avian and mammalian predation?", *Oikos*, 66, pp. 359–63.

Korpimäki, E. and Krebs, C.J. (1996) "Predation and population cycles of small mammals," *BioScience*, 46, pp. 754–64.

Korpimäki, E. *et al.* (2005) "Predator-induced synchrony in population oscillations of coexisting small mammal species," *Proceedings of the Royal Society B: Biological Sciences*, 272, pp. 193–202.

Korpimäki, E. *et al.* (2002) "Dynamic effects of predators on cyclic voles: field experimentation and model extrapolation," *Proceedings of the Royal Society B: Biological Sciences*, 269, pp. 991–7.

Krebs, C.J. (2001) "What drives the 10-year cycle of snowshoe hares?", *BioScience*, 51, pp. 25–35.

Krebs, C.J. (2002) "Two complementary paradigms for analysing population dynamics," *Philosophical Transactions of the Royal Society B: Biological Sciences*, 357, pp. 1211–19.

Krebs, C.J. *et al.* (1995) "Impact of food and predation on the snowshoe hare cycle," *Science*, 269, pp. 1112–1115.

Kruuk, H. (1972) "Surplus killing by carnivores," *Journal of Zoology*, 166, pp. 233–44.

Laundré, J.W., Hernández, L. and Altendorf, K.B. (2001) "Wolves, elk, and bison: reestablishing the 'landscape of fear' in Yellowstone National Park, U.S.A.," *Canadian Journal of Zoology*, 79, pp. 1401–9.

le Roux, E., Kerley, G.I.H. and Cromsigt, J.P.G.M. (2018) "Megaherbivores modify trophic cascades triggered by fear of predation in an African savanna ecosystem," *Current Biology*, 28, pp. 2493–9.

Lee, D.E. *et al.* (2016) "Migratory herds of wildebeests and zebras indirectly affect calf survival of giraffes," *Ecology and Evolution*, 6, pp. 8402–11.

Letnic, M. and Ripple, W.J. (2017) "Large-scale responses of herbivore prey to canid predators and primary productivity," *Global Ecology and Biogeography*, 26, pp. 860–6.

Lima, S.L. (1995) "Back to the basics of anti-predatory vigilance: the group-size effect," *Animal Behaviour*, 49, pp. 11–20.

Lima, S.L. (2002) "Putting predators back into behavioral predator–prey interactions," *Trends in Ecology and Evolution*, 17, pp. 70–5.

Lima, S.L. and Dill, L.M. (1990) "Behavioral decisions made under the risk of predation: a review and prospectus," *Canadian Journal of Zoology*, 68, pp. 619–40.

Lindström, E.R. *et al.* (1994) "Disease reveals the predator: sarcoptic mange, red fox predation, and prey populations," *Ecology*, 75, pp. 1042–9.

Linnell, J.D.C., Aanes, R. and Andersen, R. (1995) "Who killed Bambi? The role of predation in the neonatal mortality of temperate ungulates," *Wildlife Biology*, 1, pp. 209–23.

Litvaitis, J.A., Sherburne, J.A. and Bissonette, J.A. (1985) "Influence of understory characteristics on snowshoe hare habitat use and density," *Journal of Wildlife Management*, 49, pp. 866–73.

Losey, R.J. *et al.* (2014) "Craniomandibular trauma and tooth loss in northern dogs and wolves: implications for the archeological study of dog husbandry and domestication," *PLoS ONE*, 9, p. e99746.

MacLeod, K.J. *et al.* (2017) "Fear and lethality in snowshoe hares: the deadly effects of non-consumptive predation risk," *Oikos*, 127, pp. 375–80.

Matias, R. and Catry, P. (2008) "The diet of feral cats at New Island, Falkland Islands, and impact on breeding seabirds," *Polar Biology*, 31, pp. 609–16.

McArthur, C. *et al.* (2014) "The dilemma of foraging herbivores: dealing with food and fear," *Oecologia*, 176, pp. 677–89.

McLellan, B.N. *et al.* (2010) "Predator-mediated Allee effects in multi-prey systems," *Ecology*, 91, pp. 286–92.

Mech, L.D. (1970) *The wolf: ecology and behavior of an endangered species*. New York: Natural History Press.

Melis, C. *et al.* (2009) "Predation has a greater impact in less productive environments: variation in roe deer, *Capreolus capreolus*, population density across Europe," *Global Ecology and Biogeography*, 18, pp. 724–34.

Mertz, D.B. and Wade, M.J. (1976) "The prudent prey and the prudent predator," *American Naturalist*, 110, pp. 489–96.

Metz, M.C. *et al.* (2018) "Predation shapes the evolutionary traits of cervid weapons," *Nature Ecology and Evolution*, 2, pp. 1619–25.

Middleton, A.D. *et al.* (2013) "Linking anti-predator behaviour to prey demography reveals limited risk effects of an actively hunting large carnivore," *Ecology Letters*, 16, pp. 1023–30.

Miller, F.L., Gunn, A. and Broughton, E. (1985) "Surplus killing as exemplified by wolf predation on newborn caribou," *Canadian Journal of Zoology*, 63, pp. 295–300.

Mori, E., Maggini, I. and Menchetti, M. (2014) "When quills kill: the defense strategy of the crested porcupine *Hystrix cristata* L., 1758," *Mammalia*, 78, pp. 229–34.

Mukherjee, S. and Heithaus, M.R. (2013) "Dangerous prey and daring predators: a review," *Biological Reviews*, 88, pp. 550–63.

Ng'weno, C.C. *et al.* (2017) "Lions influence the decline and habitat shift of hartebeest in a semiarid savanna," *Journal of Mammalogy*, 98, pp. 1078–87.

Ng'weno, C.C. *et al.* (2019a) "Apparent competition, lion predation, and managed livestock grazing: can conservation value be enhanced?", *Frontiers in Ecology and Evolution*, 7, p. 123.

Ng'weno, C.C. *et al.* (2019b) "Interspecific prey neighborhoods shape risk of predation in a savanna ecosystem," *Ecology*, 100, p. e02698.

Nielsen, S.E. *et al.* (2017) "Complementary food resources of carnivory and frugivory affect local abundance of an omnivorous carnivore," *Oikos*, 126, pp. 369–80.

Norrdahl, K. and Korpimäki, E. (2000) "The impact of predation risk from small mustelids on prey populations," *Mammal Review*, 30, pp. 147–56.

O'Mahony, D. *et al.* (1999) "Fox predation on cyclic field vole populations in Britain," *Ecography*, 22, pp. 575–81.

Oates, B.A. *et al.* (2019) "Antipredator response diminishes during periods of resource deficit for a large herbivore," *Ecology*, 100, p. e02618.

Ostfeld, R.S. (1982) "Foraging strategies and prey switching in the California sea otter," *Oecologia*, 53, pp. 170–8.

Palmer, M.S. *et al.* (2017) "A 'dynamic' landscape of fear: prey responses to spatiotemporal variations in predation risk across the lunar cycle," *Ecology Letters*, 20, pp. 1364–73.

Patterson, B.R., Benjamin L.K. and Messier, F. (1998) "Prey switching and feeding habits of eastern coyotes in relation to snowshoe hare and white-tailed deer densities," *Canadian Journal of Zoology*, 76, pp. 1885–97.

Peers, M.J.L. *et al.* (2018) "Quantifying fear effects on prey demography in nature," *Ecology*, 99, pp. 1716–23.

Person, D.K., Bowyer, R.T. and Van Ballenberghe, V. (2001) "Density dependence of ungulates and functional responses of wolves: effects on predator–prey ratios," *Alces*, 37, pp. 253–73.

Pierce, B.M. *et al.* (2012) "Top-down versus bottom-up forcing: evidence from mountain lions and mule deer," *Journal of Mammalogy*, 93, pp. 977–88.

Polanowski, A. *et al.* (1997) "Proteinase inhibitors in the nonvenomous defensive secretion of grasshoppers: antiproteolytic range and possible significance," *Comparative Biochemistry and Physiology*, 117, pp. 525–9.

Post, E. *et al.* (2002) "Phase dependence and population cycles in a large-mammal predator–prey system," *Ecology*, 83, pp. 2997–3002.

Preisser, E.L., Bolnick, D.I. and Benard, M.F. (2005) "Scared to death? The effects of intimidation and consumption in predator–prey interactions," *Ecology*, 86, pp. 501–9.

Proffitt, K.M. *et al.* (2014) "Bottom-up and top-down influences on pregnancy rates and recruitment of northern

Yellowstone elk," *Journal of Wildlife Management*, 78, pp. 1383–93.

Prugh, L.R. *et al.* (2019) "Designing studies of predation risk for improved inference in carnivore–ungulate systems," *Biological Conservation*, 232, pp. 194–207.

Quick, H.F. (1953) "Occurrence of porcupine quills in carnivorous mammals," *Journal of Mammalogy*, 34, pp. 256–9.

Salo, P. *et al.* (2007) "Alien predators are more dangerous than native predators to prey populations," *Proceedings of the Royal Society B: Biological Sciences*, 274, pp. 1237–43.

Schaefer, J.A. and Mahoney, S.P. (2001) "Antlers on female caribou: biogeography of the bones of contention," *Ecology*, 82, pp. 3556–60.

Schmidt, K.A. and Whelan, C.J. (1998) "Predator-mediated interactions between and within guilds of nesting songbirds: experimental and observational evidence," *American Naturalist*, 152, pp. 393–402.

Schmitt, M.H. *et al.* (2014) "Determining the relative importance of dilution and detection for zebra foraging in mixed-species herds," *Animal Behaviour*, 96, pp. 151–8.

Seldal, T., Andersen, K.-J. and Högstedt, G. (1994) "Grazing-induced proteinase inhibitors: a possible cause for lemming population cycles," *Oikos*, 70, pp. 3–11.

Shapira, I., Sultan, H. and Shanas, U. (2008) "Agricultural farming alters predator–prey interactions in nearby natural habitats," *Animal Conservation*, 11, pp. 1–8.

Sheriff, M.J. *et al.* (2015) "Predator-induced maternal stress and population demography in snowshoe hares: the more severe the risk, the longer the generational effect," *Journal of Zoology*, 296, pp. 305–10.

Sheriff, M.J., Krebs, C.J. and Boonstra, R. (2009) "The sensitive hare: sublethal effects of predator stress on reproduction in snowshoe hares," *Journal of Animal Ecology*, 78, pp. 1249–58.

Sheriff, M.J., Krebs, C.J. and Boonstra, R. (2010) "The ghosts of predators past: population cycles and the role of maternal programming under fluctuating predation risk," *Ecology*, 91, pp. 2983–94.

Sheriff, M.J., Krebs, C.J. and Boonstra, R. (2011) "From process to pattern: how fluctuating predation risk impacts the stress axis of snowshoe hares during the 10-year cycle," *Oecologia*, 166, pp. 593–605.

Short, J., Kinnear, J.E. and Robley, A. (2002) "Surplus killing by introduced predators in Australia—evidence for ineffective anti-predator adaptations in native prey species?", *Biological Conservation*, 103, pp. 283–301.

Simberloff, D. (2004) "Community ecology: is it time to move on?", *American Naturalist*, 163, pp. 787–99.

Sinclair, A.R.E. *et al.* (2003) "Mammal population cycles: evidence for intrinsic differences during snowshoe hare cycles," *Canadian Journal of Zoology*, 81, pp. 216–20.

Sinclair, A.R.E. and Krebs, C.J. (2002) "Complex numerical responses to top-down and bottom-up processes in vertebrate populations," *Philosophical Transactions of the Royal Society B*, 357, pp. 1221–31.

Sinclair, A.R.E. and Pech, R.P. (1996) "Density dependence, stochasticity, compensation and predator regulation," *Oikos*, 75, pp. 164–73.

Skogland, T. (1991) "What are the effects of predators on large ungulate populations?", *Oikos*, 61, pp. 401–11.

Slobodkin, L.B. (1968) "How to be a predator," *American Zoologist*, 8, pp. 43–51.

Smith, J.A. *et al.* (2019) "Integrating temporal refugia into landscapes of fear: prey exploit predator downtimes to forage in risky places," *Oecologia*, 189, pp. 883–90.

Stankowich, T. and Caro, T. (2009) "Evolution of weaponry in female bovids," *Proceedings of the Royal Society B: Biological Sciences*, 276, pp. 4329–34.

Stensland, E., Angerbjörn, A. and Berggren, P. (2003) "Mixed species groups in mammals," *Mammal Review*, 33, pp. 205–23.

Suarez-Bregua, P., Guerreiro, P.M. and Rotllant, J. (2018) "Stress, glucocorticoids and bone: a review from mammals to fish," *Frontiers in Endocrinology*, 9, p. 526.

Taylor, R.H. (1979) "How the Macquarie Island parakeet became extinct," *New Zealand Journal of Ecology*, 2, pp. 42–5.

Thorne, E.T., Bogan, M.A. and Anderson, S.H. (eds.) (1989) *Conservation biology and the black-footed ferret*. New Haven: Yale University Press.

Twining, J.P. and Mills, C. (2021) "Cooperative hunting in the yellow-throated marten (*Martes flavigula*): evidence for the not-so-solitary marten?", *Ecosphere*, 12, p. e03398.

Van Ballenberghe, V. and Ballard, W.B. (1994) "Limitation and regulation of moose populations: the role of predation," *Canadian Journal of Zoology*, 72, pp. 2071–7.

Van Duyne, C.V. *et al.* (2009) "Wolf predation among reintroduced Przewalski horses in Hustai National Park, Mongolia," *Journal of Wildlife Management*, 73, pp. 836–43.

Wayne, A.F. *et al.* (2017) "Recoveries and cascading declines of native mammals associated with control of an introduced predator," *Journal of Mammalogy*, 98, pp. 489–501.

Welch R.J. *et al.* (2017) "Hunter or hunted? Perceptions of risk and reward in a small mesopredator," *Journal of Mammalogy*, 98, pp. 1531–7.

White, P.J., Proffitt, K.M. and Lemke, T.O. (2012) "Changes in elk distribution and group sizes after wolf restoration," *American Midland Naturalist*, 167, pp. 174–87.

Whitten, K.R. (1995) "Antler loss and udder distention in relation to parturition in caribou," *Journal of Wildlife Management*, 59, pp. 273–7.

Widga, C. *et al.* (2017) "Late Pleistocene proboscidean population dynamics in the North American Midcontinent," *Boreas*, 46, pp. 772–82.

Wright, S.M. (1993) "Observations of the behaviour of male eastern grey kangaroos when attacked by dingoes," *Wildlife Research*, 20, pp. 845–9.

Ylönen, H. (1989) "Weasels *Mustela nivalis* suppress reproduction in cyclic bank voles *Clethrionomys glareolus*," *Oikos*, 55, pp. 138–40.

Ylönen, H. *et al.* (2006) "Is the antipredatory response in behaviour reflected in stress measured in faecal corticosteroids in a small rodent?", *Behavioral Ecology and Sociobiology*, 60, pp. 350–8.

Zalewski, A., Jędrzejewski, W. and Jędrzejewska, B. (1995) "Pine marten home ranges, numbers and predation on vertebrates in a deciduous forest (Białowieża National Park, Poland)," *Annales Zoologici Fennici*, 32, pp. 131–44.

CHAPTER 9

Cascades

"Cascade" commonly is used to describe a sequence of effects in which the interaction between two species affects other community members. The metaphor connotes that an ecological effect descends trophic levels like water flows downhill. When more than two trophic levels are involved, ecologists have tended to invoke "trophic cascade" as a descriptor, in spite of the fact that some of these effects ascend, rather than descend, trophic levels and in some cases operate via non-trophic interactions.

9.1 Ecological cascades

This term refers to a wide range of community interactions—trophic or otherwise—among guilds and species. The interactions may involve competition, predation, pollination, dispersal of plant propagules, nutrient transport, or facilitation of scavenging, the sole criterion being a causally linked chain of events with a single forcing process. One well-documented and especially lengthy cascade, involving multiple taxa and trophic levels, features domestic dogs, northern raccoons, and intertidal invertebrate prey of raccoons on coastal islands of British Columbia, Canada (Figure 9.1). Dogs harass and kill raccoons in the intertidal zone and by playing recordings of dog barks—and using vocalizations of pinnipeds as controls—researchers showed that raccoons feared dogs and avoided intertidal foraging. After one month of treatment, the intertidal prey of raccoons increased dramatically, one species of intertidal crab by 97%, one species of intertidal fish by 81%, and an intertidal annelid worm by 59%. These effects extended into the subtidal zone via the intertidal crab, which competed with a subtidal fish and preyed on

a subtidal mollusk, both of which responded to changes initiated by the simulated dog harassment of raccoons (Suraci *et al.*, 2016). This experimental study showed how a competitive interaction (dog–raccoon) extended to three lower trophic levels, and from land into the subtidal zone, without reported effects on herbivory or plants. While striking in their effect sizes and penetration into the near-shore ocean, these interactions do not meet one criterion for a trophic cascade: predation affecting plants via herbivory.

Another reported cascade involved the strong influence of arctic fox predation on nutrient transporters. Foxes were introduced to many of the Aleutian Islands in the eighteenth century by Russian fur hunters, who primarily sought sea otter pelts. Prior to the arrival of foxes, abundant sea birds had nested and defecated on the treeless islands, and native vegetation was dominated by tall, lush graminoids (Figure 9.2). The invasive foxes preyed on nesting seabirds, driving them to ecological extinction on most islands. With more than 150 years of seabird exclusion, vegetation shifted from graminoids to a mix of low-lying forbs and dwarf shrubs. In the absence of seabirds, soils became poor in phosphorus and nitrogen, and plants were low in tissue nitrogen (Croll *et al.*, 2005; Maron *et al.*, 2006). Experimental fertilization of some seabird-poor islands showed that these effects were reversible, and consistent with seabirds having transported the marine-derived nutrients. With the eradication of foxes from most islands beginning in the 1980s, seabirds reoccupied historical nesting islands, and vegetation gradually recovered its pre-fox form. In this case, an introduced carnivoran interrupted a nutrient transport pathway by which marine-derived nutrients had reached land, with

Carnivoran Ecology. Steven W. Buskirk, Oxford University Press. © Steven W. Buskirk (2023). DOI: 10.1093/oso/9780192863249.003.0009

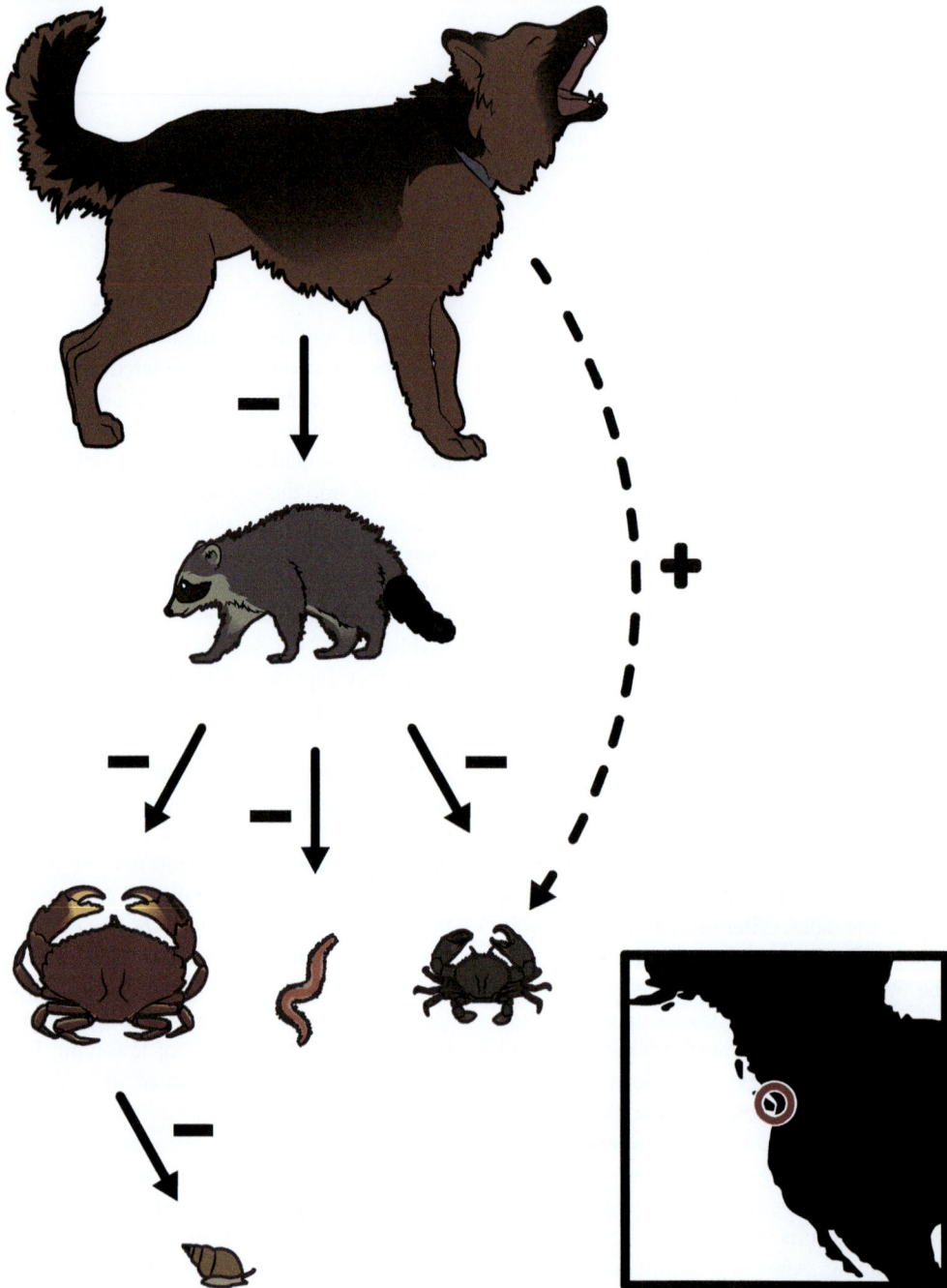

Figure 9.1 An ecological cascade features one community interaction, in this case involving a carnivoran, driving other interactions within and across trophic levels. The interactions can be trophic or competitive in nature or involve nutrient transport or another mechanism. Recorded vocalizations of domestic dogs, but not local pinnipeds, caused northern raccoons to reduce foraging in the intertidal zone of a British Columbia, Canada island. This allowed intertidal crabs, fish, and annelid worms to recover. Subtidal fish and mollusks responded to changes in intertidal crab populations via changes in competition and predation (Suraci et al., 2016).

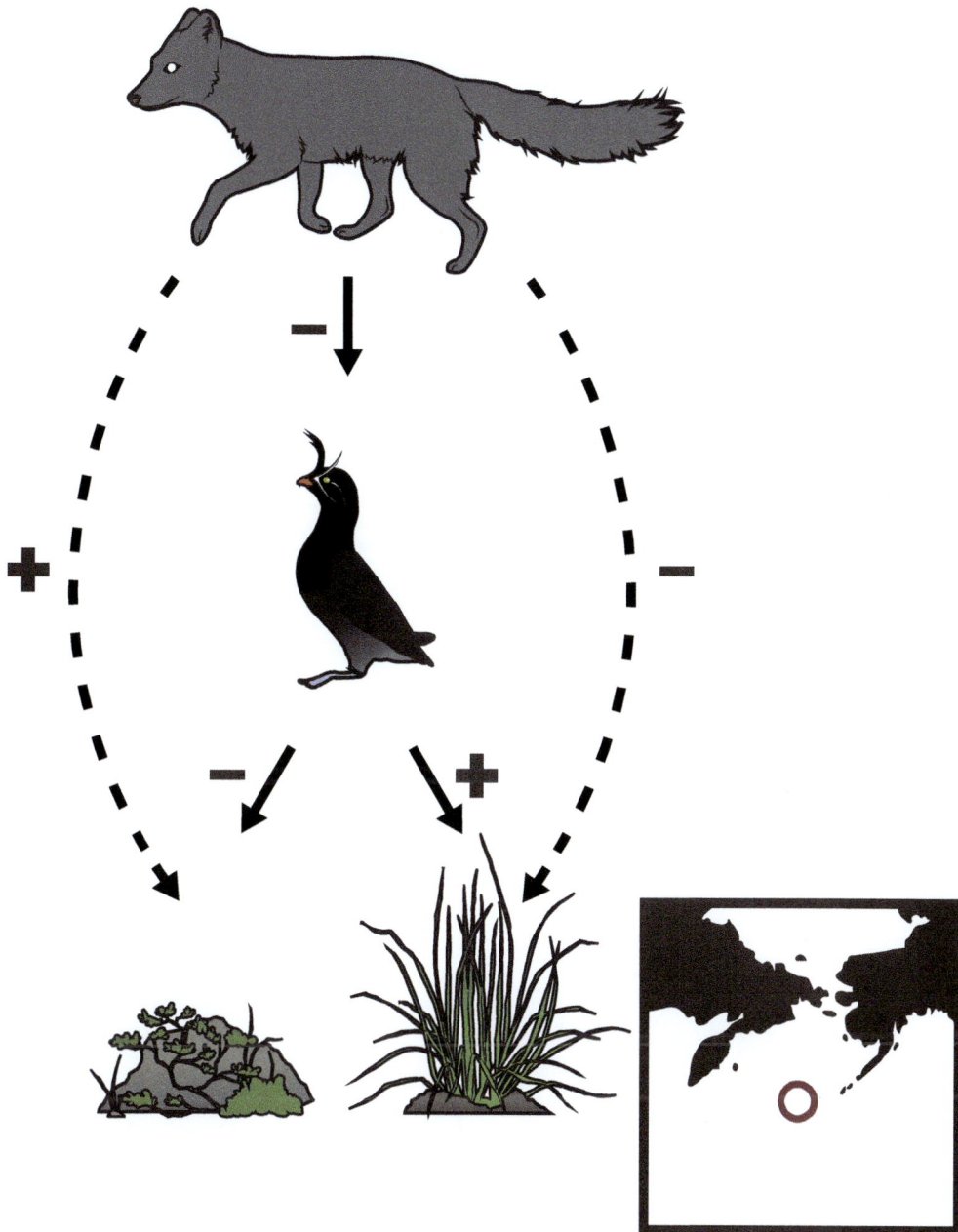

Figure 9.2 Arctic foxes, introduced to the Aleutian Islands during the eighteenth century, preyed heavily on nesting and resting seabirds, which previously had transported large quantities of marine-derived nutrients in urine and feces to land. Lacking fertilization, vegetation shifted from lush graminoids to small forbs and subshrubs. With management extirpation of arctic foxes from most islands during 1980–2015, sea birds returned, and vegetation recovered its condition before arctic foxes were introduced (Croll *et al.*, 2005).

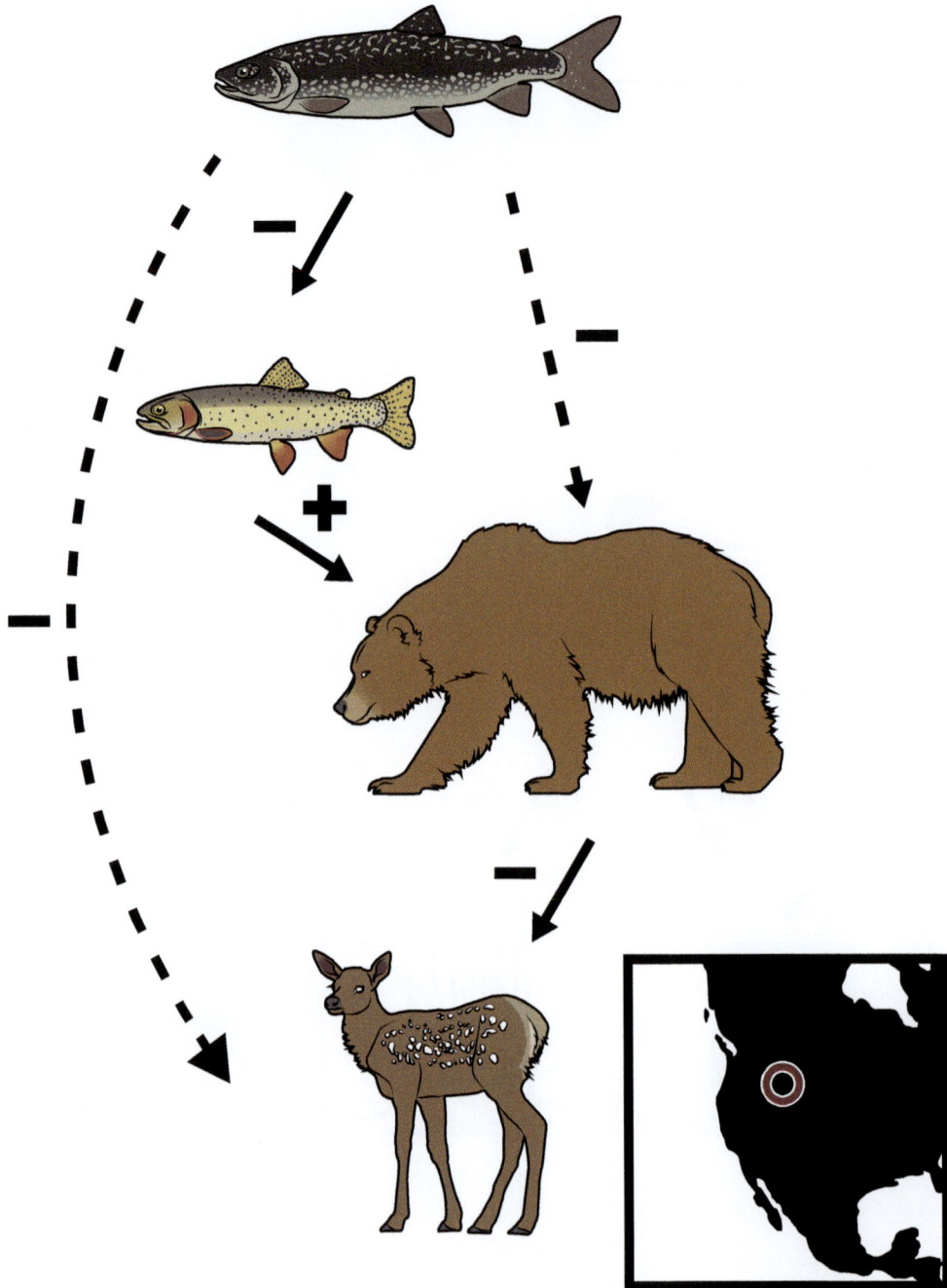

Figure 9.3 The cascade postulated for the trout–bear–elk system in Yellowstone National Park, US. Bears historically preyed on stream-spawning cutthroat trout during early summer. However, introduced lake trout drastically reduced the abundance of cutthroat trout, with the result that bears shifted their diets to elk calves at that season. Recruitment and population growth of elk have declined (Middleton *et al.*, 2013a). Lake trout are putative exploitation competitors with brown bears for cutthroat trout, with a cascading effect to elk.

a top-down process (predation) mediating bottom-up effects (fertilization) on vegetation. Herbivory on land played no role in the observed effect on vegetation.

Middleton and colleagues (2013a) and Marshall and colleagues (2015) postulated another intriguing ecological cascade involving carnivorans in Yellowstone National Park. Brown bears (and secondarily black bears) traditionally fed on native cutthroat trout during early summer spawning runs in shallow streams that drain into Yellowstone Lake (Figure 9.3). The arrival, likely via illegal translocation, of non-native lake trout to Yellowstone Lake led to a more than 90% decline in cutthroat trout numbers because of predation by lake trout, although other factors may have contributed. Lake trout did not enter streams, and therefore did not compensate for the loss of cutthroat trout as brown bear food (Koel *et al.*, 2005). The loss to brown bears of an important food source—spawning cutthroat trout—caused them to switch to preying on elk calves in other areas at that time of summer, contributing to a more than threefold increase in elk calf mortality attributable to bears. This unusual cascade postulates an invasive salmonid triggering carnivoran prey switching, with regional implications for the most abundant ungulate. Again, herbivory is not a mechanism underlying the cascade, and effects on vegetation have not been proposed.

9.2 Trophic cascades

Trophic cascades, first formally defined by Paine (1980), are a subset of ecological cascades in which predation on herbivores affects herbivory and therefore plants. Ripple and colleagues (2016) defined them less restrictively as originating with predators, and having indirect, downward effects. The earliest understandings of these cascades were not mechanistic and lacked comparative or experimental data. For example, Leopold (1949) metaphorically suggested that deer feared their predators, whereas "mountains" (i.e. plants) feared their herbivores. Trophic cascades and plant defenses were the implicit mechanisms underlying the central premise of Hairston and colleagues (1960)—the world is green despite the presence of herbivores that eat green plants. Therefore, predators and plant defensive compounds must limit herbivores, allowing green plants to exist. Trophic cascades fall into two categories—density-mediated (DMTC) and behaviorally mediated (BMTC)—a distinction made by Schmitz and colleagues (1997). A DMTC occurs where predation affects vegetation via differing or changed abundance of herbivores. DMTCs are described for areas large enough to enable estimates of herbivore density. BMTCs are described for areas with predation risk gradients, across which prey behaviors vary. In both categories of trophic cascade, vegetative responses can involve production, standing biomass, reproduction, lengths of defensive thorns, or concentrations of plant defensive compounds.

9.2.1 Density-mediated trophic cascades

In spite of the presumed omnipresence of density-mediated trophic cascades initiated by carnivorans, the number of well-documented cases is modest. The best-documented one occurs in nearshore waters of the North Pacific, where the sea otter, which reduces densities of sea urchins and other herbivores, has cascading effects on kelp (Figure 9.4). Where sea otters are common, herbivorous invertebrates are rare and kelp is abundant (Estes and Duggins, 1995). Over a dozen spatially and temporally replicated studies conducted by various investigators from central California to Russia (Estes and Duggins, 1995, Table 1) have documented this phenomenon. Although not all of these studies demonstrate all elements of a trophic cascade, the replication and consistency of response show that sea otters are strong forcing factors in North Pacific kelp forest communities. An extension of the sea otter example to include killer whale predation on sea otters has been more controversial (Estes *et al.*, 1998; Wade *et al.*, 2007). Why the sea otter–urchin–kelp system consistently shows such strong cascading effects—serving as the poster child for trophic cascades—is not clear. In part it reflects trophic simplicity—the sea otter is the dominant predator on sea urchins and sea urchins dominate kelp herbivory. Further, sea otters have such high metabolic demands (Section 4.3.1) that they consume, on a per-animal basis, several times the prey biomass expected from their body weight. The

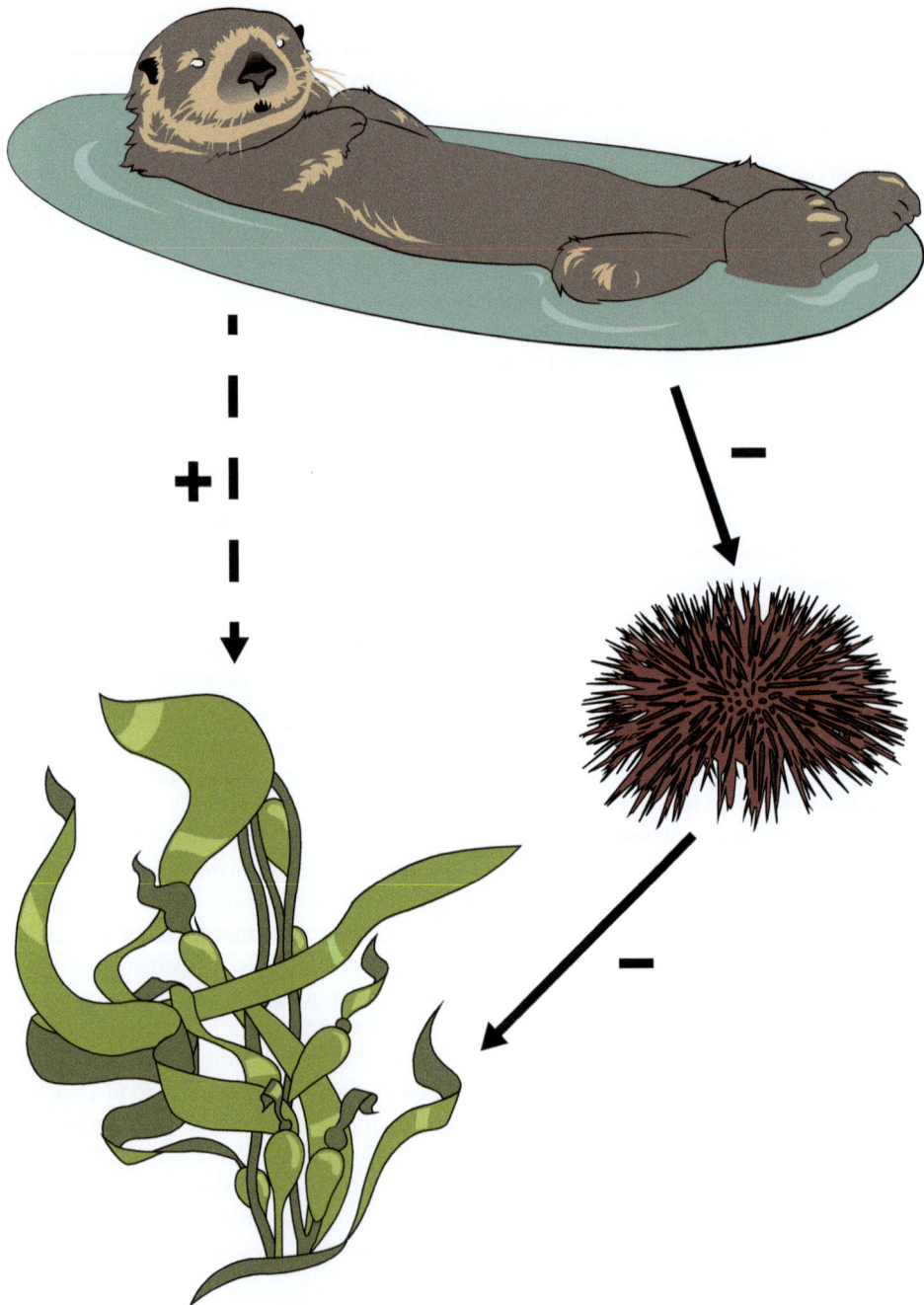

Figure 9.4 The well-documented trophic cascade involving sea otters, invertebrate herbivores, and kelp in North Pacific coastal waters. This system produces the strongest and most consistently demonstrated examples of trophic cascades involving carnivorans.

predator has an especially high metabolic rate and high local population densities, and the prey is an ectotherm living in cool water.

A second well-studied system involves Australian dingoes, kangaroos (or wallabies), and plants. Fencing excludes dingoes from large areas of southern Australia, so that dingo abundance differs by a factor of over 80 across the fence (Figure 9.5). The abundance of macropodids was much higher where dingoes were absent than where present (Morris and Letnic, 2017). Vegetation biomass was reduced where kangaroos were abundant, but not where they were rare. Soil phosphorus tracked the same trends observed in vegetation. Another study conducted by Colman and colleagues (2014) in southern Australian forests showed that dingoes suppressed the abundance of macropodids and introduced red foxes, which increased understory vegetation and small mammal abundance (Figure 9.6). In this case, competition between carnivorans cascaded to the herbivorous prey of the smaller carnivoran, then to vegetation—a trophic cascade.

One report describes an intriguing trophic cascade on a near-shore archipelago in the Baltic Sea (Fey *et al.*, 2009). Introduced American minks preyed on two species of voles, which fed on sparse graminoid and shrubby plants in this subarctic setting. The mink was the only important mammalian predator of voles, and mink removal was conducted for 12 to 15 years on treatment islands; control islands retained mink at moderate or high densities. Eradication of mink resulted in two- to fivefold increases in vole abundance, which reduced diversity and evenness of plant communities. Plant communities were more diverse and individual plants were more evenly distributed across plant species where mink were removed, showing that herbivory maintained some community traits. On the other hand, several plant species were either uncommon or absent on mink-free islands, showing that predation was protecting some plant species. The strength of this cascade can be attributed to the simplicity of the subarctic community and novelty of the introduced carnivoran predator, which had not evolved with the native herbivores.

In contrast to these demonstrated strong DMTCs, other case studies have demonstrated no—or only

weak—effects of predation cascading to vegetation. African wild dogs recolonized the Laikipia Plateau, Kenya after a two-decade absence, causing a 33% reduction in abundance of their primary prey, Günther's dik-dik. In turn, dik-dik browsing affected the abundance of three species of acacias, with the strongest effect on *Acacia mellifera*. This suggested that wild dogs might moderate the effect of dik-dik browsing on acacias; however, no such effect was observed. The effect of herbivory by small browsers on acacias in the presence of wild dogs was unchanged or increased. Ford and colleagues (2015) postulated that absence of the expected effect could have resulted from other small antelopes that were not important prey of wild dogs but browsed the same acacia species. A further factor could have been increased rainfall (by 36%) from before to after wild dog recovery (a bottom-up effect), which may have confounded the effects of herbivory. In another example in a Finnish grassland, voles (mostly *Microtus*) were the most abundant small mammals, and their primary predators were least weasels and ermines. European kestrels and small owls also killed some voles. Predator-proof enclosures allowed vole numbers to rise to levels twenty times as high as without fencing. Surprisingly, this high vole abundance did not cascade to vegetation, because severe winters limited vole abundance, which was limited by population growth during a single summer. Vole herbivory was held to low or moderate levels across years (Klemola *et al.*, 2000; Norrdahl *et al.*, 2002).

A third case study shows that cascading effects of carnivoran predation to vegetation are not universal. The study was conducted in a grassland in Montana, US, where the carnivoran guild comprised the coyote, ermine, and American badger, the primary predators of small mammals: deer mice, Columbian ground squirrels, and voles (cervids were also present). Predator (carnivoran and avian) exclusion, combined with some rodent exclusion within plots, showed that predators had no measurable effect on primary production. Rodents reduced aboveground biomass of graminoids, but not forbs, and predators did not alter these interactions. Predators did not affect ground squirrel or deer mouse densities, so these species would not be expected to transmit effects to vegetation. Predators did affect

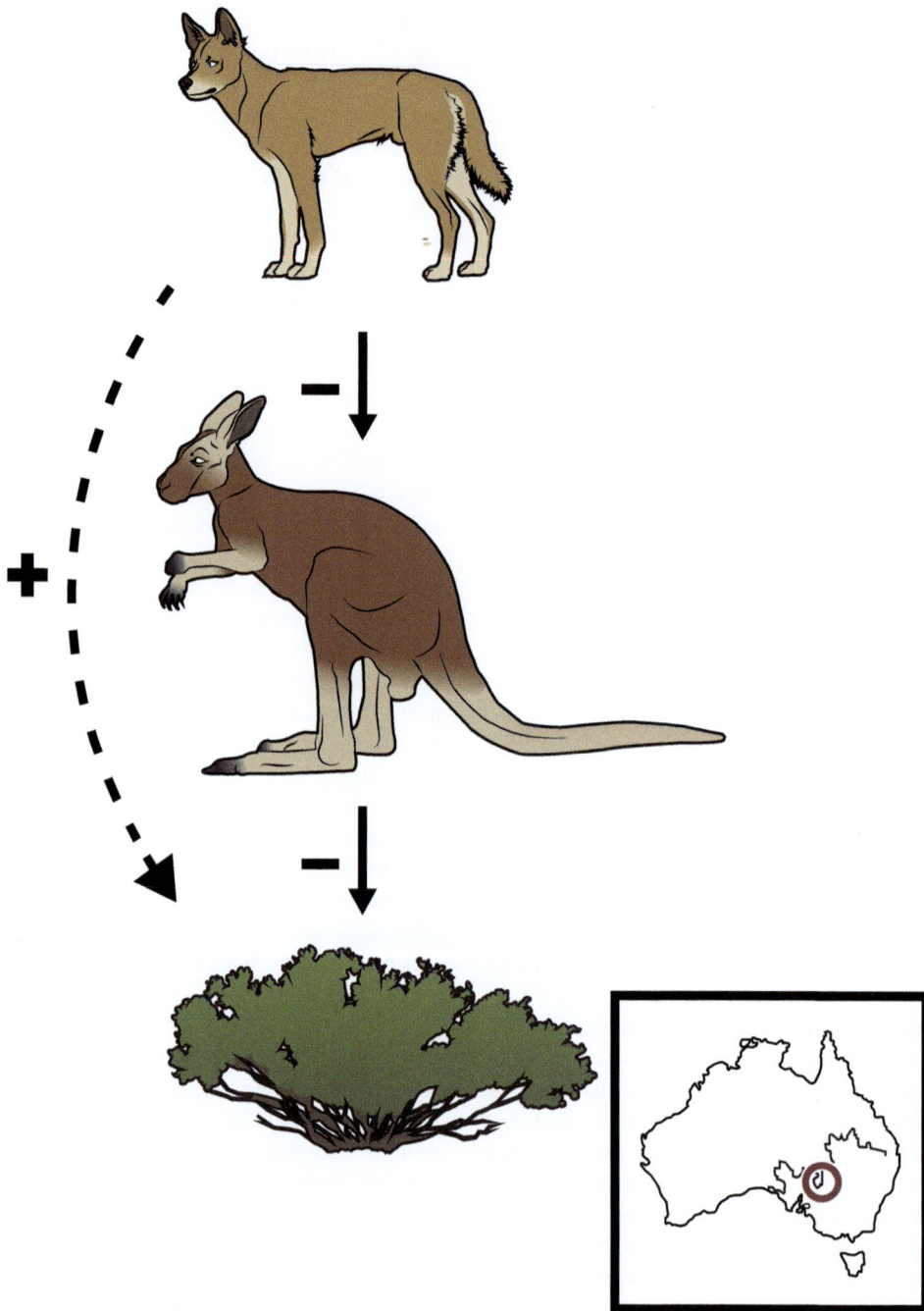

Figure 9.5 A density-mediated trophic cascade in which the Australian dingo lowers the abundance of kangaroos, which leads to increases in plant abundance (Morris and Letnic, 2017).

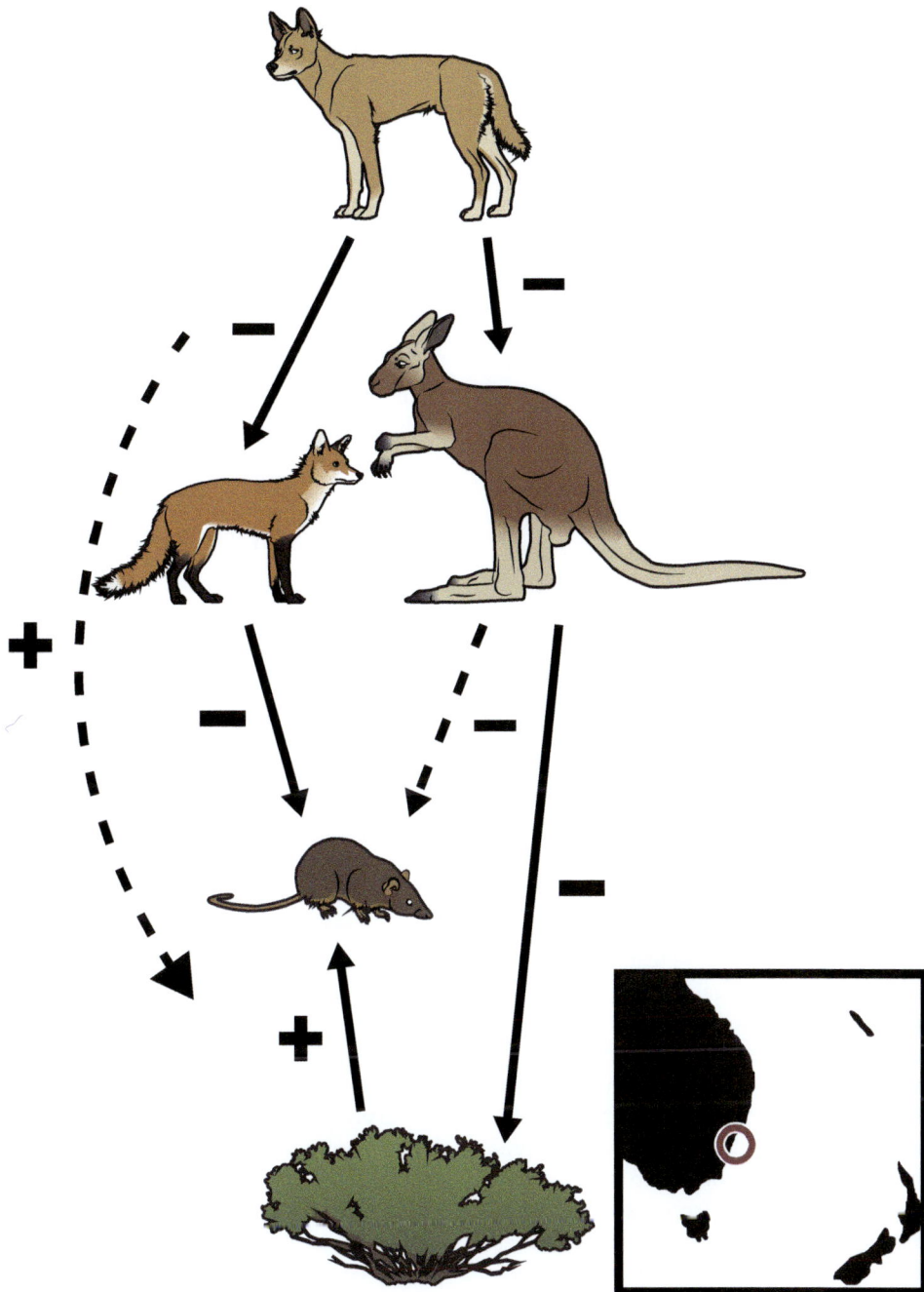

Figure 9.6 A density-mediated trophic cascade in which dingoes affect both herbivores (kangaroos) via predation and red foxes via competition in Australia (Colman *et al.*, 2014). Both of those prey taxa affect small herbivores; vegetation is affected via kangaroo herbivory.

densities of voles, a potential prerequisite for cas-cading effects to vegetation, but the latter effects were not observed. In this case, the reticulate nature of the carnivoran and rodent trophic interactions dampened the effect of any individual trophic path-way (Maron and Pearson, 2011).

9.2.2 Strength of density-mediated trophic cascades

The strength of a DMTC reflects two sequential processes—predation on herbivores, and herbivory on plants—both of which must be strong for a DMTC to occur. If the effect is entirely density mediated, the herbivore–plant effect will be attenu-ated relative to the predator–herbivore one (Shurin et al., 2002, Figure 1) and some of the factors that affect DMTC strength can be parsed between these two processes. First, trophic cascades tend to be stronger in aquatic or marine systems than in ter-restrial systems, although this finding may reflect the singularly strong effect sizes reported for the sea otter–urchin–kelp cascade (Shurin et al., 2002). This is so because herbivores that convert plant biomass efficiently but are highly vulnerable to predation tend to generate strong DMTCs (Polis and Strong, 1996). Those herbivores tend to occur in aquatic or marine systems. Second, community complexity appears to influence both the predatory and her-bivorous parts of the cascade; reticulate food webs reduce the likelihood of strong cascades involving a single species. Alternate trophic pathways obscure effect sizes along focal pathways (e.g. Singer et al., 2014). As an example, megaherbivores (e.g. ele-phants, rhinos, hippos) too large to be vulnerable to predation will tend to exert herbivory without respect to predation risk. In this way they dampen any possible trophic cascade on co-occurring her-bivores (le Roux et al., 2018). Third, variable pre-cipitation events confound bottom-up vs. top-down comparisons. Fourth, strong seasonality tends to obscure multi-year changes in herbivore responses to top-down or bottom-up forces; multi-year sys-tem changes are precluded by intra-annual density dependence. Fifth, trophic cascades tend to be tran-sient, occurring when favorable conditions arise. The moose–wolf–tree and lynx–hare–shrub systems are carnivoran examples of effects that depend

on relative carnivoran and herbivore abundance (Piovia-Scott et al., 2017).

Effect sizes for DMTCs are expressed as the pro-portional change or difference in plant attributes divided by the proportional change or difference in herbivore abundance attributable to predators. Schmitz and colleagues (2000, Table 1) reviewed DMTCs for avian, ectotherm vertebrate, and arthro-pod predators. Most of the systems reviewed showed a stronger response in plants (mean effect size = 5.6) than had occurred in in herbivore den-sity (mean effect size = –0.18), which suggests some other, non-density factor in operation. For example, behaviorally mediated effects of predation could result in functional, but not numerical, responses in herbivores, leading to stronger effects on vegetation than just those of herbivore densities.

9.2.3 Behaviorally mediated trophic cascades

A BMTC, by contrast, results when predation risk elicits behavioral responses that cascade to vegeta-tion. In its most commonly reported form, hetero-geneous predation risk creates spatial variation in herbivore behavior, which affects plant traits vari-ably across the risk gradient. A prey animal might spend more total time in a habitat patch where the predation risk is lower or spend less time vigilant or more time foraging there. The key difference between DMTCs and BMTCs is scalar. In a BMTC, a single population or subpopulation of herbivores affects vegetation differentially across a spatially explicit gradient of predation risk. The same ani-mals behave differently under different threat envi-ronments, and this is shown to affect vegetation. By contrast, in a DMTC, two or more herbivore populations or subpopulations differ in their expo-sure to predation (or a single population can receive differing predation treatments over time), and the altered herbivory regime affects plant traits. A few well-documented BMTCs involve carnivorans.

In an East African savanna, impalas avoided woody cover, especially at times when leopards (which accounted for 52% of impala predation) and African wild dogs (31% of impala predation) were active (Ford et al., 2014). Risk of encounter-ing and being killed by those carnivorans increased with woody cover. Two important browse species

(*Acacia* spp.) of impalas differed in their thorniness, and the presence of woody cover allowed the *Acacia* with short thorns to increase there (Figure 9.7). As a result, the short-thorned *Acacia* out-competed the long-thorned *Acacia* in habitats impalas perceived as risky, and the long-thorned *Acacia* was better able to persist in open habitats that impalas perceived as safe.

Another example comes from simple, high-elevation grassland of the Argentinean Andes, where an ambush predator (puma) was responsible for almost all predation on its nearly exclusive prey (vicuña) in a landscape of three distinct habitat types (Figure 9.8) (Donadio and Buskirk, 2016). Flat, open plains provided good visibility for vicuñas and little ambush cover for pumas. Here, vicuñas spent less than 10% of their time vigilant, and grazed *Stipa* and other grasses intensively. By contrast, canyons provided physical complexity—rock outcroppings, cliffs, and shrubs—that pumas used to ambush vicunas. Pumas were ten times as common in canyons as in plains and vicuñas were 3.7 times as vigilant in canyons as in plains. The effect of grazing on seed production was greater in plains than in canyons (by a factor of 16) and its effect on grass biomass was greater by a factor of 4 in the same comparison. So, puma presence had a strong protective effect on plants. Importantly, the effect was confounded by differences between the habitats in primary production, so that the apparent effects of herbivory were greater where production was lower. The puma–vicuña interaction is the primary source of food for avian scavengers (Perrig *et al.*, 2017).

Yovovich and colleagues (2021) provided an especially clear example of fine-scale spatial changes in carnivoran–prey interactions. They studied pumas, deer, and vegetation in a California landscape strongly modified by humans and found that pumas selectively killed deer more than 340 m from roads or human structures. Deer doubled their activity within 70 m of human development, where they browsed more heavily than farther away (Figure 9.9). This case illustrates the dominant influence of humans in some BMTCs.

A number of studies have tested for suspected BMTCs initiated by carnivorans but found only weak effects. A system in which BMTCs have been

sweepingly alleged—but frequently disputed—is the Yellowstone system of wolves, elk, and plants. Although wolf reintroduction to Yellowstone in 1995 was widely expected to trigger a trophic cascade that would release aspen trees from decades of heavy elk browsing, by the 2010s no evidence was observed of a trophic cascade of any type (Kauffman *et al.*, 2010), or of a BMTC specifically (Winnie, 2012). Subsequent studies (e.g. Beschta *et al.*, 2018) have likewise failed to demonstrate the predicted cascade (Fleming, 2019). Reasons for controversy about this issue among ecologists include suboptimal experimental conditions, differing evidential standards, the number of species considered in the analysis, and possibly biased sampling methods. The suboptimal experimental conditions resulted from several factors that confounded the wolf reintroduction experiment, which was essentially a before–after comparison:

1. The Yellowstone reintroduction area was affected by extensive wildfires in 1988, covering 3213 km^2, or 36% of the area within the national park boundary. Forests and shrublands burned by the fires were in early successional stages when wolves were reintroduced, potentially changing the availability of winter browse for elk (Boyce, 2018).

2. The region experienced a drought during the time of early wolf colonization, similarly confounding before–after comparisons of vegetative conditions (Middleton *et al.*, 2013b).

3. The brown bear population of the Yellowstone area was recovering while wolves became established. Brown bear numbers increased by a factor of 3–5 from the late 1970s to 2015 (White *et al.*, 2017, p. 6). In addition, over the period 1995–2018, spawning cutthroat trout, an important mid-summer food of bears, declined by over 90%, causing bears to prey more heavily on elk calves during that period (Section 9.1). These factors together increased bear predation on elk calves markedly during 1990–2015 (Van Manen *et al.*, 2013).

4. In some cases, the vegetative changes to browse species caused by high elk densities were not reversible by merely lowering elk densities via wolf predation. Instead, browsing needed to be

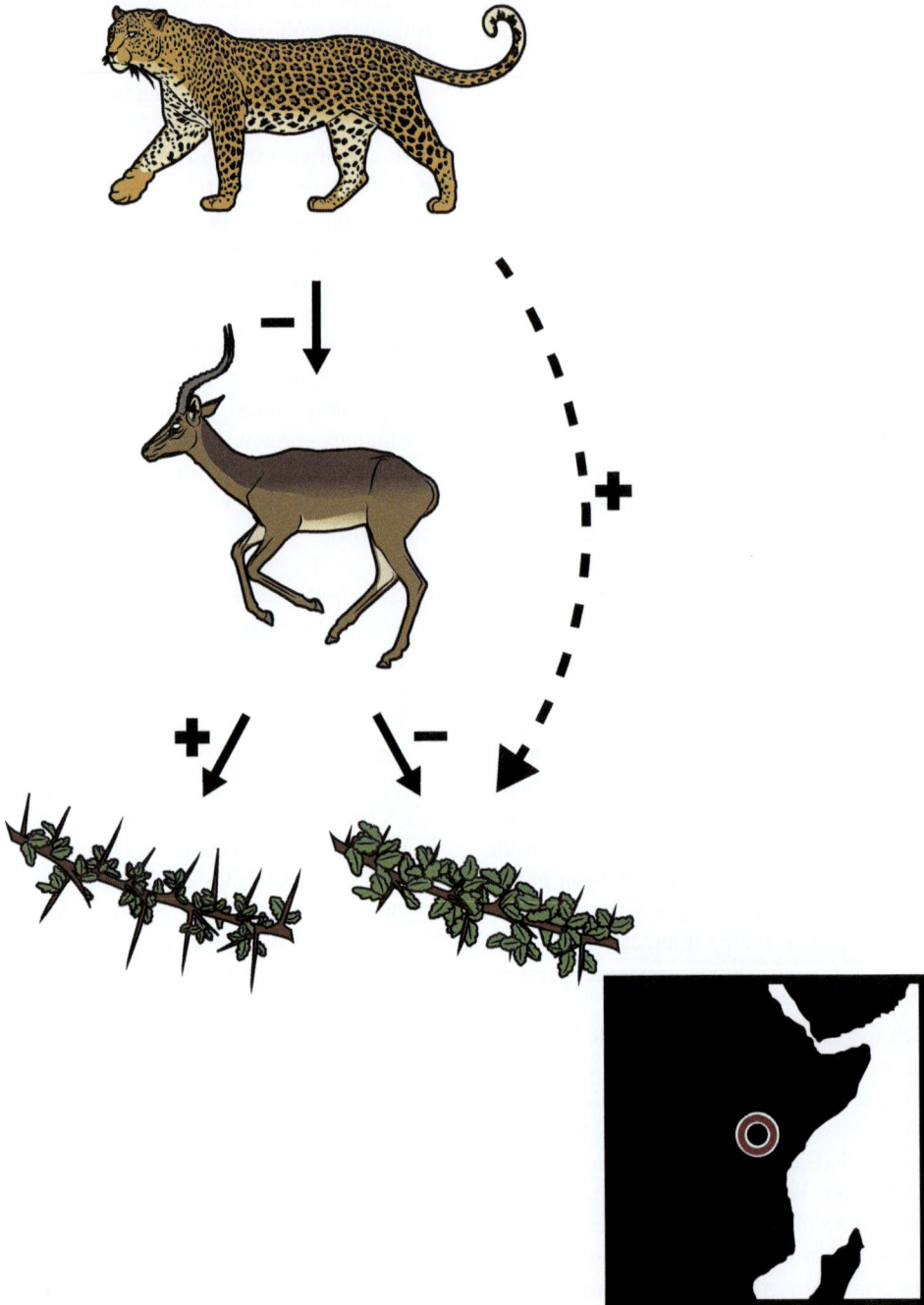

Figure 9.7 A behaviorally mediated trophic cascade in which increased impala browsing in areas of Kenya with lower risk of leopard predation increased anti-herbivory defenses, in this case thorns, in plants (Ford *et al.*, 2014). Woody cover was associated with increased predation risk and allowed a short-thorned acacia to outcompete the long-thorned species.

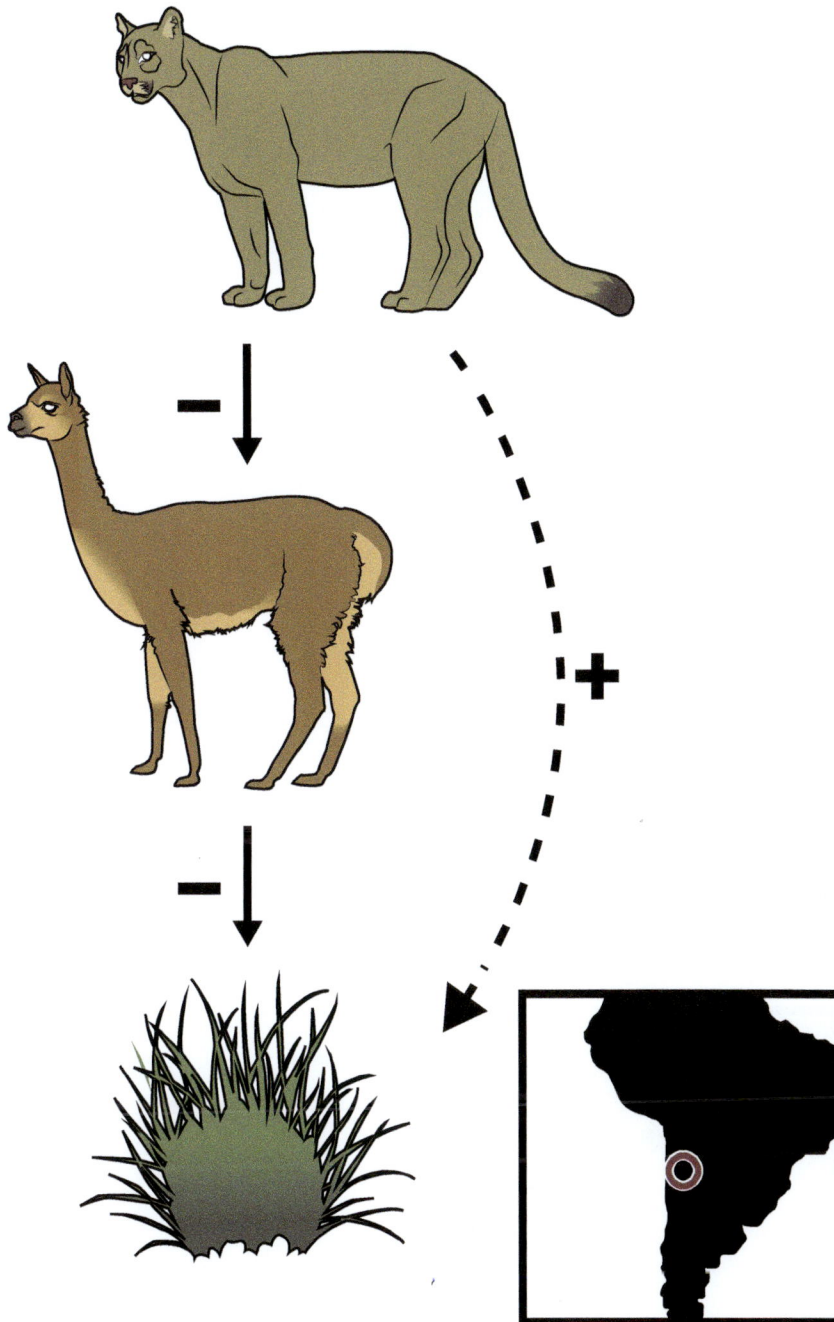

Figure 9.8 A behaviorally mediated trophic cascade involving pumas, vicuñas, and three distinct vegetation types in the Argentinean Andes. Flat, open plains provided pumas with little ambush cover; vicuñas could detect them at long distances and grazed without pausing to scan for predators. Herbaceous plants were heavily grazed. Canyon habitats were characterized by rocks, shrubs, and much better ambushing cover. Plant biomass was much higher than on plains, but vicuñas spent much more time vigilant (Donadio and Buskirk, 2016).

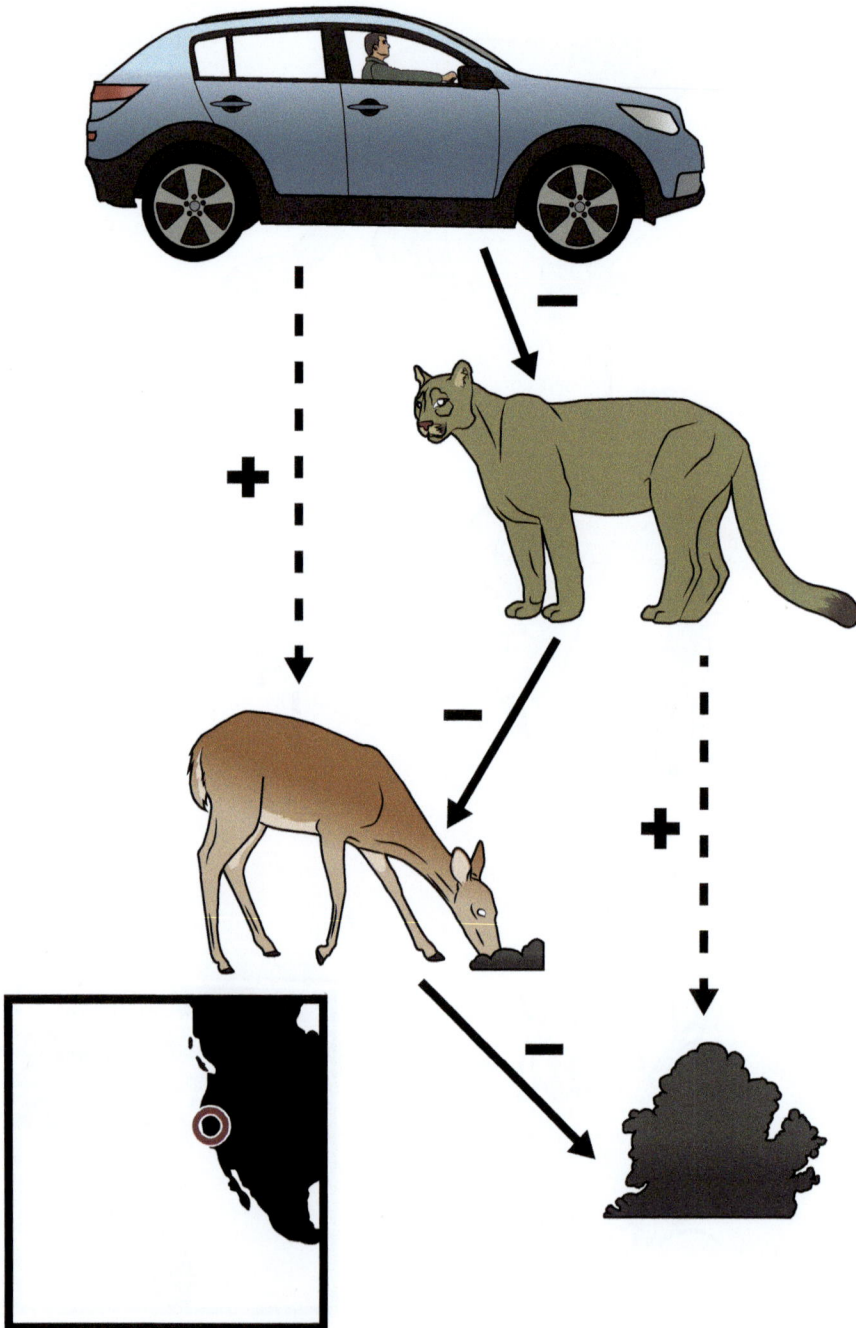

Figure 9.9 A behaviorally mediated trophic cascade in which human development in California affected fine-scale behaviors of pumas and deer, leading to a behaviorally mediated trophic cascade in which the protection provided by pumas to browse plants is cancelled by human presence (Yovovich *et al.*, 2021).

excluded (in this case via exclosures) to allow browse species to grow above browsing height. Such exclusion is not practicable at a landscape scale (Wolf *et al.*, 2007). Recovery of browse plants in some settings may require many decades, or another major perturbation that allows woody plants to escape elk browsing.

Brice and colleagues (2022) described a case of sampling bias in the Yellowstone trophic cascade literature, showing that using the customary method of sampling only the tallest regenerating aspen stems overestimated regeneration by a factor of 3-44, compared with random sampling of stems. Similarly, Boyce (2018) speculated that GPS sampling interval for elk could have affected inferences about habitat use for foraging. The Yellowstone system arguably has generated more published examples of postulated or documented trophic cascades than any other comparable geographic area on Earth. However, it is likewise remarkable for the number of ecologists who attribute the inferred cascades to intellectual biases, methodological weaknesses, or confounding influences.

9.2.4 Strength of behaviorally mediated trophic cascades

Across diverse aquatic, marine, and terrestrial systems, BMTCs involving carnivorans have tended to show effect sizes comparable to those of DMTCs, recognizing that BMTCs and DMTCs cannot be directly compared. Donadio and Buskirk (2016) showed a mean difference between plains and both canyons and meadows—the most strongly differing habitats in terms of predation risk—of two- to sixteenfold, depending on vegetative response. Other studies have tended not to measure differences in herbivory across a risk gradient while adjusting for variation in primary production across the same gradient. With few data available for carnivorans, this is an area of critical need if we are to understand the ecological importance of trophic cascades (e.g. Schmitz *et al.*, 2004). Commonly, several habitat features mediate predation risk: visual cover and escape structures, the latter including trees, burrows, and water bodies. Additionally, the prey species must sense those features and behave

adaptively. For example, moose in Wyoming, US responded to the proximity of wolves in early winter by increasing their rate of movement. However, over winter, responses to wolves within 0.5 km became attenuated; the deteriorated nutritional condition of moose in late winter caused them to alter their assessment of predation risk level (Oates *et al.*, 2019). If a BMTC results from a mix of mechanisms (reduced time spent by an herbivore in risky habitat, increased time vigilant, reduced bite size, reduced bite number), calculating the composite effect size becomes problematic outside of a modeling framework. The obvious trade-off made by an herbivore involved in a BMTC involves foraging vs. predation risk, but comparing these dissimilar currencies is challenging.

Among carnivorans, hunting mode affects the strength of BMTCs (Preisser *et al.*, 2007). Ambushing predators elicit stronger avoidance reactions in prey than do coursing predators that detect prey while moving. This results from the advantage that a certain habitat feature—typically, concealment—confers upon a sit-and-wait predator. This, in turn, affects herbivore presence, vigilance, forage intake, and plant attributes. Thaker and colleagues (2011) tested this idea for African ungulates and predators and found that small ungulates (up to the size of kudu) avoided areas used by all large carnivorans. Large ungulates (from the size of wildebeest to giraffe) avoided areas used by ambush predators, specifically lions and leopards. Ambush habitats were specifically avoided by prey species most vulnerable there. Coursing predators (e.g. cheetahs and wild dogs) were less selective in hunting habitats.

Lastly, some field studies show firm evidence of a trophic cascade, but the underlying mechanism of the cascade (behavior or demography) is unclear. Hebblewhite and colleagues (2005) described a system in Banff National Park, Canada in which recolonizing wolves caused elk densities to diverge between the town of Banff, where wolf densities remained low in spite of recolonization, and an adjacent high-wolf-density area. The wolf densities were inversely related to elk densities, with cascading effects on browse species. For this cascade, it is not clear the extent to which the effect reflects elk behavior (elk moving into the wolf-scarce town, BMTC), or differences in elk density across distinct subpopulations (DMTC). Similarly, Atkins and colleagues (2019) described an African cascade involving the near eradication of large and medium-sized predators during a military conflict in Mozambique. This loss of predators permitted bushbuck, ordinarily closed-habitat specialists, to occupy flat, treeless landscapes. One plant species of open habitats that was eaten almost exclusively by bushbuck showed strong browsing effects. In this case, a single population may have subdivided into open-country and closed-habitat subpopulations. The only behavior invoked by the authors to qualify this as a BMTC is habitat selection by bushbucks (Atkins *et al.*, 2019). Lacking reported behavioral differences between habitats, DMTC seems an equally or more apt descriptor of the relationship. Both of these studies point out the nuanced nature of any DMTC–BMTC dichotomy. If the herbivores in a landscape showing a change in the heterogeneity of risk modify their perception of risk and alter their uses of habitats accordingly, densities of prey subpopulations may differ, and the behavioral-density distinction becomes moot.

Kuijper and colleagues (2013) raised this issue in their study of wolf predation on red deer in Białowieża Primeval Forest, Poland. They compared predation rates, browsing rates, and sapling regeneration inside wolf territory core areas vs. non-core areas of the same territories. Predation rates were higher inside core areas (by a factor of 5), where browsing intensity was 15% lower than outside core areas. They did not report relative deer densities. Saplings were 13% taller inside wolf core areas. The abundance of coarse woody debris (CWD), a presumed impediment to escaping from wolves, strongly influenced the browsing effect. Near (less than one meter away from) CWD inside core areas, browsing was curtailed, whereas outside core areas the effect of CWD on browsing was slight. The authors interpreted these results as reflecting two effects: demographic (core area vs. outside) and behavioral (avoiding of CWD vs. not), although no demographic closure relative to the core area boundary was demonstrated. Deer avoidance of risky sites at two different scales could account for both results. Again, the scalar aspects of DMTC vs. BMTC confuse how effects are classified. If simple avoidance of an area is considered

the behavior in BMTC, then most DMTCs involving heterogeneous predation risk might be considered BMTCs.

9.2.5 Evidence required for trophic cascades

The evidence presented to document trophic cascades differs between DMTCs and BMTCs, and ranges from controlled, mechanistic, and spatially replicated, to uncontrolled, non-mechanistic, and lacking replication. Ford and Goheen (2015) characterized five approaches to test for DMTCs involving large carnivorans. The least rigorous approach assumes but does not measure a carnivoran–herbivore or herbivore–plant interaction, does not measure the overall effect of the carnivoran on vegetation, treats predation as a binary variable, and lacks replication. The most rigorous approach shows that carnivoran and herbivore vital rates and abundances are negatively correlated, herbivore and plant abundances are negatively correlated, carnivoran and plant abundances are positively correlated, and includes temporal breadth and spatial replication.

Evidence for BMTCs overlaps somewhat with that for DMTCs. A BMTC should demonstrate spatial heterogeneity in predation risk within the home ranges of the prey species; herbivores can discern safe vs. risky habitats. Predators should be shown to spend more total time hunting in, or to cause higher rates of predation in, habitats that herbivores perceive as risky. Herbivores must display behaviors consistent with perceiving risk gradients, and their foraging decisions should cause vegetative responses. Lastly, for both DMTCs and BMTCs, rates of primary production should not differ between treatment and control plots, or should be controlled for in analyses of apparent herbivory (Donadio and Buskirk, 2016; Grinath, 2018). Lacking this correction, differences in primary production across risk environments (in the case of BMTCs) or across populations or time periods (in the case of DMTCs) may be attributed to or mask differences in herbivory.

Regardless of the semantic conventions used to classify cascades, some community interactions will remain difficult to classify. Monk and colleagues (2022) described the effects of a sarcoptic mange epizootic in the puma–vicuña–plant system described earlier. Mange mites nearly eradicated the vicuña (herbivore) population, affecting pumas (predators), Andean condors (scavengers), and vegetation. Only a small amount of vicuña biomass is converted to mite biomass, and the interactions ascend and descend trophic levels. This case study illustrates the problem of assigning simple terms that imply top and bottom to interactions with complex dimensionality.

Key points

- Cascades are community responses linked to a single interaction between two species. When a third species—typically at a lower trophic level—responds to an interaction at a higher trophic level, a "trophic cascade" commonly is invoked.
- Other interactions may ascend trophic levels or be non-trophic in nature—"ecological cascades." These processes can be triggered by competition, transport of nutrients, pollination, or other processes, in addition to trophic ones. They can ascend or descend trophic levels.
- Trophic cascades can be density-mediated or behaviorally mediated. The former are caused by predators reducing herbivore densities, eliciting vegetative responses, which are typically increases in plant production, biomass, or reduced plant defenses. Demonstrating them requires comparing traits of populations or subpopulations of herbivores.
- Behaviorally mediated trophic cascades are caused by differences in herbivore behaviors across risk regimes within a subpopulation—altered foraging times, vigilance times, or other responses that link risk to vegetative traits. These responses are smaller in scale than for density-mediated cascades—typically across habitat types.
- Cascades of both kinds often are attributed to carnivoran predation, but some studies fail to show strong effects where expected. Others show effects without evidence for the underlying mechanisms. Strong cascades are not universal features of communities with carnivoran predation.
- Density-mediated trophic cascades tend to be more commonly reported in aquatic systems, in simple communities (single predator, single prey, single food plant), and where variable weather, including seasonality, does not confound top-down effects.
- Behaviorally mediated trophic cascades are more commonly reported where ambush predators consume a single prey species and account for most predation on that species, and where ambushing cover is heterogenous across the landscape.
- Several lines of evidence are needed to demonstrate compellingly a trophic cascade.

References

Atkins, J.L. *et al.* (2019) "Cascading impacts of large-carnivore extirpation in an African ecosystem," *Science*, 364, pp. 173–7.

Beschta, R.L., Painter, L.E. and Ripple, W.J. (2018) "Trophic cascades at multiple spatial scales shape recovery of young aspen in Yellowstone," *Forest Ecology and Management*, 413, pp. 62–9.

Boyce, M.S. (2018) "Wolves for Yellowstone: dynamics in time and space," *Journal of Mammalogy*, 99, pp. 1021–31.

Brice, E.M., Larsen, E.J. and MacNulty, D.R. (2022) "Sampling bias exaggerates a textbook example of a trophic cascade," *Ecology Letters*, 25, pp. 177–88.

Colman, N.J. *et al.* (2014) "Lethal control of an apex predator has unintended cascading effects on forest mammal assemblages," *Proceedings of the Royal Society B: Biological Sciences*, 281, p. 20133094.

Croll, D.A. *et al.* (2005) "Introduced predators transform subarctic islands from grassland to tundra," *Science*, 307, pp. 1959–61.

Donadio, E. and Buskirk, S.W. (2016) "Linking predation risk, ungulate antipredator responses, and patterns of vegetation in the high Andes," *Journal of Mammalogy*, 97, pp. 966–77.

Estes, J.A. and Duggins, D.O. (1995) "Sea otters and kelp forests in Alaska: generality and variation in a community ecological paradigm," *Ecological Monographs*, 65, pp. 75–100.

Estes, J.A. *et al.* (1998) "Killer whale predation on sea otters linking oceanic and nearshore ecosystems," *Science*, 282, pp. 473–6.

Fey, K. *et al.* (2009) "Does removal of an alien predator from small islands in the Baltic Sea induce a trophic cascade?", *Ecography*, 32, pp. 546–52.

Fleming, P.J.S. (2019) "They might be right, but Beschta *et al.* (2018) give no strong evidence that 'trophic cascades shape recovery of young aspen in Yellowstone

National Park': a fundamental critique of methods," *Forest Ecology and Management*, 454, p. 117283.

Ford, A.T. and Goheen, J.R. (2015) "Trophic cascades by large carnivores: a case for strong inference and mechanism," *Trends in Ecology and Evolution*, 30, pp. 725–35.

Ford, A.T. et al. (2015) "Recovery of African wild dogs suppresses prey but does not trigger a trophic cascade," *Ecology*, 96, pp. 2705–14.

Ford, A.T. et al. (2014). "Large carnivores make savanna tree communities less thorny," *Science*, 346, pp. 346–9.

Grinath, J.B. (2018). "Short-term, low-level nitrogen deposition dampens a trophic cascade between bears and plants," *Ecology and Evolution*, 8, pp. 11213–23.

Hairston, N.G., Smith, F.E. and Slobodkin, L.B. (1960) "Community structure, population control, and competition," *The American Naturalist*, 94, pp. 421–5.

Hebblewhite, M. et al. (2005) "Human activity mediates a trophic cascade caused by wolves," *Ecology*, 86, pp. 2135–44.

Kauffman, M.J., Brodie, J.F. and Jules, E.S. (2010) "Are wolves saving Yellowstone's aspen? A landscape-level test of a behaviorally mediated trophic cascade," *Ecology*, 91, pp. 2742–55.

Klemola, T. et al. (2000) "Experimental tests of predation and food hypotheses for population cycles of voles," *Proceedings of the Royal Society B: Biological Sciences*, 267, pp. 351–6.

Koel, T.M. et al. (2005) "Nonnative lake trout result in Yellowstone cutthroat trout decline and impacts to bears and anglers," *Fisheries*, 30, pp. 10–9.

Kuijper, D.P.J. et al. (2013) "Landscape of fear in Europe: wolves affect spatial patterns of ungulate browsing in Białowieża Primeval Forest, Poland," *Ecography*, 36, pp. 1263–75.

le Roux, E., Kerley, G.I.H. and Cromsigt, J.P.G.M. (2018) "Megaherbivores modify trophic cascades triggered by fear of predation in an African savanna ecosystem," *Current Biology*, 28, pp. 2493–9.

Leopold, A. (1949) *A Sand County almanac*. Oxford: Oxford University Press.

Maron, J.L. et al. (2006) "An introduced predator alters Aleutian Island plant communities by thwarting nutrient subsidies," *Ecological Monographs*, 76, pp. 3–24.

Maron, J.L. and Pearson, D.E. (2011) "Vertebrate predators have minimal cascading effects on plant production or seed predation in an intact grassland ecosystem," *Ecology Letters*, 14, pp. 661–9.

Marshall, K.N. et al. (2015) "Conservation challenges of predator recovery," *Conservation Letters*, 9, pp. 70–8.

Middleton, A.D. et al. (2013b) "Animal migration amid shifting patterns of phenology and predation: lessons from a Yellowstone elk herd," *Ecology*, 94, pp. 1245–56.

Middleton, A.D. et al. (2013a) "Grizzly bear predation links the loss of native trout to the demography of migratory elk in Yellowstone," *Proceedings of the Royal Society B: Biological Sciences*, 280, p. 20130870.

Monk, J.D. et al. (2022) "Cascading effects of a disease outbreak in a remote protected area," *Ecology Letters*, 25, pp. 1152–63.

Morris, T. and Letnic, M. (2017) "Removal of an apex predator initiates a trophic cascade that extends from herbivores to vegetation and the soil nutrient pool," *Proceedings of the Royal Society B: Biological Sciences*, 284, p. 20170111.

Norrdahl, K. et al. (2002) "Strong seasonality may attenuate trophic cascades: vertebrate predator exclusion in boreal grassland," *Oikos*, 99, pp. 419–30.

Oates, B.A. et al. (2019) "Antipredator response diminishes during periods of resource deficit for a large herbivore," *Ecology*, 100, p. e02618.

Paine, R.T. (1980) "Food webs: linkage, interaction strength and community infrastructure," *Journal of Animal Ecology*, 49, pp. 667–85.

Perrig, P.L. et al. (2017) "Puma predation subsidizes an obligate scavenger in the high Andes," *Journal of Applied Ecology*, 54, pp. 846–53.

Piovia-Scott, J., Yang, L.H. and Wright, A.N. (2017) "Temporal variation in trophic cascades," *Annual Review of Ecology, Evolution, and Systematics*, 48, pp. 281–300.

Polis, G.A. and Strong, D.R. (1996) "Food web complexity and community dynamics," *American Naturalist*, 147, pp. 813–46.

Preisser, E.L., Orrock, J.L. and Schmitz, O.J. (2007) "Predator hunting mode and habitat domain alter nonconsumptive effects in predator-prey interactions," *Ecology*, 88, pp. 2744–51.

Ripple, W.J. (2016) "What is a trophic cascade?", *Trends in Ecology & Evolution*, 31, pp. 842–9.

Schmitz, O.J., Beckerman, A.P. and O'Brien, K.M. (1997) "Behaviorally mediated trophic cascades: effects of predation risk on food web interactions," *Ecology*, 78, pp. 1388–99.

Schmitz, O.J., Hambäk, P.A. and Beckerman, A.P. (2000) "Trophic cascades in terrestrial systems: a review of the effects of carnivore removals on plants," *American Naturalist*, 155, pp. 141–53.

Schmitz, O.J., Krivan, V. and Ovadia, O. (2004) "Trophic cascades: the primacy of trait-mediated indirect interactions," *Ecology Letters*, 7, pp. 153–63.

Shurin, J.B. et al. (2002) "A cross-ecosystem comparison of the strength of trophic cascades," *Ecology Letters*, 5, pp. 785–91.

Singer, M.S. et al. (2014) "Herbivore diet breadth mediates the cascading effects of carnivores in food webs,"

Proceedings of the National Academy of Sciences of the United States of America, 111, pp. 9521–6.

Suraci, J.P. *et al.* (2016) "Fear of large carnivores causes a trophic cascade," *Nature Communications*, 7, p. 10698.

Thaker, M., Vanak, A.T., Owen, C.R., Ogden, M.B., Niemann, S.M., and Slotow, R. 2011. Minimizing predation risk in a landscape of multiple predators: effects on the spatial distribution of African ungulates. *Ecology* 92: 398–407.

Van Manen, F.T. *et al.* (eds.) (2013). *Yellowstone grizzly bear investigations: annual report of the Interagency Grizzly Bear Study Team.* Bozeman: US Geological Survey.

Wade, P.R. *et al.* (2007) "Killer whales and marine mammal trends in the North Pacific—a re-examination of evidence for sequential megafaunal collapse and the prey switching hypothesis," *Marine Mammal Science*, 23, pp. 766–802.

White, P.J., Gunther, K.A. and Wyman, T.C. (2017) "The population—attributes, behavior, genetics, nutrition, and status," in White, P.J., Gunther, K.A. and van Manen, F.T. (eds) *Yellowstone grizzly bears: ecology and conservation of an icon of wildness.* Bozeman: Yellowstone Forever, pp. 1–11. https://www.nps.gov/yell/learn/nature/bearbook.htm (Accessed: November 12, 2021).

Winnie, J.A., Jr. (2012) "Predation risk, elk, and aspen: tests of a behaviorally mediated trophic cascade in the Greater Yellowstone Ecosystem," *Ecology*, 93, pp. 2600–14.

Wolf, E.C., Cooper, D.J. and Hobbs, N.T. (2007) "Hydrologic regime and herbivory stabilize an alternate state in Yellowstone National Park," *Ecological Applications*, 17, pp. 1572–87.

Yovovich, V., Thomsen, M. and Wilmers, C.C. (2021) "Pumas' fear of humans precipitates changes in plant architecture," *Ecosphere*, 12, p. e03309.

CHAPTER 10

Population ecology

In its narrow sense, a population comprises collections of interbreeding organisms, and population processes include births, deaths, and dispersal. More broadly, population biology includes movement ecology and the distribution and movement of genes between populations and across landscapes and seascapes. Selection, genetic drift, and genetic exchange link population biology and evolution. These within-species processes are called microevolution, in contrast to macroevolution, or heritable changes above the species level. Phenotypic or genetic differences among populations tell us whether they are of a single interbreeding species, distinct species that share alleles at a zone of contact, or separate species that have not shared genes over evolutionary time spans. Decades ago, these differences were inferred entirely from morphological features, including size, shape, and color. Since the 1970s, zoologists have used genetic information—heritable protein polymorphisms, DNA markers, or entire genomes—to infer among-population genetic differentiation, and much more.

Movements over corridors connect populations, and barriers to movement isolate them. Humans have studied carnivoran movements for millennia by following tracks in soil or snow. Modern biologists do so by attaching unique marks or tags to individuals, by affixing transmitters that allow researchers to follow animals, and by using remote cameras to photograph them. These methods can be combined: carnivorans can be photographed, the presence of a tag or electronic chip detected, and hair collected for its DNA, all at a single station. DNA permits human observers to build encounter histories for individual animals and infer habitat use and population size, which links the fields of behavior, molecular biology, population biology,

biogeography, and microevolution, both conceptually and in practice.

10.1 How carnivorans die

Although we lack analyses of known-cause mortality across mammalian orders, carnivorans appear to die from quantitatively different causes. Importantly, researchers tend to report proximate causes of death, which may not reflect underlying factors. For example, an animal with a deleterious mutation might struggle to feed itself, causing it to take more risks than it would otherwise, which could lead to it being killed by another carnivoran. At necropsy, it might exhibit poor body condition and trauma, but the underlying genetic condition likely would go undetected. Still, we learn much from proximate causes of death, and our best understandings of them derive from telemetry studies using mortality-activated transmitters, which lead biologists to animals that have stopped moving. Other data sources include carcasses found by biologists or the public and necropsied. These latter data can be biased; for example, animals killed by vehicles on roads are more visible to motorists than animals dying of other causes, away from roads, leading to a false inference that vehicle collisions cause a high proportion of deaths.

Carnivorans die from a wide range of species- and site-specific factors, including interspecific killing (82% of deaths of Pacific martens; Bull and Heater, 2001), human hunting or trapping (90% of deaths of American martens; Hodgman *et al.*, 1994), and disease (39% of urban red foxes; Gosselink *et al.*, 2007). Of European otters collected from eastern Germany over several decades, 71% were killed by vehicles (Hauer *et al.*, 2002). Collins and Kays (2011)

Carnivoran Ecology. Steven W. Buskirk, Oxford University Press. © Steven W. Buskirk (2023). DOI: 10.1093/oso/9780192863249.003.0010

reviewed twenty-seven species of medium- and large-sized North American mammals and found that carnivorans were less likely than other orders (artiodactyls, lagomorphs, and rodents) to die of predation, but more likely to die of anthropogenic factors, including hunting and trapping. As shown earlier, small-bodied carnivorans are especially vulnerable to interspecific killing.

By contrast, herbivorous mammals, particularly neonates and those in protected areas, tend to die of predation more than any other factor, with disease, interference competition, human harvest (for species smaller than ungulates), and vehicle collisions fairly unimportant. Deaths caused by predation for thirteen species of herbivores ranging in size from voles to moose, but emphasizing neonatal ungulates, were 78% of all known-cause deaths (mean of study means, Boutin et al., 1986; Keith and Bloomer, 1993; Linnell et al., 1995; Norrdahl and Korpimaki, 1995; Van Vuren, 2001; DelGiudice et al., 2002; Barber-Meyer et al., 2008; Boland and Litvaitis, 2008; Donadio et al., 2012).

In addition to background mortality patterns, some carnivoran populations experience episodes of mass mortality. This is especially true of pinnipeds, which can suffer losses of more than 50% over small areas and brief periods. The proximate causes of these events tend to be disease, toxins, and trampling on rookeries. The deaths of tens of thousands of harbor seals on the coast of Europe in 1988 were attributed to a paramyxovirus now called phocine distemper virus (Harwood and Hall, 1990). A similar outbreak of a toxic diatom was suspected to have caused the deaths of several hundred California sea lions in central California in 1998 (Scholin et al., 2000). Rabies and canine distemper likewise can cause high rates of mortality among terrestrial species. Rabies of various strains is carried by many carnivoran species, and canine distemper has caused nearly 100% mortality among exposed black-footed ferrets and African wild dogs (Williams et al., 1988; van de Bildt et al., 2002). Sarcoptic mange infects a wide range of mammals worldwide, including carnivorans. It is especially prevalent in canids and had dramatic effects on red fox, coyote, and wolf populations in North America in the early 2000s. Typically, these canid populations decline, recover from epizootics, and

show slight or no long-term effects (Pence and Ueckermann, 2002).

10.2 Demography

A few carnivorans of conservation concern are among the best-studied free-ranging mammals. However, population attributes of carnivorans generally are less well known than those of some other mammalian orders for two reasons. First, predators are inherently rarer than herbivores of similar body size, so that sample sizes can be small. Second, many carnivorans are secretive, nocturnal, or both, so that observing them directly is difficult. Carnivorans give birth to altricial young, mostly below ground, in dense woody vegetation, or in other protected sites, so that recording birth events is challenging. Dates of parturition, litter sizes, and neonatal survival are less observable than for ungulates. Even today, many studies of carnivoran demography focus on methodology—how to infer vital rates in lieu of direct data.

Early understandings of carnivoran demography were limited to such traits as minimum number alive, litter size after leaving the den, age at first reproduction, and inter-birth interval. Data that enabled the construction of life tables—schedules of reproductive rate, mortality rate, and life expectancy by age—for extant carnivoran populations emerged after the mid-twentieth century. For example, Kenyon and colleagues (1954) documented juvenile mortality in northern fur seals, and Kurtén (1976) used tooth measurements to estimate age-specific survival and life expectancy for extinct cave bears from various European sites. Although the cave bear data derive from caves distributed across Europe, with thousands of years between sample collections, they provide remarkable understandings of a long-extinct taxon that had a life history similar to that of extant brown bears (Craighead et al., 1974).

The next major analytical step was to matrix demographic modeling, which uses Leslie matrices to estimate demographic traits. Leslie matrices require age-specific fecundity and mortality, and Lefkovich matrices are the corresponding tool for stage-specific data, in which individuals can advance in stage (e.g. juvenile, subadult) with increasing age or remain in the same stage. Fecundity is sequential (typically annual) reproductive output and combines inter-birth interval with litter size. Survival estimates are probabilities of animals of one age (or stage) living to the next. Matrix models are particularly useful in demonstrating which life stage transitions best explain variation in population growth or decline. If small proportional variation in a single transition rate—for example, that from birth to one year of age—results in a large response in population growth, the population shows high elasticity to that transition. Elasticity analyses are the clearest means to visualize how populations respond to perturbations and are important for estimating sustainable rates of off-take and for analyzing population viability.

Of carnivoran species, few are known well enough to have been modeled using Leslie matrices; van de Kerk and colleagues (2013) reviewed reports of thirty-five studies conducted on twenty-seven carnivoran taxa, dominated taxonomically by canids and ursids and geographically by North America. Species of conservation concern were especially well represented, and those species have tended to be large bodied. The studies illustrated the difficulty of parameterizing demographic models with data from scarce and elusive carnivorans.

10.3 Spatial aspects of population organization

Because carnivoran home ranges are often territorial, the spatial features of populations can serve as surrogates for the animals themselves. Home range size and packing (proportion of the landscape filled with territories) have been used to estimate population size using models that consider occupancy rates and other variables (e.g. Spencer *et al.*, 2011). When territory holders die, vacancies are filled by resident, subdominant animals. When populations increase, they do so by range expansion into less-optimal, but undefended, habitat; these are the ideal free distribution and ideal despotic distribution proposed by Fretwell and Lucas (1970). Territories tend to prevail where resources are concentrated or defensible, and most terrestrial carnivorans occupy territories as adults. Marine carnivorans do not defend resources in the pelagic environment, but do

so on breeding rookeries, where space is limited and critical to breeding opportunities.

10.3.1 Dispersal and homing

Carnivorous carnivorans disperse longer distances than herbivorous-omnivorous mammals, by a factor of 1.2–4.5, and large-bodied species disperse farther than small ones (Sutherland *et al.*, 2000). The scaling of dispersal distance to body mass differs across analyses; Sutherland and colleagues (2000) reported a scaling exponent of 0.89 for carnivores compared to 0.54 for non-carnivores; carnivores likewise had a higher y-intercept than did non-carnivores. The slope reported by Bowman and colleagues (2002) was 0.62 for all mammals. Similar to Sutherland and colleagues (2000), Lindstedt and colleagues (1986) found that as body size increases, carnivorans show larger increases in home range size than do herbivorous mammals, which was attributed to the confounding effect of latitude. Estimates of dispersal distance and home range size for large carnivorans come mostly from high geographic latitudes, where productivity is low, requiring larger home ranges. This confounds the influence of body size and latitude. The median distance dispersed by a mammalian species, carnivorans included, was about seven times the home range radius (Bowman *et al.*, 2002). Some of the variation in this relationship among carnivorans is due to the habitat niche breadth. For example, swift foxes and kit foxes, weighing 1.5–3 kg, are steppe and shrub-steppe dwellers and poor dispersers across mountain ranges in western North America, primarily because of forests that separate desert and steppe basins (Mercure *et al.*, (1993). The similarly sized arctic fox readily crosses long expanses of ice-covered ocean and snow-covered land (Box 10.1), including steep terrain, and likewise avoids forests.

Homing movements are those made by animals after translocation away from their home ranges and include some extraordinary carnivoran examples of navigation and endurance. Bowman and colleagues (2002) documented the maximum documented distances for some North American carnivorans that returned to their home ranges, for example, 282 km for the wolf, 56 km for the red fox,

---

Box 10.1 Arctic fox mariners

Arctic foxes are the most accomplished travelers on sea ice (Figure 10.1). The most dramatic reported case was that of a fox that traveled from Svalbard Island, Norway to Ellesmere Island, Canada in 76 days. The female covered 1789 km straight-line, 4415 km between successive GPS points. The greatest straight-line distance moved in a single day was 155 km while traversing the Green-land ice cap (Fuglei and Tarroux, 2019). Comparable dispersal distances are traversed by arctic foxes riding on wind-driven sea ice but require longer times (Fay and Rausch, 1992). It is not surprising that with such exceptional dispersal abilities, arctic foxes show little spatial land-scape genetic structure (Geffen *et al.*, 2007; Norén *et al.*, 2011).

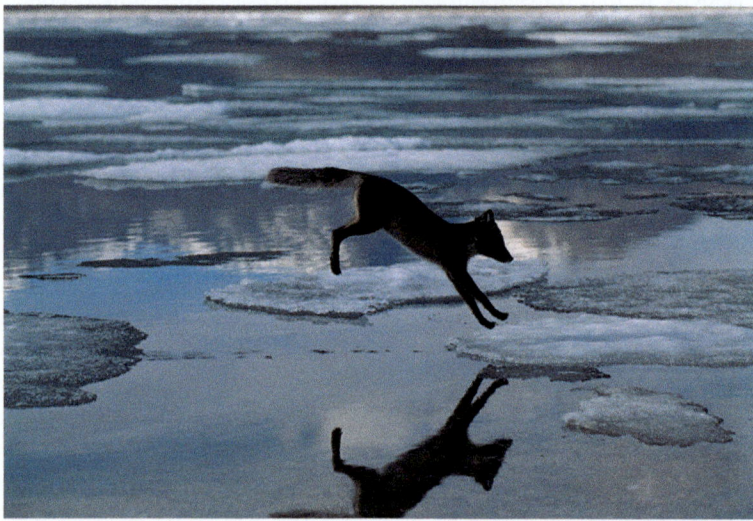

Figure 10.1 An arctic fox negotiates an open lead on sea ice. Arctic foxes can traverse long distances on contiguous sea ice, and ride wind-blown ice, but small sections of open water can be complete barriers.
Photo: M. Watson/Science Source.

---

and 258 km for brown bear. These values scale to body size with an exponent that approaches that for the linear dimension of the home range, maximum homing distance being about forty times the home range radius. Figures such as these are of applied as well as conceptual interest; planning translocations of problem carnivorans requires knowing from how far they can return.

10.3.2 Migration

Long seasonal movements are relatively uncommon in terrestrial carnivorans because abandoning a defended territory for part of a year interrupts site tenure. Further, the carnivoran reproductive pattern

of raising altricial neonates in protected sites limits the mobility of adults and young during the reproductive season. Land carnivorans that do migrate traverse greater distances than other mammalian species, adjusting for body size. Wolves are particularly mobile, traversing total yearly distances of up to 7200 km in Mongolia, and 5900 km in northern Canada. Most of the other longest-migrating land mammals are ungulates several times larger than wolves (Joly *et al.*, 2019). Some pinnipeds migrate much farther; Northern elephant seals migrate each year along a precise great-circle route from Southern California to the North Pacific Ocean, covering 18,000–20,000 km round-trip, not including underwater movements (Brillinger and Stewart,

1998). Clearly, in terms of size-specific movement rates, carnivorans surpass all other mammalian orders, but migration and dispersal should not be conflated. Dispersal involves movement away from an animal's birthplace, in search of vacant habitat, whereas migration represents periodic, typically seasonal movements that are retraced. Some carnivorans are highly migratory but give birth near where they were born. For example, the Steller sea lion forages widely across the Gulf of Alaska during winter, but gives birth near where mothers were born, and nearly always in the same place from year to year (Hastings et al., 2017).

10.3.3 Population genetic structure

The genetic structure of populations links biogeography, evolution, and ecology, and searches for evidence of corridors, barriers, and dispersal mechanisms. Modern molecular tools allow us to quantify dispersal rates across landscapes without tracking individual animal movements. The same tools also allow us to identify genetic events at various scales of time and relatedness—whether two individuals are first-order relatives (parent–offspring, full sibling) or related more distantly. Other methods can be used to estimate when two lineages diverged millions of years in the past. Genetic variation across long distances and heterogeneous habitat—so-called landscape genetics—is useful in understanding how carnivorans perceive corridors and barriers differently from other mammals. Inasmuch as carnivorans disperse farther than other mammals, do they show correspondingly less landscape genetic structure? Several basic patterns of landscape genetic variation are possible, and at least some carnivoran species illustrate each one.

The absence of genetic structure (panmixia) occurs in species that disperse readily over the distances at issue, in some cases across the entire geographic range of a taxon. Apparent panmixia can also be shown for large populations, even lacking dispersal, because genetic drift occurs so slowly. The carnivoran species that lack spatial genetic structure tend to be large-bodied, habitat generalists, or pelagic marine species (Mercure et al., 1993). Northern fur seals illustrate this pattern, showing little genetic structure across the 7500 km from

California, US to Sakhalin Island, Russia (Dickerson et al., 2010). Hawaiian monk seals are similarly panmictic across the 2500-km Hawaiian archipelago (Schultz et al., 2010). These patterns occur because ocean water is a homogenous dispersal corridor for a large-bodied mammal. As do marine mammals, polar bears share this lack of spatial genetic structure across their vast Holarctic range, showing only minor genetic structure that could reflect recent changes in sea ice conditions (Paetkau et al., 1999). On land, various carnivoran species have been shown to lack genetic structure over major parts of the geographic ranges; examples include the red fox, coyote, Canada lynx, and jaguar (Lehman et al., 1991; Eizirik et al., 2001; Teacher et al., 2011; Row et al., 2012).

Isolation by distance—increased differentiation with distance—is the most commonly reported spatial genetic pattern among terrestrial carnivorans (Figure 10.2A). No analyses have considered covariates affecting isolation by distance in Carnivora; however, body size is positively correlated with both home range size and dispersal distance (Bowman et al., 2002). Dispersal distance is negatively correlated with degree of population genetic structure in non-carnivoran mammals (Burns and Broders, 2014). Therefore, large carnivorans should exhibit less spatial structure than small species, inferred from a low slope of the isolation-on-distance regression. Degree of habitat specialization also affects landscape genetic structure; species with narrow tolerances (e.g. martens, kit foxes, and swift foxes) seem to show stronger isolation-by-distance effects than habitat generalists (e.g. jackals, coyote). It has not been possible to test for phylogenetic effects in rates of dispersal across long distances because of limited numbers of studies, varying metrics of genetic structure, and the confounding effect of body size.

Isolation by barrier describes abrupt discontinuities in gene flow where barriers bisect corridors (Figure 10.2B). These tend to be bodies of water (for terrestrial species), land masses (for aquatic species), and habitats that the focal species does not tolerate. Ocean expanses are the most frequently identified natural barriers to carnivoran dispersal, evidenced in the traditional taxonomic literature by the designation of island populations

(a)   (b)   (c)

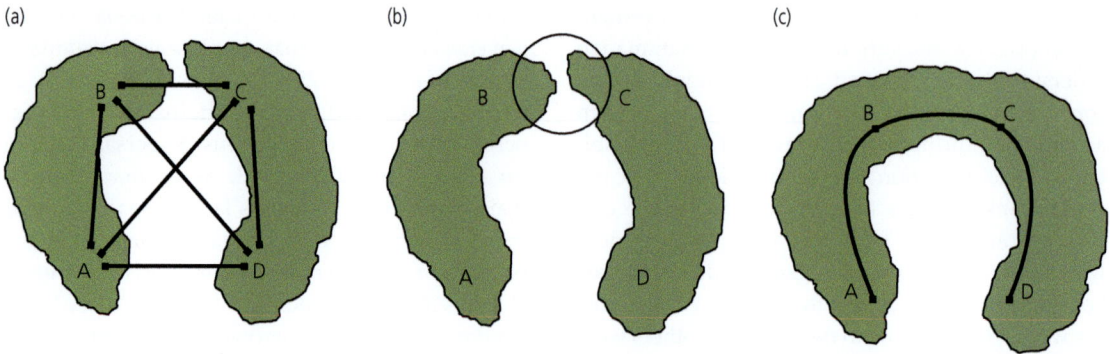

Figure 10.2 Differences between isolation-by-distance, isolation-by-barrier, and isolation-by-ecology patterns in landscape genetics. Points A–D represent populations of a single animal species, separated by distances that could result in genetic divergence. A: Isolation by distance. The green area represents habitat used by the species equally to the surrounding white matrix. The straight-line distance between populations best explains spatial genetic structure. B: Isolation by barrier. The green area represents a habitat type preferred by the species; it seldom disperses across the white matrix. The width and other attributes of the white gap separating the preferred habitat best explain spatial genetic structure. C: Isolation by ecology. The species prefers green habitat and avoids white. The distance between populations within the obligate habitat best explains spatial genetic structure.

as discrete taxa—species or subspecies. Examples include a Canada lynx subspecies on Newfoundland Island (Row *et al.*, 2012), and a subspecies of *Meles* badger in Japan (Marmi *et al.*, 2006). The *Mustela erminea* complex provides the most dramatic example of insular differentiation in the Carnivora. The complex was formerly considered a single Holarctic species comprising thirty-four named subspecies, fourteen of them insular endemics. Based on genomic and morphological traits, Colella and colleagues (2021) found that the complex holds at least three species, with some speciation events explained by water barriers, whereas others are not.

Likewise, the St. Lawrence River in eastern Canada poses a partial barrier to dispersal of lynx, as evidenced from genetic studies (Koen *et al.*, 2015). Brown bears of coastal mainland Alaska were traditionally regarded as differentiated from interior "grizzly bears" at the subspecific or specific level based on body size or skull shape, but microsatellite DNA has shown that mainland bears near the coast differ only weakly from those in the interior. Instead, strong differentiation occurs between mainland bears and those on islands a surprisingly short distance offshore; islands farther offshore (>4 km for females, and >7 km for males) show marked genetic isolation (Paetkau *et al.*, 1998). Wide rivers are barriers to dispersal for European

badgers in western England, raccoons in Ontario, Canada, and striped skunks in southern Quebec, Canada (Cullingham *et al.*, 2009; Frantz *et al.*, 2010; Talbot *et al.*, 2012).

These corridor-barrier distinctions can be intermittent. For example, surface ice connects land bodies seasonally or less frequently, so that carnivorans can episodically colonize via water routes too far to swim. Coyotes colonized Newfoundland Island, Canada by an anomalous surface-ice connection, and in an extreme example, red foxes colonized St. Matthew Island in the Bering Sea, which lies more than 400 km from the Siberian and Alaskan mainlands, in the last sixty years. By 2005, a red fox population of at least three mitochondrial (maternal) lineages had completely displaced arctic foxes, formerly the only native canids (Colson *et al.*, 2017).

Other carnivorans are also strong dispersers across polar landscapes and seascapes. Hoekstra and Fagan (1998) examined the native nonvolant mammalian fauna of the Kuril Islands. This archipelago extends from northernmost Hokkaido, Japan to southernmost Kamchatka, Russia, separating the Sea of Okhotsk from the North Pacific Ocean. Individual islands present a range of land areas and distances from the Japanese and Siberian mainlands. Carnivorans, not including pinnipeds, accounted for thirty of the ninety-four species-island occupancies, much higher than predicted from general ecological patterns. Medium- and large-bodied carnivorans—particularly red fox and brown bear—were more likely to occur on some remote islands than nearer the mainlands. The remotest mammal-occupied islands had the largest-bodied mammals, which were carnivorans, representing compositional disharmony—a pattern inconsistent with predictions based on island area and water distances. The authors attributed this pattern to large carnivorans being more resistant to high-latitude temperature extremes, which facilitated their dispersal while crossing Pleistocene ice bridges or while swimming in sea water. Hoekstra and Fagan (1998) also proposed that carnivorans are better able to subsist on the scarce food resources of high-latitude islands than are other mammals of the region, which include rodents, shrews, lagomorphs, and artiodactyls. Foxes and bears can more efficiently use rich nutrients in marine-derived carrion, nesting sea birds, and spawning salmonids. One counterintuitive example of isolation by barrier exists in the marine otter, which might seem well-suited to disperse along coastlines. However, the marine otter requires rocky haul-out areas where it can rest underground. This species limits its time in water, being poorly adapted to the cold water along the Chilean coast, and rests and recovers in burrows not found on long stretches of sandy beach. Because of these physiological limitations, its dispersal is restricted by unbroken sandy beaches, discernable as genetic discontinuities (Vianna *et al.*, 2010).

Human-created linear features are often recognized as barriers to dispersal, shaping genetic structure in carnivorans. European badgers, red foxes in Japan, American black bears in the Rocky Mountains, and pumas in southern California show genetic discontinuities at highways or railway routes (Frantz *et al.*, 2010; Bull *et al.*, 2001; Ernest *et al.*, 2014; Kato *et al.*, 2017). The barrier effect tends to be greater for highways that have many lanes or incorporate fences that block animal movements. Water canals similarly affect genetic connectivity in European pine martens in France (Mergey *et al.*, 2017). In this case, the barrier effect results from the banks of 20-meter-wide canals being covered with corrugated sheeting, which hinders swimming animals from coming ashore. In summary, carnivorans

show many of the same human-created barrier effects that have been reported for other land vertebrates (Holderegger and Di Giulio, 2010), with effect sizes sometimes greater than reported for the latter.

Isolation by ecology—synonymous with "isolation by resistance," combines elements of isolation by distance and isolation by barrier, recognizing that the corridor-barrier variable is continuous, rather than categorical. It describes a pattern in which physical distance via connected suitable habitats explains genetic distance better than straight-line distance (Figure 10.2C). Examples include semi-aquatic species that disperse via water routes more readily than across land. This is the case for European otters in southern France, where populations are connected by watercourses but separated by topographic ridges. The steeper and longer the slope separating aquatic habitats, the greater the barrier to dispersal (Janssens *et al.*, 2008). American minks in Scotland show the same pattern—their movements are restricted primarily by mountain ranges, reflected in spatial genetic structure (Zalewski *et al.*, 2009). Another example involves the preference by the American marten for moving through unharvested vs. harvested forest. Working in Ontario, Canada, Broquet and colleagues (2006) found that connectivity of unharvested forest interacted with movement behaviors by martens to best explain spatial genetic structure. Koen and colleagues (2012) studied American martens in the same region but found that the effective resistance of partially harvested landscapes was less important than linear distance. Distance was more important than resistance in explaining genetic structure, although both effects could be detected. These patterns illustrate that carnivorans show variable facility of dispersal with diverse dispersal modes. The Carnivora includes some of the best dispersers with the least amount of spatial genetic structure of land and aquatic mammals. It also includes dispersal-sensitive species that seldom traverse certain habitat features and show strong landscape genetic structure.

10.3.4 Hybridization and introgression

Population genetics also informs inferences about hybridization and introgression, within or between biological species. Hybridization refers to all lineage crosses including first generation (F1) crosses that produce sterile offspring, whereas introgression describes deeper lineage mixing, including repeated backcrosses. Both phenomena are fairly common across nominal carnivoran species. A number of congeneric species hybridize in zones of sympatry; examples include European pine martens and sables in Fennoscandia, which produce sterile crosses called kidus (Bakeyev and Sinitsyn, 1994). Steppe polecats, European polecats, European minks, and domestic ferrets can produce fertile hybrids in captivity, but do so rarely in the wild (Davison *et al.*, 2001). Polecat males mate with European mink females, but not vice versa, and backcrossing is rare (Cabria *et al.*, 2011), maintaining species barriers.

The tendency to introgress is taxon and context specific. The spotted skunks of North America comprise two genomic and biological species at their northern distributional limits, where eastern and western forms, long-recognized, have non-overlapping breeding seasons and only the western form exhibits embryonic diapause (Mead 1968a,b). Farther south, some pairs of lineages exhibit discordance between introgression inferred from mitochondrial vs. nuclear genomes. The Sonoran Clade of far western Sonora, Mexico and southernmost Arizona (recently identified solely from its mitochondrial genome) showed mitochondrial introgression with *Spilogale leucoparia* in southern Arizona. By contrast, *S. leucoparia* showed no evidence of introgression in its zone of contact with *S. interrupta* in west Texas, where sampling was similarly intense (McDonough *et al.*, 2022). Figure 10.2C illustrates a scenario in which populations B and C introgress freely, but A and D become allopatric and evolve barriers to gene flow.

Introgression is a particular problem where rare wild carnivorans encounter introduced species or domestic strains. Examples include domestic cats introgressing with European wildcats, European pine martens introgressing with introduced American martens, and Australian dingoes mixing with domestic dogs (Kyle *et al.*, 2003; Hertwig *et al.*, 2009; Jones, 2009). Introgression can involve more than two species; the Geoffroy's cat, oncilla, and colocolo are members of Genus *Leopardus*, which radiated

in South America beginning around 2.9 Ma. The first two species are morphologically similar and broadly allopatric, with a narrow zone of sympatry. The colocolo overlaps the distributions of the other two species. Genetic analyses show that the Geoffroy's cat and oncilla produce a hybrid zone, with introgression dependent on distance from the range boundary. Sequence segments from the colocolo have also been found in oncilla-like animals, as have introgressive changes in the other direction. So, these three neotropical felids form complex admixtures that vary with distance from the zones of contact (Trigo et al., 2008).

One of the most closely studied examples of wild–domestic lineage introgression involves American mink in eastern Canada. In this region, the spatial density of trapped wild mink is inversely related to the density of ranch mink trapped in the same area (Bowman et al., 2007), the latter of which are often recognizable from pelage color (Figure 10.3). This finding alone would be consistent with plausible competition between wild and escaped animals, or with other processes. Kidd and colleagues (2009) used molecular markers to show that nearly two-thirds of free-ranging mink near mink farms were either escapees or their descendants. Over a broader area of eastern Canada, Bowman and colleagues (2017) found that one-third of wild-caught mink were escapees or introgressed animals, with the highest frequencies of such animals found where mink farms were most common. Further, Morris and colleagues (2020) considered how introgression affected genetic variability and demonstrated that free-ranging mink near mink farms had lower functional genetic diversity than those distant from mink farms. This reflects the strong human-applied selection on ranch mink for fur characteristics or other desirable traits. Introgression with wild mink has led to outbreeding depression among free-ranging mink, so that wild mink are at risk of decline across regions where ranching occurs.

Introgression in wild carnivorans has been influenced by a history of human exploitation. Fur seals of multiple unrecorded species were harvested to extinction on subantarctic Macquarie Island during the nineteenth century. Recolonization by three species of *Arctocephalus* (New Zealand fur seal,

Figure 10.3 An adult male hybrid of a black-pelage domestic mink and a wild American mink in the Niagara region of Ontario, Canada. Hybrid and introgressed domestic-wild mink are common in this region, where genetic mixing threatens wild American mink populations. Photo: © Larissa Nituch.

Antarctic fur seal, and subantarctic fur seal) began around 1950, and by 1955, a New Zealand fur seal pup was observed with its mother, the first reproduction by the species there in over 130 years. Fur seals reestablished breeding rookeries asynchronously across sexes, so that sex ratios of the recolonizing populations were highly skewed. The first New Zealand fur seals to return in numbers were non-breeding males, and that species still had not established a strong breeding population by 2010. Prior to 1991, heterospecific territories (females of one species in a territory held by a male of a different species) and interspecific matings were common. Across all species, 17–30% of pups were hybrids, and backcrossing and introgression were fairly common (Lancaster *et al.*, 2006; Goldsworthy *et al.*, 2009). In this system, the combination of human eradication of native populations, the close relatedness of the three fur seal species, and slow and sex-asynchronous recolonization from distant breeding rookeries has exacerbated introgression. The problem is expected to resolve over time as pure-line individuals become more common, and matings with like individuals become more common (Lancaster *et al.*, 2007).

The Genus *Canis* represents another special case of introgression. The extant Canidae are closely related, having diverged over the last 10,000,000 years (Wayne *et al.*, 1997). In the even more closely related *Canis*, broad sympatry of nominal species is common, as is hybridization (Figure 10.4) (Lindblad-Toh *et al.*, 2005; Gaubert *et al.*, 2012; Galov *et al.*, 2015; Gopalakrishnan *et al.*, 2018). This has important implications for intraspecific evolutionary change because rather than *Canis* species evolving independently, they have shared substantial parts of their genomes over long time spans. Introgression has occurred directly as well as through intermediary lineages. For example, genes from the ancestral coyote lineage can be traced to extant African *Canis*. The domestic dog, the most abundant and widespread carnivoran in the world, has been an important intermediary, crossing with most *Canis* species except the side-striped jackal and black-backed jackal (Figure 10.4). *Canis* hybridization events do not always lead to introgression and are sexually asymmetrical. For example, crosses of male dogs and female wolves are common, but female wolves seldom mate with male dogs, as shown by wolf mitochondrial haplotypes. This is so because of asynchronous breeding between the two lineages. Most wild female *Canis* have one estrous period per year, and males are sexually active only then. By contrast, female dogs typically have two estrous periods yearly, and males can impregnate females year-round. As a result, female wolves in estrus that encounter male dogs are readily impregnated, whereas female dogs that encounter male wolves while in estrous are not (Leonard *et al.*, 2014).

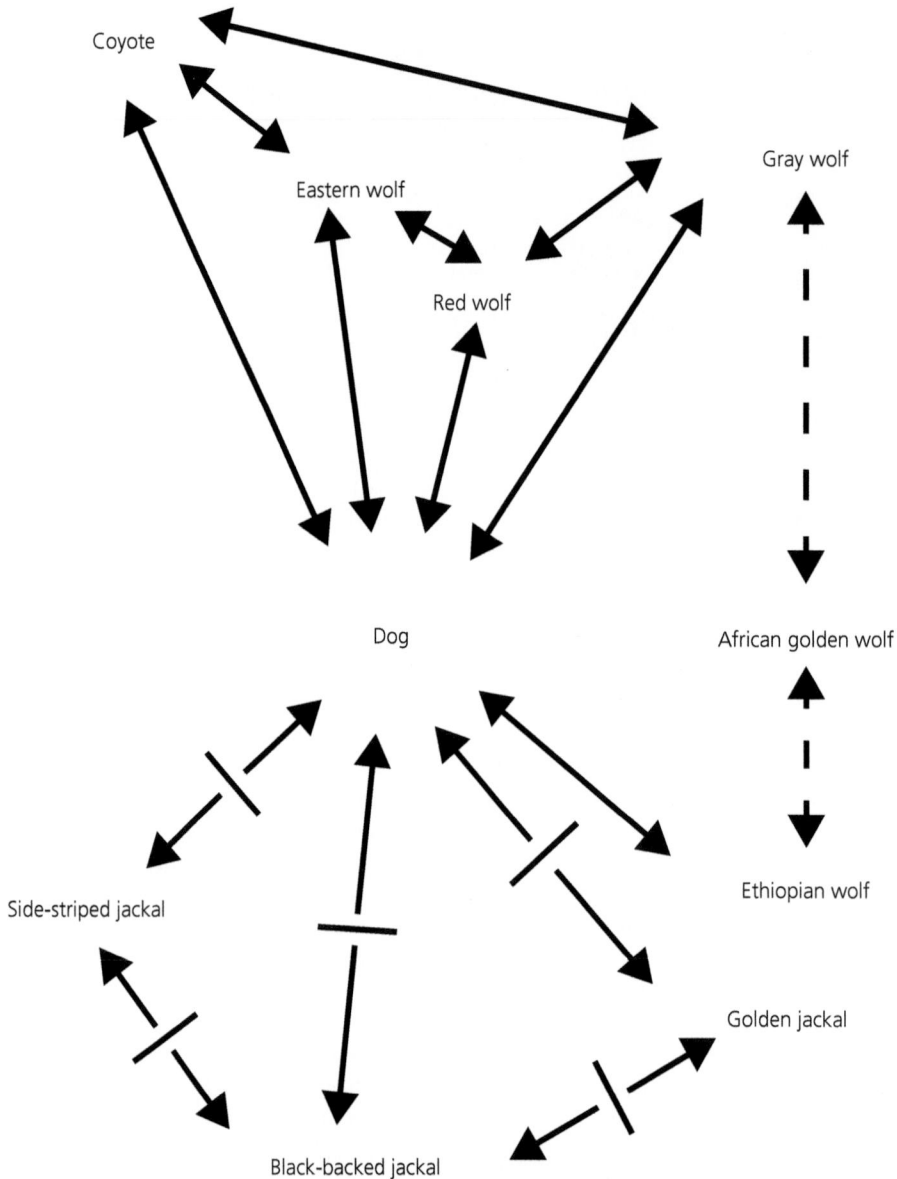

Figure 10.4 Observed modern and inferred past introgression between nominal species of *Canis*. Dogs include various dingo varieties from Indo–Australia. Known modern introgression between extant species is denoted by solid arrows, inferred ancient introgression is denoted by dashed arrows. Mitochondrial data show that the coyote is basal to all other species except the black-backed jackal, side-striped jackal, and golden jackal. These three species diverged from other *Canis* several million years ago, hybridize less than other *Canis*, and likely warrant generic rank. A lineage ancestral to the modern dhole and African hunting dog is basal to coyotes. Hybridization with dogs tends to be asymmetrical, with female wild *Canis*–male dog hybrids the more common outcome.

10.4 Carnivoran life histories

At a basic level, an organism's life history comprises its pattern of growth to adult size, reproductive maturation, breeding, and producing and caring for young. These traits occur on a time scale that is fundamentally constrained by metabolic rate and longevity, both of which are correlated with body size (Lindstedt and Calder, 1981). Life history traits

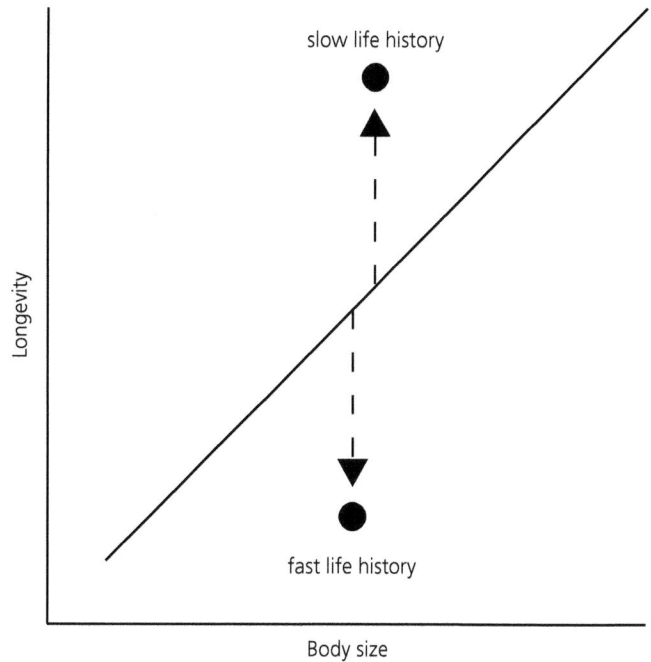

Figure 10.5 Life history traits can be interpreted from body-size-dependent traits or defined on a body-size-adjusted basis. The latter approach considers the residual variation of a species' trait from the trait predicted from its body size, based on comparator organisms. Here, the trait is longevity, which across mammalian species is positively correlated with body size. Species with longer lives than predicted from body size are considered to have slow life histories, and vice versa. Life history traits occur in suites; species with short lives tend to have high mortality rates, early ages of first reproduction, large litters, and short inter-birth intervals.

tend to occur in suites; small-bodied mammals tend to grow and sexually mature quickly, produce large litters, care for them briefly, and die young—a "fast" life history. By contrast, large-bodied species do the reverse—a "slow" life history. This body-size-mediated life "speed" varies strongly across the Carnivora because they display so much variation (five orders of magnitude) in body size, from that of a 40-g least weasel to that of a 4000-kg southern elephant seal.

Considering a species' departure from the life history predicted by its body size, another view of life history emerges; for its body size, a species is faster or slower than expected (Figure 10.5) (Dobson, 2007). For example, the five species of boreal forest martens (European marten, sable, Japanese marten, Pacific marten, and American marten) and the black-footed ferret all weigh around 1.2 kg, live in the north temperate or boreal zone, produce one litter per year, and eat primarily mammals that they kill on land near the soil surface. The martens exhibit delayed implantation, produce first litters at the age of two years with litter sizes of three to four, and can survive in the wild for over ten years.

Population growth is most sensitive to adult survival rate (Buskirk et al. 2012), so that martens have slow life histories for their body size. By contrast, the black-footed ferret implants embryos without delay and produces its first litters at one year. Litters in the wild are slightly larger than those of martens but reach as high as seven (Eads and Marsh, 2020) and can reach nine in captivity. However, female fecundity declines sharply after about four years of age. Population growth is most sensitive to first-year survival and fertility; the black-footed ferret has a fast life history for its body size, atypical of a critically endangered species (Grenier et al., 2007). The environments occupied by the two species may account for this difference. Boreal forest martens occur in mesic mid- to late successional forests with predictable disturbance dynamics. Resources are stable through time. Black-footed ferrets occupy prairie dog colonies, where bison grazing, frequent wildfire, and drought cause unpredictable collapses in prey abundance. Alternatively, the differences may reflect the phylogenetic traits of the two lineages, which diverged about 12 Ma (Koepfli et al., 2008).

The Carnivora, occupying trophic positions differing from those of most community members, might be expected to have life histories distinct from their prey. However, comparative analyses do not support such a view. Van de Kerk and colleagues (2013) showed that carnivoran life histories are as slow and fast as those of other mammals, whether considering body size or not. The sexual size dimorphism common among carnivorans complicates this comparison, with males-larger dimorphism more pronounced in larger-bodied carnivorans (Frynta *et al.*, 2012), a pattern found in various mammalian lineages (Rensch's rule). However, among mustelid dentitions the pattern is reversed; canine tooth size is more sexually dimorphic in the smallest-bodied species (Gittleman and Van Valkenburgh 1997). Males-larger size dimorphism commonly is inferred to reflect sexual selection—males with the largest bodies (or teeth in the case of mustelids) enjoy fitness advantages during aggressive encounters with conspecifics of the same sex. The intraspecific pattern observed in domestic dog breeds is another curious exception. These breeds result from artificial selection, much of it applied over the last few centuries. Humans surely did not mimic sexual selection in creating dog breeds, yet sexual size dimorphism is pronounced across breeds ranging from chihuahuas to mastiffs. However, the life history traits that scale with body size across wild canids differ across dog breeds. For example, across fourteen breeds spanning a tenfold range of body sizes, gestation period is a nearly invariant sixty-two days, whereas among mammals generally, it scales to the ¼-power of body mass (Kirkwood, 1985). Even more remarkably, longevity, which also scales positively to the ¼-power of body mass across mammalian species, scales negatively across dog breeds (Lindstedt and Calder, 1981). Chihuahuas live longer than mastiffs. Apparently, selective pressures for body size are coupled to those for sexual size dimorphism across carnivorans but decoupled from those for other life history traits.

10.5 Social and mating systems

More broadly, life history embraces social (e.g. group vs. solitary living) and mating systems (e.g. reproductive monogamy vs. polygamy)

associated with reproductive patterns. It considers the developmental state of neonates at birth (altriciality vs. precociality), and parental contributions to care of young (paternal care of young vs. none). Paternal care is common where extra-pair matings are rare. Again, carnivorans are highly diverse. Considering only neonatal development, they equal or surpass all other mammalian orders in the diversity of their strategies (Derrickson, 1992).

These broader aspects of life history strategy, with strong links to behavior, also relate to population density, resource abundance, and vulnerability to intra- or interspecific aggression. The strategies employed by carnivorans include:

1. Solitary living, intra-sexual territoriality, unrestricted mating, no paternal care—many mustelid species. Males exclude other males from territories, but range beyond territory boundaries to seek mating opportunities. Sexual size dimorphism is high.
2. Kin group living, group territoriality, mating restricted by dominance, paternal care of young—wolf, African wild dog, and African lion. Participation in mating depends on resource abundance. Females jointly defend against infanticide, and large groups facilitate territorial defense (Smith *et al.*, 2017). Sexual size dimorphism is modest.
3. Mated pair living, territoriality across mated pairs, social monogamy with some extra-pair matings—island gray fox (Roemer *et al.*, 2001). This strategy is uncommon among carnivorans and may reflect the limitations on dispersal imposed by the restricted land area available to this species. Sexual size dimorphism is reduced.
4. Solitary living, non-territorial foraging, territorial breeding rookeries, brief maternal care, no paternal care—many pinnipeds. Adults forage at sea, but assemble on shore, where sexual selection is manifest: females join harems, males compete for control of females, and highly polygynous matings result (Figure 10.6). Postweaning care by the mother, especially joint foraging, is modest or nil. Sexual size dimorphism is the highest of all mammals, with a mean male/female body mass ratio of 3 across the Otariidae (Weckerly, 1998).

Figure 10.6 Female northern fur seals assembled in a harem on the Komandorski Islands, Russia. They are attended by a male, which defends the harem against interlopers.
Photo: pilipenkoD/CanStock.

5. Same-sex group living, group territoriality, unrestricted mating—narrow-striped mongoose. Diet does not differ by sex, and the sexual segregation is hypothesized to result from frequent male harassment of females, eliciting female defense of group members. This is an apparent by-product of polygyny (Schneider and Kappeler, 2016). Sexual size dimorphism is nil.

6. Kin-group living, territoriality, female philopatry, male dispersal, female dominance over males after dispersal, female-choice mating, kin-based within-sex dominance—spotted hyena. Foraging may be solitary or in groups. This complex social system is unique among carnivorans, resembling the systems of cercopithecine primates (Holekamp *et al.*, 2007).

Polygyny is a recurring feature among carnivoran life histories and raises the question of multiple paternity of single litters. Although few species have been tested using molecular tools (Yamaguchi *et al.*, 2004), wide variation is apparent within and among carnivoran species (Table 10.1). Palomares and colleagues (2017) found that 98% of matings in Iberian lynx were with the territory-holding male, whereas Nielsen and Nielsen (2007) reported that no more than 12% of raccoon litters were single-fathered. Even socially monogamous species, notably canids, are now understood to produce many multiply fathered litters (e.g. 52% of litters for swift fox; Kitchen *et al.*, 2006). If females conceive after matings with multiple males followed by induced ovulation, it raises the possibility of superfetation, whereby females ovulate in response to each mating, even if separated by days. Superfetation has been reported for several carnivoran species (Corner *et al.*, 2015) and allows females to ovulate in response to the relative quality of copulations with multiple males. Likewise, they could employ *in utero* sexual selection via termination of inferior embryos (Larivière and Ferguson, 2003).

Some reproductive aspects of life history relate to energetics, to habitat, or to maternal defense of young. Female phocids are an example, coming ashore with all the somatic stores needed to give birth and nurse a pup to weaning; mothers do not forage during the entire pup-rearing period. By contrast, female otariids fast for only a few days surrounding parturition, returning to the sea for days at a time to restore somatic reserves (Trillmich, 1996). They forego defending neonates during these absences and therefore must whelp at predator-free sites. The consequence is that phocids can separate foraging from parturition temporally and spatially, whereas otariids must give birth near food resources.

Table 10.1 Frequency of extra-pair paternities reported for carnivoran species. Other species, including American mink and ermine, have been reported to display EPP, without frequency data

Species	Proportion of litters with multiple fathers	Proportion of young fathered by male not associated with mother	Reference
Arctic fox		≥ 0.31	Cameron *et al.* (2011)
Swift fox		0.52	Kitchen *et al.* (2006)
Bat-eared fox		0.1	Wright *et al.* (2010)
Iberian lynx		0.02	Palomares *et al.* (2017)
Northern raccoon	≥ .88		Nielsen and Nielsen (2007)
Brown bear	0.14		Bellemain *et al.* (2006)
European badger	0.45		Dugdale *et al.* (2007)

10.6 Adaptation in populations

Two kinds of adaptation are closely linked to community processes: character displacement and its reciprocal, character (sometimes "competitive") release. Character displacement describes phenotypic—most commonly morphological—change in response to the sympatry or arrival of a species of similar phenotype or niche characteristics. Because body size is so important in predatory and competitive interactions and is so well studied in the Carnivora, many examples of character displacement and release in this group use body size as the trait of interest. Studies come from three main settings: paleoecological studies of co-occurring fossil remains, studies of similar species within vs. outside of zones of sympatry, and studies of how introduced (or eradicated) species affect sympatric competitors. The most consistently useful body size metrics in these comparisons have been cranial or dental structures, especially the sizes of one or both carnassial teeth (Davies *et al.*, 2007).

Paleoecological studies have shown evidence of character displacement for Nearctic canids and Old World hyaenids (Dayan and Simberloff, 1998); these groups have evolved divergent morphologies to reduce niche overlap between them. Geographical comparisons of morphology have been confounded in some cases by clinal variation in body size of carnivorans or their prey (e.g. Meiri *et al.*, 2011). For example, the coyote, where it recently colonized northeastern North America, is larger-bodied than in the pre-settlement western range. The difference is related to the shift from primarily lagomorph prey in the West to ungulate prey in the Northeast. Common garden experiments have suggested that coyotes from northern New England differ in heritable traits from those from the West (Silver and Silver, 1969); these body-size differences seem to have arisen quickly. The mechanism underlying this change could be selection for larger body size to subdue larger prey, or introgression with wolves, which were more common in the Northeast than the intermountain West when these changes occurred (Lariviere and Crête, 1993). Searches for other evidence of displacement or release resulting from carnivoran introductions have been inconclusive. When body sizes were compared between small Asian mongooses introduced to coastal islands of Croatia with and without stone martens, few salient patterns were found (Barun *et al.*, 2015). Skull lengths were similar across the native range of the mongoose, islands with martens, and islands without martens. Character displacement was not supported, although the study did show that mongooses on Croatian islands were much smaller than those on other subtropical sites of introduction (e.g. Oahu, Mauritius). So, while carnivoran communities have seemed to be ideal places to document character displacement and release, few clearly supportive examples have emerged.

Character displacement in body size is further complicated by the multiple selective forces operating on sexual size dimorphism. Especially in the Mustelidae and Herpestidae, a nominal species tends to comprise two morphotypes—male and female—either of which may approach the body sizes of interspecific competitors. Sexual selection was long regarded as the primary driver of sexual size dimorphism in mustelids (Moors, 1980); however, niche divergence in response to competition has gained favor as an explanation. Males and females of different sizes can utilize different food sources, especially if the species is solitary and highly predaceous (Law, 2019). Andreasen and colleagues (2021) showed sex differences in prey size selection by pumas in Nevada, US, but they were opposite to those expected; females were more inclined to kill larger prey. Sunde and Kvam (1997) showed sexual differences in prey size selection by Eurasian lynx in Norway but attributed them to factors other than body weight. *Mustela* spp. are prominent examples of sexual size dimorphism resulting from sexual selection interacting with selection for niche divergence (Giery and Layman, 2019). Herpestids provide additional examples; the small Asian mongoose occurs sympatrically with two slightly larger congeneric species as well as allopatrically from them. Where allopatric with the two congenerics, male small Asian mongooses are larger and sexual size dimorphism is greater than where they are sympatric (Simberloff *et al.*, 2000).

Key points

- Carnivorans die from causes that differ from those for other mammals. Near humans and their activities, human-inflicted causes, including vehicle collisions, trapping, and shooting, tend to be primary mortality causes. Non-carnivoran mammals are more likely to die of predation, with hunting a major cause for hunted ungulates. Away from human influences, interspecific killing is a major—or the primary—cause of carnivoran deaths.
- Mass mortality events may affect carnivorans more than other mammals. These events are typically caused by epizootics of rabies, canine distemper, sarcoptic mange, or other infectious diseases. Other mass mortalities are caused by toxins and by trampling on pinniped haul-outs or rookeries.
- Our understandings of carnivoran demography have been slow to develop, except for some high-profile species of conservation concern, mostly in developed countries.
- Carnivorans have lower population densities and larger home ranges, disperse longer distances, and undergo some of the longest mammalian annual migrations of all mammals except bats and whales.
- A wide range of patterns of population genetic structure is exhibited by carnivorans, depending on body size, habitat specialization, and dispersal distances. Several pinniped species exhibit panmixia over long distances. On land, various strongly dispersing carnivoran species show little genetic structure over geographic ranges that span continents. Habitat-specialist species show strong genetic discontinuities at dispersal barriers. Many carnivoran examples show isolation by distance, isolation by barrier, or isolation by ecology.
- Barriers to dispersal can be intermittent, exemplified by seasonal ice on water bodies.
- Carnivorans are especially strong dispersers along ocean archipelagos, on a geological time scale, because of their ability to travel on ice, withstand cold, and feed on intertidal food resources.
- Human structures strongly affect dispersal by some carnivorans, commonly impeding gene flow. These structures include highways, fences, and canals.
- Hybridization is common among named carnivoran species, and introgression somewhat common. Genetic mixing is a conservation concern where wild carnivorans encounter introduced or domestic varieties. Genus *Canis* is a particularly interesting complex of named species, with many identified cases of introgression, some of them via domestic dogs.
- In spite of occupying niches different from those of mammalian herbivores, carnivorans share no special life history traits. They include strategies that span the slow–fast continuum, even accounting for variation in body size.

- Diverse life histories result in large differences in carnivoran social and mating strategies, which include solitary, social, mated-pair, and same-sex group living.
- A range of paternity patterns of litters has been reported, including nearly exclusive paternity of all young born within an adult male's home range to multiply fathered litters in species formerly regarded as having exclusively single-paternity litters.
- Adaptation in carnivoran populations has been studied mostly for body size in the context of community interactions—character displacement and release. Displacement has been confirmed in paleocommunities of canids and hyaenids, but evidence has been inconclusive for most modern communities studied. The effects of selection for niche separation and for mating opportunities are easily confused. Introgression among *Canis* spp., which is especially common, also can affect relative body sizes of named species.

References

Andreasen, A.M., Stewart, K.M, Longland, W.S., and Beckmann, J.P. 2021. Prey specialization by cougars on feral horses in a desert environment. *Journal of Wildlife Management* 85:1104–20.

Bakeyev, N.N. and Sinitsyn, A.A. (1994) "Status and conservation of sables in the Commonwealth of Independent States," in Buskirk, S.W. et al. (eds.) *Martens, sables, and fishers: biology and conservation.* Ithaca: Cornell University Press, pp. 246–54.

Barber-Meyer, S.M., Mech, L.D. and White, P.J. (2008) "Elk calf survival and mortality following wolf restoration to Yellowstone National Park," *Wildlife Monographs,* 169, pp. 1–30.

Barun, A. et al. (2015) "Possible character displacement of an introduced mongoose and native marten on Adriatic Islands, Croatia," *Journal of Biogeography,* 42, pp. 2257–69.

Bellemain, E., Swenson, J.E. and Taberlet, P. (2006) "Mating strategies in relation to sexually selected infanticide in a non-social carnivore: the brown bear," *Ethology,* 112, pp. 238–46.

Boland, K.M. and Litvaitis, J.A. (2008) "Role of predation and hunting on eastern cottontail mortality at Cape Cod National Seashore, Massachusetts," *Canadian Journal of Zoology,* 86, pp. 918–27.

Boutin, S. et al. (1986) "Proximate causes of losses in a snowshoe hare population," *Canadian Journal of Zoology,* 64, pp. 606–10.

Bowman, J. et al. (2017) "Hybridization of domestic mink with wild American mink (*Neovison vison*) in eastern Canada," *Canadian Journal of Zoology,* 95, pp. 443–51.

Bowman, J., Jaeger, J.A.G. and Fahrig, L. (2002) "Dispersal distance of mammals is proportional to home range size," *Ecology,* 83, pp. 2049–55.

Bowman, J. et al. (2007) "Assessing the potential for impacts by feral mink on wild mink in Canada," *Biological Conservation,* 139, pp. 12–8.

Brillinger, D.R. and Stewart, D.S. (1998) "Elephant-seal movements: modelling migration," *Canadian Journal of Statistics,* 26, pp. 431–43.

Broquet, T. et al. (2006) "Genetic isolation by distance and landscape connectivity in the American marten (*Martes americana*)," *Landscape Ecology,* 21, pp. 877–89.

Bull, E.L. and Heater, T.W. (2001) "Survival, causes of mortality, and reproduction in the American marten in northeastern Oregon," *Northwestern Naturalist,* 82, pp. 1–6.

Burns, L.E. and Broders, H.G. (2014) "Correlates of dispersal extent predict the degree of population genetic structuring in bats," *Conservation Genetics,* 15, pp. 1371–9.

Buskirk, S.W., Bowman, J. and Gilbert, J.H. (2012) "Population biology and matrix demographic modeling of American martens and fishers," in Aubry, K.B. et al. (eds.) *Biology and conservation of martens, sables, and fishers: a new synthesis.* Ithaca: Cornell University Press, pp. 77–92.

Cabria, M.T. et al. (2011) "Bayesian analysis of hybridization and introgression between the endangered European mink (*Mustela lutreola*) and the polecat (*Mustela putorius*)," *Molecular Ecology,* 20, pp. 1176–90.

Cameron, C., Berteaux, D. and Dufresne, F. (2011) "Spatial variation in food availability predicts extrapair paternity in the arctic fox," *Behavioral Ecology,* 22, pp. 1364–73.

Colella, J.P. et al. (2021) "Extrinsically reinforced hybrid speciation within Holarctic ermine (*Mustela* spp.) produces an insular endemic," *Diversity and Distributions,* 27, pp. 747–62.

Collins, C. and Kays, R. (2011) "Causes of mortality in North American populations of large and medium-sized mammals," *Animal Conservation,* 14, pp. 474–83.

Colson, K.E., Smith J.D. and Hundertmark, K.J. (2017) "St. Matthew Island colonized through multiple long-distance red fox (*Vulpes vulpes*) dispersal events," *Canadian Journal of Zoology,* 96, pp. 607–9.

Corner, L.A.L. *et al.* (2015) "Reproductive biology including evidence for superfetation in the European badger *Meles meles* (Carnivora: Mustelidae)," *PLoS ONE*, 10, p. e0138093.

Craighead, J.J., Varney, J.R. and Craighead, F.C. (1974) "A population analysis of the Yellowstone grizzly bears," *Bulletin of the Montana Forest and Conservation Experiment Station*, 40, pp. 1–20.

Cullingham, C.I. *et al.* (2009) "Differential permeability of rivers to raccoon gene flow corresponds to rabies incidence in Ontario, Canada," *Molecular Ecology*, 18, pp. 43–53.

Davies, T.J. *et al.* (2007) "Species co-existence and character divergence across carnivores," *Ecology Letters*, 10, pp. 146–52.

Davison, A. *et al.* (2001) "Mitochondrial phylogeography and population history of pine martens Martes martes compared with polecats Mustela putorius," *Molecular Ecology*, 10, pp. 2479–88.

Dayan, T. and Simberloff, D. (1998) "Size patterns among competitors: ecological character displacement and character release in mammals, with special reference to island populations," *Mammal Review*, 28, pp. 99–124.

Delgiudice, G.D. *et al.* (2002) "Winter severity, survival, and cause-specific mortality of female white-tailed deer in north-central Minnesota," *Journal of Wildlife Management*, 66, pp. 698–717.

Derrickson, E.M. (1992) "Comparative reproductive strategies of altricial and precocial eutherian mammals," *Functional Ecology*, 6, pp. 57–65.

Dickerson, B.R. *et al.* (2010) "Population structure as revealed by mtDNA and microsatellites in northern fur seals, *Callorhinus ursinus*, throughout their range," *PLoS ONE*, 5, p. e10671.

Dobson, F.S. (2007) "A lifestyle view of life-history evolution," *Proceedings of the National Academy of Sciences of the United States of America*, 104, pp. 17565–6.

Donadio, E., Buskirk, S.W. and Novaro, A.J. (2012) "Juvenile and adult mortality patterns in a vicuña (*Vicugna vicugna*) population," *Journal of Mammalogy*, 93, pp. 1536–44.

Dugdale, H.L. *et al.* (2007) "Polygynandry, extra-group paternity and multiple-paternity litters in European badger (*Meles meles*) social groups," *Molecular Ecology*, 16, pp. 5294–306.

Eads, D.A. and Marsh, D. (2020) "Possible litter of seven wild black-footed ferret kits," *Western North American Naturalist*, 80, pp. 543–6.

Eizirik, E. *et al.* (2001) "Phylogeography, population history and conservation genetics of jaguars (*Panthera* onca, Mammalia, Felidae)," *Molecular Ecology*, 10, pp. 65–79.

Ernest, H.B., Vickers, T.W., Morrison, S.A., Buchalski, M.R., and Boyce, W.M. 2014. Fractured genetic connectivity threatens a Southern California puma (Puma concolor) population. *PLoS ONE* 9(10):e107985.

Fay, F.H. and Rausch, R.L. (1992) "Dynamics of the arctic fox population on St. Lawrence Island, Bering Sea," *Arctic*, 45, pp. 393–7.

Frantz, A.C. *et al.* (2010) "Using isolation-by-distance-based approaches to assess the barrier effect of linear landscape elements on badger (*Meles meles*) dispersal," *Molecular Ecology*, 19, pp. 1663–74.

Fretwell, S.D. and Lucas, H.L., Jr. (1970) "On territorial behavior and other factors influencing habitat distribution in birds. I. Theoretical development," *Acta Biotheoretica*, 19, pp. 16–36.

Frynta, D. *et al.* (2012) "Allometry of sexual size dimorphism in domestic dog," *PLoS ONE*, 7, p. e46125.

Fuglei, E. and Tarroux, A. (2019) "Arctic fox dispersal from Svalbard to Canada: one female's long run across sea ice," *Polar Research*, 38, p. 3512.

Galov, A. *et al.* (2015) "First evidence of hybridization between golden jackal (*Canis aureus*) and domestic dog (*Canis familiaris*) as revealed by genetic markers," *Royal Society Open Science*, 2, p. 150450.

Gaubert, P. *et al.* (2012) "Reviving the African wolf *Canis lupus lupaster* in North and West Africa: a mitochondrial lineage ranging more than 6,000 km wide," *PLoS ONE*, 7, p. e42740.

Geffen, E. *et al.* (2007) "Sea ice occurrence predicts genetic isolation in the Arctic fox," *Molecular Ecology*, 16, pp. 4241–55.

Giery, S.T. and Layman, C.A. (2019) "Ecological consequences of sexually selected traits: an eco-evolutionary perspective," *Quarterly Review of Biology*, 94, pp. 29–74.

Gittleman, J.L., and Van Valkenburgh, B. 1997. Sexual dimorphism in the canines and skulls of carnivores: effects of size, phylogeny, and behavioural ecology. *Journal of Zoology*, London 242: 97–117.

Goldsworthy, S.D. *et al.* (2009) "Fur seals at Macquarie Island: post-sealing colonization trends in abundance and hybridisation of three species," *Polar Biology*, 32, pp. 1473–86.

Gopalakrishnan, S. *et al.* (2018) "Interspecific gene flow shaped the evolution of the Genus Canis," *Current Biology*, 28, pp. 3441–9.

Gosselink, T.E. *et al.* (2007) "Survival and cause-specific mortality of red foxes in agricultural and urban areas of Illinois," *Journal of Wildlife Management*, 71, pp. 1862–73.

Grenier, M.B., McDonald, D.B. and Buskirk, S.W. (2007) "Rapid population growth of a critically endangered carnivore," *Science*, 317, p. 779.

Harwood, J. and Hall, A. (1990) "Mass mortality in marine mammals: its implications for population dynamics and genetics," *Trends in Ecology and Evolution*, 8, pp. 254–7.

Hastings, K.K. *et al.* (2017) "Natal and breeding philopatry of female steller sea lions in southeastern Alaska," *PLoS ONE*, 12, p. e0176840.

Hauer, S., Ansorge, H. and Zinke, O. (2002) "Mortality patterns of otters (*Lutra lutra*) from eastern Germany," *Journal of Zoology*, 256, pp. 361–8.

Hertwig, S.T. *et al.* (2009) "Regionally high rates of hybridization and introgression in German wildcat populations (*Felis silvestris*, Carnivora, Felidae)," *Journal of Zoological Systematics and Evolutionary Research*, 47, pp. 283–97.

Hodgman, T.P. *et al.* (1994) "Survival in an intensively trapped marten population in Maine," *Journal of Wildlife Management*, 58, pp. 593–600.

Hoekstra, H.E. and Fagan, W.F. (1998) "Body size, dispersal ability and compositional disharmony: the carnivore-dominated fauna of the Kuril Islands," *Diversity and Distributions*, 4, pp. 135–49.

Holderegger, R. and Di Giulio, M. (2010) "The genetic effects of roads: a review of empirical evidence," *Basic and Applied Ecology*, 11, pp. 522–31.

Holekamp, K.E., Sakai, S.T. and Lundrigan, B.L. (2007) "Social intelligence in the spotted hyena (*Crocuta crocuta*)," *Transactions of the Royal Society B: Biological Sciences*, 362, pp. 523–38.

Janssens, X. *et al.* (2008) "Genetic pattern of the recent recovery of European otters in southern France," *Ecography*, 31, pp. 176–86.

Joly, K. *et al.* (2019) "Longest terrestrial migrations and movements around the world," *Scientific Reports*, 9, p. 15333.

Jones, E. (2009) "Hybridisation between the dingo, *Canis lupus dingo*, and the domestic dog, *Canis lupus familiaris*, in Victoria: a critical review," *Australian Mammalogy*, 31, pp. 1–7.

Kato, Y. *et al.* (2017) "Population genetic structure of the urban fox in Sapporo, northern Japan," *Journal of Zoology*, 301, pp. 118–24.

Keith, L.B. and Bloomer, S.E.M. (1993) "Differential mortality of sympatric snowshoe hares and cottontail rabbits in central Wisconsin," *Canadian Journal of Zoology*, 71, pp. 1694–7.

Kenyon, K.W., Scheffer, V.B. and Chapman, D.G. (1954) "A population study of the Alaska fur seal herd," US Fish and Wildlife Service Wildlife, Special Scientific Report, No. 12, pp. 1–77.

Kidd, A.G. *et al.* (2009) "Hybridization between escaped domestic and wild American mink (*Neovison vison*)," *Molecular Ecology*, 18, pp. 1175–86.

Kirkwood, J.K. (1985) "The influence of size on the biology of the dog," *Journal of Small Animal Practice*, 26, pp. 97–110.

Kitchen, A.M. *et al.* (2006) "Multiple breeding strategies in the swift fox, Vulpes velox," *Animal Behaviour*, 71, pp. 1029–38.

Koen, E.L. *et al.* (2012) "Landscape resistance and American marten gene flow," *Landscape Ecology*, 27, pp. 29–43.

Koen, E.L., Bowman, J. and Wilson, P.J. (2015) "Isolation of peripheral populations of Canada lynx (*Lynx canadensis*)," *Canadian Journal of Zoology*, 93, pp. 521–30.

Koepfli, K.-P. *et al.* (2008) "Multigene phylogeny of the Mustelidae: Resolving relationships, tempo and biogeographic history of a mammalian adaptive radiation," *BMC Biology*, 6, p. 10.

Kurtén, B. (1976) *The cave bear story.* New York: Columbia University Press.

Kyle, C.J., Davison, A. and Strobeck, C. (2003) "Genetic structure of European pine martens (*Martes martes*), and evidence for introgression with *M. americana* in England," *Conservation Genetics*, 4, pp. 179–88.

Lancaster, M.L. *et al.* (2007) "Lower reproductive success in hybrid fur seal males indicates fitness costs to hybridization," *Molecular Ecology*, 16, pp. 3187–97.

Lancaster, M.L. *et al.* (2006) "Ménage à trois on Macquarie Island: hybridization among three species of fur seal (*Arctocephalus* spp.) following historical population extinction," *Molecular Ecology*, 15, pp. 3681–92.

Larivière, S. and Crête, M. (1993) "The size of eastern coyotes (*Canis latrans*): a comment," *Journal of Mammalogy*, 74, pp. 1072–4.

Larivière, S. and Ferguson, S.H. (2003) "Evolution of induced ovulation in North American carnivores," *Journal of Mammalogy*, 84, pp. 937–47.

Law, C.J. (2019) "Solitary meat-eaters: solitary, carnivorous carnivorans exhibit the highest degree of sexual size dimorphism," *Scientific Reports*, 9, p. 15344.

Lehman, N. *et al.* (1991) "Introgression of coyote mitochondrial DNA into sympatric North American gray wolf populations," *Evolution*, 45, pp. 104–19.

Leonard, J.A. *et al.* (2014) "Impact of hybridization with domestic dogs on the conservation of wild canids," in Gompper, M. E. (ed.) *Free-ranging dogs and wildlife conservation.* Oxford: Oxford University Press, pp. 170–84.

Lindblad-Toh, K., Wade, C.M., Mikkelsen, T.S, et al. (2005). Genome sequence, comparative analysis and haplotype structure of the domestic dog. *Nature* 438:803–819.

Lindstedt, S.L. and Calder, W.A., III. (1981) "Body size, physiological time, and longevity of homeothermic animals," *Quarterly Review of Biology*, 56, pp. 1–16.

Lindstedt, S.L., Miller, B.J. and Buskirk, S.W. (1986) "Home range, time, and body size in mammals," *Ecology*, 67, pp. 413–8.

Linnell, J.D.C., Aanes, R. and Andersen, R. (1995) "Who killed Bambi? The role of predation in the neonatal mortality of temperate ungulates," *Wildlife Biology*, 1, pp. 209–23.

Marmi, J. *et al.* (2006) "Mitochondrial DNA reveals a strong phylogeographic structure in the badger across Eurasia," *Molecular Ecology*, 15, pp. 1007–20.

McDonough, M.M. *et al.* (2022) "Phylogenomic systematics of the spotted skunks (Carnivora, Mephitidae, Spilogale): additional species diversity and Pleistocene climate change as a major driver of diversification," *Molecular Phylogenetics and Evolution*, 167, p. 107266.

Mead, R.A. (1968a) "Reproduction in eastern forms of the spotted skunk (genus *Spilogale*)," *Journal of Zoology*, 156, pp. 119–36.

Mead, R.A. (1968b) "Reproduction in western forms of the spotted skunk (genus *Spilogale*)," *Journal of Mammalogy*, 49, pp. 373–90.

Meiri, S., Simberloff, D. and Dayan, T. (2011) "Community-wide character displacement in the presence of clines: a test of Holarctic weasel guilds," *Journal of Animal Ecology*, 80, pp. 824–34.

Mercure, A. *et al.* (1993) "Genetic subdivisions among small canids: mitochondrial DNA differentiation of swift, kit, and arctic foxes," *Evolution*, 47, pp. 1313–28.

Mergey, M. *et al.* (2017) "Identifying environmental drivers of spatial genetic structure of the European pine marten (*Martes martes*)," *Landscape Ecology*, 32, pp. 2261–79.

Moors, P.J. (1980) "Sexual dimorphism in the body size of mustelids (Carnivora): the roles of food habits and breeding systems," *Oikos*, 34, pp. 147–58.

Morris, K.Y. *et al.* (2020) "Functional genetic diversity of domestic and wild American mink (*Neovison vison*)," *Evolutionary Applications*, 13, pp. 2610–29

Nielsen, C.L.R. and Nielsen, C.K. (2007) "Multiple paternity and relatedness in southern Illinois raccoons (*Procyon lotor*)," *Journal of Mammalogy*, 88, pp. 441–7.

Norén, K. *et al.* (2011) "Pulses of movement across the sea ice: population connectivity and temporal genetic structure in the arctic fox," *Oecologia*, 166, pp. 973–84.

Norrdahl, K. and Korpimaki, E. (1995) "Mortality factors in a cyclic vole population," *Proceedings of the Royal Society B: Biological Sciences*, 261, pp. 49–53.

Paetkau, D. *et al.* (1999) "Genetic structure of the world's polar bear populations," *Molecular Ecology*, 8, pp. 1571–84.

Paetkau, D., Shields, G.F. and Strobeck, C. (1998) "Gene flow between insular, coastal and interior populations of brown bears in Alaska," *Molecular Ecology*, 7, pp. 1283–92.

Palomares, F. *et al.* (2017) "Territoriality ensures paternity in a solitary carnivore mammal," *Scientific Reports*, 7, p. 4494.

Pence, D.B. and Ueckermann, E. (2002) "Sarcoptic mange in wildlife," *Revue Scientifique et Technique*, 21, pp. 385–98.

Roemer, G.W. *et al.* (2001) "The behavioural ecology of the island fox (*Urocyon littoralis*)," *Journal of Zoology*, 255, pp. 1–14.

Row, J.R. *et al.* (2012) "Dispersal promotes high gene flow among Canada lynx populations across mainland North America," *Conservation Genetics*, 13, pp. 1259–68.

Schneider, T.C. and Kappeler, P.M. (2016) "Gregarious sexual segregation: the unusual social organization of the Malagasy narrow-striped mongoose (*Mungotictis decemlineata*)," *Behavioral Ecology and Sociobiology*, 70, pp. 913–26.

Scholin, C.A. *et al.* (2000) "Mortality of sea lions along the central California coast linked to a toxic diatom bloom," *Nature*, 403, pp. 80–4.

Schultz, J.K. *et al.* (2010) "Range-wide genetic connectivity of the Hawaiian monk seal and implications for translocation," *Conservation Biology*, 25, pp. 124–32.

Silver, H. and Silver, W.T. (1969) "Growth and behavior of the coyote-like canid of northern New England with observations on canid hybrids," *Wildlife Monographs*, 17, pp. 3–41.

Simberloff, D. *et al.* (2000) "Character displacement and release in the small Indian mongoose, Herpestes javanicus," *Ecology*, 81, pp. 2086–99.

Smith, J.E. *et al.* (2017) "Insights from long-term field studies of mammalian carnivores," *Journal of Mammalogy*, 98, pp. 631–41.

Spencer, W. *et al.* (2011) "Using occupancy and population models to assess habitat conservation opportunities for an isolated carnivore population," *Biological Conservation*, 144, pp. 788–803.

Sunde, P. and Kvam, T. (1997) "Diet patterns of Eurasian lynx *Lynx lynx*: what causes sexually determined prey size segregation?", *Acta Theriologica*, 42, pp. 189–201.

Sutherland, G.D. *et al.* (2000) "Scaling of natal dispersal distances in terrestrial birds and mammals," *Conservation Ecology*, 4, p. 16.

Talbot, B. *et al.* (2012) "Lack of genetic structure and female-specific effect of dispersal barriers in a rabies vector, the striped skunk (*Mephitis mephitis*)," *PLoS ONE*, 7, p. e49736.

Teacher, A.G.F., Thomas, J.A. and Barnes, I. (2011) "Modern and ancient red fox (*Vulpes vulpes*) in Europe show an unusual lack of geographical and temporal structuring, and differing responses within the carnivores to historical climatic change," *BMC Evolutionary Biology*, 11, p. 214.

Trigo, T.C. *et al.* (2008) "Inter-species hybridization among Neotropical cats of the genus *Leopardus*, and evidence for an introgressive hybrid zone between *L. geoffroyi* and *L. tigrinus* in southern Brazil," *Molecular Ecology*, 17, pp. 4317–33.

Trillmich, F. (1996) "Parental investment in pinnipeds," *Advances in the Study of Behavior*, 25, pp. 533–77.

Van de Bildt, M.W.G. *et al.* (2002) "Distemper outbreak and its effect on African wild dog conservation," *Emerging Infectious Diseases*, 8, pp. 211–13.

van de Kerk, M. *et al.* (2013) "Carnivora population dynamics are as slow and as fast as those of other mammals: implications for their conservation," *PLoS ONE*, 8, p. e70354.

Van Vuren, D.H. (2001) "Predation on yellow-bellied marmots (*Marmota flaviventris*)," *American Midland Naturalist*, 145, pp. 94–100.

Vianna, J.A. *et al.* (2010) "Phylogeography of the marine otter (*Lontra felina*): historical and contemporary factors determining its distribution," *Journal of Heredity*, 101, pp. 676–89.

Wayne, R.K. *et al.* (1997) "Molecular systematics of the Canidae," *Systematic Biology*, 46, pp. 622–53.

Weckerly, F.W. (1998) "Sexual-size dimorphism: influence of mass and mating systems in the most dimorphic mammals," *Journal of Mammalogy*, 79, pp. 33–52.

Williams, E.S. *et al.* (1988) "Canine distemper in black-footed ferrets (*Mustela nigripes*) from Wyoming," *Journal of Wildlife Diseases*, 24, pp. 385–98.

Wright, H.W.Y. *et al.* (2010). "Mating tactics and paternity in a socially monogamous canid, the bat-eared fox (*Otocyon megalotis*)," *Journal of Mammalogy*, 91, pp. 437–46.

Yamaguchi, N. *et al.* (2004) "Multiple paternity and reproductive tactics of free-ranging American minks, Mustela vison," *Journal of Mammalogy*, 85, pp. 432–9.

Zalewski, A. *et al.* (2009) "Landscape barriers reduce gene flow in an invasive carnivore: geographical and local genetic structure of American mink in Scotland," *Molecular Ecology*, 18, pp. 1601–15.

CHAPTER 11

How carnivorans affect humans

Wild carnivorans affect human lives and livelihoods in important and diverse ways, and many of these lines of causation cause some reciprocity. Humans affect carnivorans, which change their behaviors, diets, populations, or other traits, and humans adjust their lives and economies in turn, modifying interactions further. In this chapter and in Chapter 12 I focus on wild carnivorans and humans as community interactors, mentioning domestic species where they mediate relationships.

11.1 Negative effects

11.1.1 Attacks on humans

Carnivoran attacks on humans, and human responses to those attacks, have been a force throughout human evolution (Camarós et al., 2016). These are the most dramatic and compelling of all wildlife–human interactions; the numbers of people killed or injured may be small, but the effects on human attitudes and actions are large. Of all human fatalities inflicted by vertebrates worldwide, wild carnivorans contribute a small fraction. Snake bites kill around 50,000 people each year, whereas domestic dogs, commonly infected with rabies, account for around 25,000 deaths. Crocodilians kill hundreds of people yearly across several continents, whereas hippopotami kill a comparable number, all in Africa (BBC News, 2016; Bombieri et al., 2019). Data for wild carnivorans are difficult to compile across jurisdictions, but lions kill about 100 people per year, brown bears about twenty, and wolves about ten. Additional thousands of people are injured (Packer et al., 2019). Factors predisposing humans to being attacked tend to be taxon- and context-specific; however,

several reviews have identified general patterns. These include the body size of the carnivoran and the human, with large-bodied carnivorans most attack-prone, and children disproportionately likely to be victims (Garrote et al., 2017). Groups are at less risk than solitary humans, and hunters are at greater risk than non-hunters, presumably because of their proximity to animal carcasses (Støen et al., 2018). Humans feeding wild carnivorans and Australian dingoes increases the likelihood of attack markedly, especially if the feeding involves direct contact (Burns and Howard, 2003). Goodrich and colleagues (2011) identified the health of east Asian tigers as a key predictor of attacks, with wounded tigers much more dangerous than those in good health (most wounds were inflicted by humans). Likewise, rabies infection increases risk of attack (Linnell et al., 2002). Penteriani and colleagues (2017) showed that human presence in carnivoran habitat under conditions of low light (dusk to dawn) was risky, and Bombieri and colleagues (2019) showed that the presence of cubs increased the likelihood of ursid maternal aggression. Underlying these proximate predictors, likelihood of attack is associated with economic conditions and land-use patterns; attacks are more likely in developing countries, and in areas where nature reserves adjoin areas with livestock production (Baker et al., 2008).

These deaths and injuries lead to strong societal responses, often including lethal control, both reactive and anticipatory. Perhaps ironically, the potential for attacks also increases public interest in wild carnivorans, imbuing them with symbolic power and charisma. Baynes-Rock (2013) described a remarkable example of this phenomenon following a spate of hyena attacks on livestock and humans

Carnivoran Ecology. Steven W. Buskirk, Oxford University Press. © Steven W. Buskirk (2023). DOI: 10.1093/oso/9780192863249.003.0011

in the Hararge region of Ethiopia. The attacks were localized and anomalous, and retaliatory killings were conducted, but the Harar also value hyenas. They consider them to control unseen spirits, and to act as reasonable humans would. They fear retaliation from hyena clan mates if lethal control is excessive. As a result, Harar responses were measured and coexistence with hyenas seemed stable. In addition to causing attacks on humans, wild carnivorans can mitigate them. Leopards living on the edge of Mumbai, India prey heavily on domestic dogs. Although leopards occasionally attack and kill humans, their greater effect is to reduce dog numbers, which is estimated to save ninety lives each year in the Mumbai area (Braczkowski *et al.*, 2018).

11.1.2 Attacks on domestic animals

Baker and colleagues (2008) synthesized the literature dealing with terrestrial carnivoran effects on human food production—typically herbivorous livestock—and found that they are mostly negative. Here I include other effects and extend the frame of reference to pets and hatchery fish. This topic overlaps strongly with the extensive "conflict" literature, which touches on the disparate interests of beneficiaries vs. victims of carnivoran presence. How human responses affect wild carnivorans is treated in Chapter 12. Necessarily, conservation scientists tend to deal with these various lines of causation as species- and site-specific problems, rather than aggregations of positive and negative effects, as I present them here.

Carnivoran taxa vary in some obvious ways in their tendency to affect the domesticated food supply. Dietary specialists, for example felids, tend to prey on vertebrates, whereas omnivorous bears kill livestock, eat crops, damage trees under culture, and raid apiaries. Pinnipeds consume fish near hatcheries or concentrated below dams (Figure 11.1). Larger-bodied carnivorans kill larger-bodied livestock. Depredation of livestock is the most widespread and persistent issue, but the natural histories and economics of the problems are taxon- and context-specific. Depredation involving one or two offending predators may be a critical problem for a subsistence farmer with small holdings. By contrast, the same level of damage might be a nuisance to a large landholder, especially if the land is used for wildlife-based tourism. Sympathy for the carnivoran itself affects how depredation it causes is viewed. Australian dingoes can cause high losses in sheep herds (e.g. Allen and West, 2013, 2015), but are not regarded as "wildlife" by some, and subject to a different standard. Dalerum and colleagues (2020) found that the density of wolves increased the frequency of livestock depredation, whereas densities of brown bears and Eurasian lynx did not influence losses caused by them; rather, anomalous behaviors by a few individuals accounted for most depredation incidents. Although apparent competition involving livestock and wild prey of conservation concern is a common issue (DeCesare *et al.*, 2010), the reverse effect can occur. Instead of increases in abundance of one prey species causing increased predation on another, the opposite occurs. As an example, fewer roe deer translated into higher lynx depredation on domestic sheep in Norway (Odden *et al.*, 2013). And in Pakistan, where wild prey are especially scarce, leopards ate mostly domestic goats, domestic dogs, and cattle (Shehzad *et al.*, 2014).

Approaches to reducing agricultural conflicts with wild carnivorans range from the narrowly mechanistic to the holistic (Boronyak *et al.*, 2022). Miller and colleagues (2016) evaluated the efficacy of four categories of methods, based on sixty-six published accounts from six continents. Husbandry (e.g. herding, corralling, monitoring, Ogada *et al.*, 2003) and using deterrents (e.g. noises, guard dogs) showed the greatest potential to be effective, but also showed the most variable success across applications. Removal of the offending species provided the highest and most consistent effectiveness. Indirect factors—proximity of livestock to vegetative cover, proximity to nature preserves, and the abundance of wild prey—varied in their effects and were not under the control of most livestock producers in any case. Several studies reviewed by Mukeka and colleagues (2019) showed that drought that reduced availability of native prey was a contributing factor to frequency of depredations. Financial compensation schemes seem to have merit because they spread the financial burden of depredations, but Braczkowski and colleagues (2020)

Figure 11.1 A Steller sea lion eats a salmon near a fish hatchery in Valdez, Alaska. Competition between humans and recovering pinniped populations for valuable salmonids is a growing issue. Photo: jtstewartphoto/iStock.

identified several weaknesses in practice: they tend to be unsustainable and to reduce incentives to protect livestock. Du Plessis and colleagues (2018) comprehensively reviewed the methods used to protect livestock from predators in South Africa, while van Eeden and colleagues (2018) reviewed published studies from a broader region and emphasized the importance of quantitative analyses and treatment controls, which are not generally employed.

Domestic pets commonly are killed by wild carnivorans in most parts of the world, but data are sparser for them—perhaps because economic loss is not the primary consequence and data are not aggregated. Worldwide, the primary killer of domestic dogs is the wolf, easily explained by the large zone of sympatry. The leopard is the second most frequent dog killer, following by the puma and coyote (Butler *et al.*, 2014). Leopards in Maharashtra, India preyed primarily on domestic dogs (39% of occurrences in scats) and domestic cats (15%). Other domestic species comprised another 29% of the diet (Athreya *et al.*, 2016).

Addressing conflicts involving carnivorans and livestock requires recognizing that they may have more basis in sociology and psychology than in economics. The empirical basis of one putative conflict was evaluated by Berger's (2006) analysis of the efficacy of coyote control across the United States in relation to socio-economic and other factors. The US

sheep industry declined by nearly 90% from 1940 to the early 2000s, which was loosely attributed to coyote depredation. Berger tested for various possible correlations and found that yearly changes in numbers of sheep under production were positively associated with lamb prices, but marginally associated with coyote control efforts. Sheep numbers declined with increasing hay prices (the best predictor) and wages. These results suggest that coyote depredations on domestic sheep, a high-profile carnivoran-centered conflict in the western US for decades, had little effect on the economics of sheep production. Other studies using differing approaches have tended to corroborate that finding. Gompper (2002) reviewed reports of depredation losses to sheep in the western US during the 1970s and found that coyotes caused 96% (mean of state means) of losses attributed to predation, but that predators caused only 9% of sheep losses. Social and economic factors similarly influence reported depredation and responses to it in other parts of the world. Rust and colleagues (2016) found that a tradition of apartheid, racist treatment, and economic inequality caused Namibian workers on white-owned farms to resort to retaliatory predator killings, livestock theft, and wildlife poaching. In conflict situations with strong social or psychological bases, solutions based on biological arguments or mechanistic treatments are not likely to be effective.

In unusual cases, large-bodied carnivorans can mitigate depredation losses. Thinley and colleagues (2018) showed that the presence of tigers near villages in Bhutan caused leopards and dholes to move closer to the villages and prey on wild ungulates that caused crop damage. When no tiger lived nearby, leopards and dholes moved away from villages, crop losses to wild ungulates increased, and depredation losses on wide-ranging ungulate livestock increased. Tigers near villages protected crops, suggesting a behaviorally mediated trophic cascade.

11.1.3 Killing wild animals valued by humans

The relationship between humans, carnivorans and wild meat is multifaceted and extends back at least 2,000,000 years when early humans began consuming larger wild prey than previously. This shift represented more than a change in behavior; it affected human morphology, locomotion, and habitat use (Pobiner, 2015). This transition to a large-mammal hunting economy required usurping carnivoran-killed carcasses and facing the associated risks. In that sense, carnivorans facilitated this major dietary shift in early humans. Since then, carnivorans consuming wild meat potentially useful to humans has been the basis of innumerable conflicts and management actions (Baker *et al.*, 2008).

The list of wild foods denied to humans by carnivorans is long and cosmopolitan. Examples from marine and freshwater systems seem to be better documented than those from land carnivorans and include some examples involving rare species. The sea otter experienced a close brush with extinction along the northwest coast of North America in the nineteenth century, but upon recovering in the twentieth century featured in several competitive conflicts with fisheries. During the 1960s, otter predation on abalones in California was first perceived as a problem caused by a new invader. Over a decade of public discussion and debate shifted that perception to one of the sea otter as a natural community component—a role that predated the abalone fishery and otter decline. More recently, the sea otter has been implicated in preying on commercially important sea cucumbers and crabs

in the Gulf of Alaska (Larson *et al.*, 2013; Carswell *et al.*, 2015). Other examples include various pinniped species preying on valuable salmon (e.g. Morissette *et al.*, 2012) in North America (Chasco *et al.*, 2017) and South America (Vilata *et al.*, 2009). European river otter predation on farmed and free-living salmonids is especially well documented and can be locally severe (Ludwig *et al.*, 2002; Jacobsen 2005). Sea otter predation on recovering Dungeness crab populations in California is a looming issue, but data so far suggest minor or habitat-specific effects (e.g. Boustany *et al.*, 2021). Looking more broadly, we find few estimates of ecosystem–scale overlap between pinniped diets and fishery catches, although pinnipeds seem to consume less fish biomass than fisheries remove, and of mostly different species (Kaschner and Pauly, 2005). This is an oversimplification in that marine food webs are difficult to partition clearly between what is extracted for fisheries versus what becomes, directly or indirectly, pinniped food.

Fewer documented examples come from terrestrial systems; one that has generated controversy for over fifty years involves wolves, bears, ungulates, and humans in Alaska. By the early 1970s the state wildlife agency was advocating more aggressive wolf control to increase the number of ungulates available for human harvest. However, studies in the 1970s showed that brown bears were as important in killing moose, especially calves, as were wolves (Section 8.2.1). The state agency broadened its control efforts beyond wolves to brown bears and black bears, and the Alaska Intensive Management Law of 1994 formalized the aggressive removal of brown bears to increase human harvest of moose. This law mandated liberal hunting regulations on bears and predator control using methods that would otherwise be illegal over most of Alaska's 1,509,000-km^2 area (Ripple *et al.*, 2019). The hunter harvest of brown bears increased twofold as a result of liberalized hunting regulations, and the aggressive predator reduction became controversial in the urban centers of the state. At the request of Alaska's governor, the National Research Council undertook a review of the program and its effects on hunting opportunities. Its report (National Research Council, 1997) found that efforts to study the effects of the program had been less informative than hoped, because of the vast geographic scale of the program, the number of confounding variables involved, and budgetary limitations. An intensive study conducted by the state agency in a small area in the early 2000s translocated all but four of the ninety-six black bears that used the study area and transported or killed half of the twelve independent brown bears there. About 75% of the wolves were removed by shooting and trapping, and human hunting of moose was restricted for the duration. The moose population increased by about forty-nine per year during and just after the control actions, and predation declined as a mortality cause from 97% before the controls to 65% afterward (Keech *et al.*, 2011). In this geographically restricted study (< 0.1% of the area of Alaska), substantial increases in moose numbers, and reductions in predation-caused mortality were achieved, but the treatment was too costly to apply at scale, and the efficacy of predator control—which includes liberal hunting rules for bears—remains controversial (Miller *et al.*, 2017). For the isolated brown bear population on the Kenai Peninsula, drastic reductions in densities have raised the possibility of ecological extinction. Bears could become too rare to serve the important nutrient-transport functions on salmon spawning streams (Section 6.1).

11.1.4 Limiting species of conservation concern

In many cases, introduced mammalian predators cause severe impacts to native vertebrate biodiversity worldwide. They account for more than eighty-seven avian, forty-five mammalian, and ten reptilian species extinctions—58% of all extinctions in these groups—as well as threatening hundreds of other species or populations of conservation concern. Carnivorans make up five of the seven species implicated in this damage, with the domestic cat, domestic dog, and the small Asian mongoose the chief offenders (Doherty *et al.*, 2016). These effects are most severe on islands and where native eutherians are absent; they are locally important where dogs or cats receive human food subsidies and range freely. Home and colleagues (2017) concluded that free-ranging dogs, especially travelling in packs, adversely affected several native ungulate and carnivoran species, including some IUCN

Critically Endangered species, in India. Some severe effects of feral cats, red foxes, and dingoes were discussed in Section 8.2. Carnivorans have further effects via apparent competition (Chapter 9) involving native prey species and, commonly, introduced alternate prey (Gompper and Vanak, 2008; DeCesare *et al.*, 2010). In one example, the construction of watering places within the range of roan antelope in Kruger National Park, South Africa led to increases in zebra and wildebeest numbers, thereby increasing the abundance of lions, which preyed disproportionately on roan antelope (Harrington *et al.*, 1999). A parallel situation was described for wolves, Przewalski horses, and livestock in Mongolia. There, livestock provided over 50% of the diet of wolves, increasing wolf abundance and predation rates on the endangered horse (Van Duyne *et al.*, 2009). A lion population recovering in Laikipia, Kenya after decades of lethal control reduced hartebeest numbers via an apparent competitive effect, with plains zebras providing the alternate, more abundant, but less-preferred, prey (Ng'weno *et al.*, 2017).

11.1.5 Interfering with tourism

Although carnivorans are assets—or even foundational—to many regional industries of wildlife-based tourism, they can also interfere with these activities. In some cases they threaten public safety, or recreation can be restricted because of potential harm to carnivorans. An example of the latter is the closure of active wolf denning areas to hiking in Denali National Park, Alaska (Thiel *et al.*, 1998). This is the result of the fidelity of wolves over decadal spans to denning sites near popular hiking routes. Hikers can accidentally encounter wolf dens and want to explore them, causing pup rearing to be disturbed. Carnivorans can also compete exploitatively for prey that are the basis of hunting industries. Increases in the value of game species in South Africa were associated with more negative attitudes from hunting guides towards leopards that killed some of the animals they otherwise could hunt (Pirie *et al.*, 2017). This was expected to lead to an increase in retaliatory killings.

11.2 Positive effects

11.2.1 Direct benefits to humans

11.2.1.1 Use for food, fur, and medicine

The relative scarcity of terrestrial carnivoran biomass on the landscape limits the potential for human populations to subsist on carnivoran flesh. Nevertheless, humans have always eaten carnivorans, and continue to do so in many parts of the world. The best-documented contemporary use of wild terrestrial carnivorans for food is in Africa, where various species are shot or snared along with primates and small ungulates (Brashares *et al.*, 2004). Studies of bushmeat hunting tend to focus on how it affects wild populations rather than how it affects human diets, so the importance of carnivoran meat in human nutrition is not well understood. Of course, humans have consumed domestic dog meat for as long as the two species have coexisted.

The use of pinnipeds for food is better documented and extends from far northern countries to Namibia. Industrial sealing for oils used for lighting and other purposes ended over a century ago. The highest reliance on pinnipeds for food traditionally occurred among indigenous peoples in the Arctic and Subarctic. There, seal meat was the primary winter food into the twentieth century (Sinclair, 1953). Today, killing pinnipeds for food is confounded by killing them for their pelts or to reduce competition with fisheries. Both of the latter are widespread, and in some countries hundreds of thousands of pinnipeds are killed yearly, but only a fraction of them used for human food. This disparity results partly from the high concentration of myoglobin in seal meat, which gives it a strong livery flavor and reduces its palatability to many modern consumers. Data on subsistence hunting of seals by indigenous peoples of Alaska, Canada, Greenland, and Russia are perhaps the most relevant to human nutrition because the animals are killed specifically to be eaten, although some parts of the carcass are fed to dogs (Robards and Reeves, 2011). Although pinnipeds are valued sources of meat, in the Far North they also carry elevated levels of mercury and other metals. Contaminant levels are higher in species at high trophic levels (e.g. harbor seals), in contrast to the filter-feeding crabeater seal (Aubail

et al., 2011). Human populations are affected in North America and Eurasia, leading to warnings against consuming pinniped tissues, particularly by pregnant women (Sonne *et al.*, 2019).

Another carnivoran use is for medicine. The bile of many mammals, especially carnivorans, has been considered healing or invigorating in east Asia for centuries. Hunting of various species for their gall bladders, bile, and other body parts has been widespread from the Amur River to Indonesia and has led to population declines of Asiatic black bears, sun bears, and brown bears. In the 1970s, North Koreans devised a means of harvesting bile more sustainably from captive bears, and the bear bile farm industry was launched (Li, 2004). Bears are taken from the wild, closely confined, and their bile periodically harvested by various methods, including permanent catheterization of the bile duct. By the 1990s, hundreds of bear farms held more than 10,000 bears in China alone, and publicity about the practice reached the popular press. International reaction was swift and strong, and regulations and prohibitions reduced the size of the monitored parts of the industry. Vietnam banned the harvesting of bile from captive bears, the numbers of which declined from 4500 in 2005 to 1250 in 2016. In Laos, growth of the industry occurred later, surging from 2008 to 2012 (Livingstone and Shepherd, 2016). Synthetic substitutes for the presumptive active ingredients have been developed, which would seem to bode well for bear conservation and welfare. In Vietnam, however, bile from wild bears is more desirable than that from captive bears born in the wild, and that from bears born in captivity is the least desirable (Crudge *et al.*, 2020). This suggests that consumer acceptance of synthetic bile substitutes is likely to be problematic. The overall effect is that, while eliminating bear bile farming might seem like a desirable reform, it could lead to a resurgence of hunting of wild bears.

For centuries, a primary positive value of wild carnivorans to humans was for fur and ancillary products. Fur hunting and trapping was not only a major economic activity, but also a geopolitical force that motivated the exploration and exploitation of much of North America and northern Asia (Figure 11.2). Russian explorations of the North Pacific Ocean in the eighteenth century aimed to

Figure 11.2 After sea otter, the most highly valued fur over the historical period was that of the sable from the Bargusin region of Russia. These were wild caught before 1900, but in the twentieth century became mostly farm raised.
Photo: Kuerschner/Wikimedia CC.

satisfy the lucrative Chinese market for sea otter fur (Black, 2004). Likewise, the quest for pelts of marten and lynx (along with beaver and snowshoe hare) fueled the interior North American territorial ambitions of England, France, and the nascent United States (Dolin, 2010). Aggressive pelagic harvesting of northern fur seals in the North Pacific by northern Pacific Rim nations reduced populations sharply, resulting in the North Pacific Fur Seal Convention of 1911, the first international treaty dealing with wildlife conservation (Lund, 1980). This treaty banned pelagic sealing, but permitted harvests on breeding rookeries, which were easier to monitor and regulate (Figure 11.3). The scale and geographic foci of the fur industry shifted quickly with fashion changes and resource depletion, but as wild fur harvests declined, fur farms assumed their role, producing a range of carnivoran pelts with more diverse colors, consistent quality, and volume than wild furs afforded.

By the early twenty-first century, the global fur industry had collapsed to a degree suggested by two facts. First, the number of furriers in New York City, the largest fur garment market in the Western Hemisphere, declined from 450 in the 1970s to fourteen in 2019. Second, the global fur industry—everything from pelts to manufactured goods—declined by nearly half from 2014 to 2019

(Givhan, 2019). The Convention on International Trade in Endangered Species of Wild Fauna and Flora (CITES) reduced trade in a wide range of carnivoran parts, the animal rights movement stigmatized wearing fur garments, and synthetic substitutes improved. An increasing number of European countries have banned or strictly regulated farmed furs, trapping, or certain trap types. Atop these trends, the COVID-19 pandemic dealt severe blows to mink farming in western Europe and some of the US, forcing millions of farm mink to be euthanized (Lesté-Lasserre, 2020). The hunting, trapping, and raising of wild or domesticated carnivorans for fur have declined drastically in the past fifty years.

11.2.1.2 Importance in nature-based tourism

Interest in carnivorans stimulates various forms of tourism, both extractive and non-extractive. This economic sector is geographically dispersed and multi-faceted, although the carnivoran species involved are few in number, but expanding. What constitutes "tourism" is subjective; activities that involve travel and paying for permits, guides, or ancillary services are more likely to be included than activities that are local, self-guided, and for subsistence purposes.

Figure 11.3 Killing northern fur seals for their pelts on St. Paul Island, Alaska, 1890s. Unregulated pelagic sealing in the North Pacific led to the Fur Seal Treaty of 1911, the first international treaty dealing with wildlife. Pelagic sealing was banned but killing seals on land could be monitored and regulated and continues under governmental oversight in the twenty-first century.
Photo: Wikimedia Commons CC.

Many large carnivorans were hunted for trophies in the past: large cats, bears of all species, and walrus. However, most of those species are inherently rare, and most felids—particularly mid-sized ones—are either difficult to hunt or protected today. Regulated hunting of black bears remains common in North America (Hristienko and McDonald, 2007) and involves large numbers of participants and considerable expenditures by those being guided. Likewise, about 1000 Asiatic black bears are taken by hunters in Japan yearly (in about 2000, Huygens *et al.*, 2001), as are 1600 brown bears taken in Alaska (in 2012, Miller et al 2017). Black bears are killed for trophies and meat, whereas Alaskan brown bears are killed almost entirely for trophies. Pumas have increased markedly in abundance in western North America in recent decades; seven western states all reported statewide population estimates in the hundreds in the early 1960s, but in the thousands in the 2000s (Logan and Runge, 2020). Managed hunting is conducted in several of these states, and a small subset of dog-assisted clients observes and photographs the chase, without killing the puma.

The bears, large cats, pinnipeds, and wolves that are the foci of regional tourism industries illustrate the emotive power of predaceous behavior and large bodies. Large felids in particular can anchor regional nature-based tourism industries, as tigers have done in India (Karanth *et al.*, 2017). There, nature-based tourism is the fastest growing of all tourism sectors, and tigers are the single largest attractor of visitors. The same is true of lions and other large felids in African protected areas, where they are aggressively marketed to wildlife tourists (e.g. Charnley, 2005). In both these regions, large-carnivoran tourism boosts local and regional economies, but income and environmental impacts flow asymmetrically to various community members. Tourism activities can alter wildlife behaviors, degrade habitat important for conserving the focal wildlife species, and add to wealth inequalities across the human community. The most charismatic carnivoran species may cause uncompensated damage to pastoralists, leading to negative attitudes toward carnivorans and tourism, and in some cases lethal retaliation against offending carnivorans. Tortato and colleagues (2017) showed such a pattern for jaguar tourism in the Brazilian Pantanal, where livestock losses were valued at a small fraction of jaguar tourism revenues but were concentrated on a few pastoralists. Various authors have decried the environmental and social effects of luxury wildlife tourism, and effective solutions that protect wildlife and spread benefits across economic levels are elusive (Charnley, 2005; Buckley, 2013; Rastogi *et al.*, 2015). Large carnivores, both cetacean and carnivoran, are singularly able to broaden the geographic reach of wildlife tourism, for example, attracting brown bear tourism to Kamchatka, Russia (Figure 11.4) and elephant seal watchers to South Georgia Island.

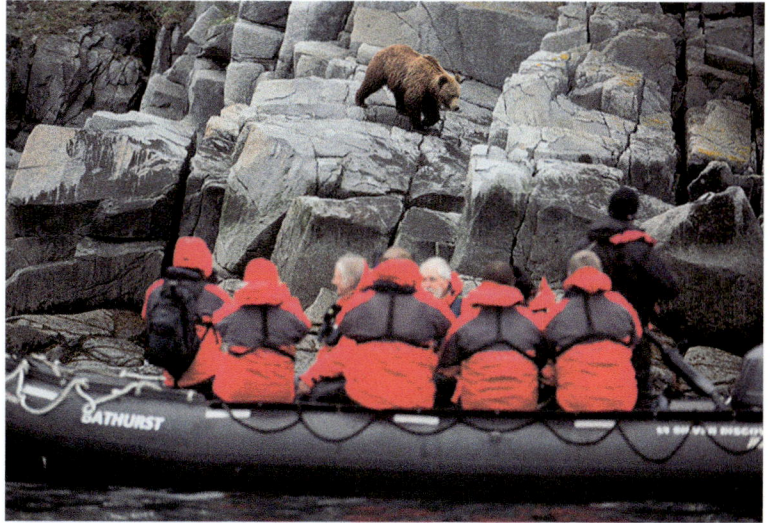

Figure 11.4 Carnivoran-based wildlife tourism is expanding beyond a few iconic destinations, distributing its economic impacts more broadly. Tourists view a brown bear from the safety of a motorized raft on the Kamchatka Peninsula, Russia.
Photo: Galaxiid/Alamy.

Figure 11.5 Tourists watch smooth-coated otters, tethered to poles, forage for fish in the mangrove bays of the Sunderbans, Bangladesh. Several hundred fishers use otters in this way and a tourism industry has grown around the activity.
Photo: Cindy Hopkins/Alamy.

Viewing of marine carnivorans represents a growing sector, especially on rookeries where large numbers of animals assemble (Bearzi, 2017). For all carnivoran-based tourism, consumptive and non-consumptive, the total impact on local economies is higher, commonly by an order of magnitude, than what tourists pay directly to participate. In the case of the seal-watching example, the costs for tickets issued by tour operators were outweighed by the costs of transportation to the site, lodging, food, and so on—leading to a multiplier of 10–20

(Kirkwood *et al.*, 2003), an amount common for this industry.

If the largest felids, canids, bears, and pinnipeds are the gold standard for carnivore-based tourism, then the mid-sized carnivores are the emeralds, sapphires, and pearls of wildlife viewing. Sightings of them can be rare and unpredictable; however, they elicit strong emotive responses from the nature enthusiasts. Martone and colleagues (2020) compared the willingness of eco-tourists to pay to see various vertebrates on boat trips primarily

---

Box 11.1 Kopi luwak

Arguably the most bizarre service provided by carnivorans to humans involves wild-captured palm civets in Indonesia. These animals are held in cages to partially digest coffee fruits (Figure 11.6), thereby producing the fecal raw material of the most highly prized coffees in the world—kopi luwak. Formerly, fecal droppings of civets were collected from the wild, but in only a few decades the extraordinary value of the coffee has led to capture, captive holding, and concerns about animal welfare and the health of the wild population (Carder et al., 2016). The price of civet coffee in 2021 was over US$500/kg, and an island-wide tourism industry centered on kopi luwak has developed on Bali.

Figure 11.6 Fecal droppings of a captive Asian palm civet containing partially digested coffee fruits. The pits will be roasted and ground to make kopi luwak—civet coffee, the costliest coffee in the world. The conditions of civet captivity raise issues of animal well-being.
Photo: kaiskynet/CanStock.

---

aimed at whale-watching in British Columbia, Canada. They found that tourists were nearly as willing to pay to see sea otters as whales, whereas tour operators considered sea otters relatively unimportant to the tourism experience. Otters can be especially entertaining and accessible tourism attractions. In and around the Sunderbans, Bangladesh captive smooth-coated otters are used to assist in fishing, and a regional tourism industry has developed based on watching and photographing otter-aided fishing from boats (Figure 11.5; Feeroz et al., 2011). The fishers that employ captive otters in this way number in the hundreds (Box 11.1).

The comparative economic value of hunting vs. viewing the same species is often a consideration that has been analyzed for American black bears in coastal British Columbia, Canada. Honey and colleagues (2016) showed that bear viewing generates more than twelve times as much economic activity and government revenue as does guided or non-guided hunting. This finding is similar to that for other viewing vs. hunting ratios for non-carnivoran species. A very different example comes from red fox hunting in Britain, a sport now banned. It employed dogs and horses and was at times justified on the basis of fox depredation on lambs. White and colleagues (2000) showed that yearly lamb losses amounted to 0.6–1% of lamb production, which is inconsequential to sheep producers. By contrast, the amount spent directly on fox hunting has been estimated at GB£176 million, not including travel and other ancillary expenses. In this case, the economic activity generated by the consumptive use vastly outweighs the value of the fox, its pelt, or the lambs it might depredate (MacMillan and Phillip, 2008).

11.2.2 Ecological services

In addition to the ecological effects of carnivorans that contribute generally to ecosystem functioning (Chapter 6), carnivorans provide services that benefit humans narrowly and specifically. One important service is limiting abundances of wild herbivores in agricultural settings. This is most commonly reported for large carnivorans that reduce exploitative competition between wild ungulates and livestock (Gordon, 2009). Allen (2015) studied a dingo–macropodid–cattle system in southeastern Australia, and his models predicted that allowing dingo presence to return would reduce kangaroo competition with cattle for forage, thereby increasing the profitability of cattle production. Considerable attention has been paid to ecological control of rodent damage to crops in Africa and Asia. In Africa, crop losses to rodents—and the potential provided by small and mid-sized carnivorans that prey on rodents—are greatest where small farms predominate (Swanepoel et al., 2017). Williams and colleagues (2018a) showed that small and mid-sized carnivorans, mostly genets and mongooses, were most diverse on smaller African farms, where livestock was present, and where domestic dogs were scarce.

The positive values of mid-sized carnivorans as waste removers are recognized in various regions. Golden jackals contribute substantially to removing organic waste from disposal sites in Serbia (Cirovic et al., 2016). That single species is estimated to remove thousands of metric tons of animal waste yearly, countrywide. Their role is enhanced by their nocturnal activity, which reduces contact with humans and dogs, but they have been mostly portrayed as competing with vultures, some species

of which are conservation concerns. A similar service was attributed to spotted hyenas across northern Ethiopia, where donkeys, cattle, and humans were the primary animal species represented in scat samples. Carnivoran consumption of human carcasses can be seen as a positive value in areas where burial or cremation are not universal (Yirga et al., 2015).

Where ungulate populations are sufficiently high, carnivoran predation on ungulates is credited with reducing ungulate–vehicle collisions. While collisions of all causes in the eastern US stabilized in the early twenty-first century, the number of reported wildlife–vehicle collisions nearly doubled (Huijser and McGowan, 2010). Carnivorans are now credited with reducing collisions in various areas. Raynor and colleagues (2021) estimated that the return of wolves to Wisconsin, US caused a 24% reduction in deer–vehicle collisions in counties where wolves were present. The reduction was not due to reduced deer densities; wolf presence/absence, not abundance, was the most important predictor, suggesting that the presence of wolves affected deer behavior related to highways. The economic benefit of the return of wolves—in terms of reduced costs of deer–vehicle collisions—was estimated to be sixty-three times greater than the verified costs of wolf depredations on livestock in those counties. Gilbert and colleagues (2017) considered the effect of restoring the puma to the eastern US and predicted that 21,000 human injuries, 155 fatalities, and US$2.1 billion in costs would be avoided in the thirty years following puma restoration.

11.3 Mixed effects

11.3.1 Roles in disease ecology

Carnivorans affect humans via several disease-related pathways. They carry and transmit pathogens that can infect humans and cause disease (zoonoses), but they also interrupt the life cycles of some pathogens by killing their hosts. They also carry and transmit diseases that infect domestic animals or wild species that humans value, including some of conservation concern (Han et al., 2021).

The question raised earlier of whether carnivoran predation facilitates or limits disease prevalence or intensity is pressing in the context of zoonoses, because they represent a large proportion of emerging infectious diseases of humans. Even disregarding the deaths caused by dog-transmitted rabies, the human toll from diseases in which wild carnivorans play some ecological role is large and increasing. Domestic carnivorans are critical links here, because 70% of pathogens of this group also infect humans, and 59% infect wild animal hosts. Domestic dogs, cats, ferrets, and farmed minks are reservoirs and transmitters between their wild counterparts and humans (Cleaveland et al., 2001). Notably, the SARS-COV-2 virus, which caused the pandemic of 2020–2022, produces its strongest symptoms, across non-human vertebrates, in carnivorans. Of all warm bloods tested, domestic ferrets, domestic cats, and civets showed the highest susceptibility, whereas ducks, chickens, rodents, and pigs exhibited lower or no susceptibility (Blanco et al., 2020). American minks were recognized early in the pandemic as contracting SARS-COV-2 from humans and showing high prevalence on mink farms. Humans introduced the disease to farms where it was transmitted to mink, the virus underwent rapid evolution and spread to adjacent farms, and then reinfected farm workers and family members in its modified form (Oude Munnink et al., 2021). These discoveries have resulted in the culling of tens of millions of captive minks in Europe, possibly endangering the entire European farmed mink industry.

The list of wild carnivorans that can serve as reservoirs or vectors for zoonotic pathogens is long and taxonomically diverse. Not surprisingly, the carnivoran families with the largest numbers of species that carry zoonotic pathogens, or the largest numbers of pathogen species, are the most speciose families. Canids and mustelids, as groups, carry more zoonotic taxa than does the monophyletic Ailuridae (Han et al., 2021). Carnivoran-borne zoonoses include those transmitted to humans directly via fecal–oral routes, those carried by arthropod vectors, and others that use carnivoran pets as intermediaries (Otranto and Deplazes, 2019). Viral diseases seem to be most likely to emerge or re-emerge as zoonotic threats, because of high rates of viral mutation (Haemig et al., 2008), but

helminth diseases of long-standing low prevalence can re-emerge because of changed ecological conditions (Deplazes et al., 2004). The guinea worm of sub-Saharan Africa has a complex life cycle, its larvae infecting aquatic copepods, which are ingested by various definitive hosts, including carnivorans, non-human primates, and humans. Carnivorans do not serve as vectors but help to maintain the disease in waters that humans may ingest (Han et al., 2021).

Lyme disease in particular has served as a focal system for how weasels, foxes, and bobcats affect the prevalence of a major emerging zoonosis. The bacterium Borrelia burgdorferi is transmitted by ticks of Genus Ixodes, which infest many vertebrate hosts, including mice and white-tailed deer. As the ticks mature, they can infect humans with Borrelia via bites. How carnivorans affect rodent densities and tick numbers is of keen interest to public health scientists. While some studies show strong effects of carnivorans on small rodent numbers (Ostfeld et al., 2018), most of these studies deal with Microtus voles, rather than the rodent taxa (e.g. Peromyscus, Apodemus) implicated as zoonotic disease reservoirs (Ostfeld and Holt, 2004). Food supply commonly emerges as the primary determinant of small rodent abundances, but the roles of specialist (e.g. weasel) vs. generalist (e.g. coyote) predators remain uncertain. Tick-borne encephalitis in Europe raises similar questions about how carnivorans might influence public health (Haemig et al., 2008).

The proximity of wild carnivorans to humans in urban and suburban settings increases the potential for zoonotic transmission. Alveolar echinococcosis is a potentially severe helminth disease of humans caused by the small hydatid tapeworm, Echinococcus multilocularis. Small rodents serve as intermediate hosts, and definitive hosts include the red fox or arctic fox, and other wild canids and northern raccoons. Where canids have become abundant near humans—for example in China and rural Alaska—prevalence of alveolar echinococcosis has spiked, and the disease is difficult to treat (Deplazes et al., 2004). Red foxes have increased in European and Japanese cities; for example, from 1985 to 1997 the number of red foxes killed or found dead in Zurich, Switzerland increased twentyfold (Gloor et al., 2001). This increased proximity to humans raises the potential for zoonotic transmission, including via domestic carnivoran hosts, which has led to experimental use of anti-helminthic baiting of foxes in peri-urban areas (Hegglin *et al.*, 2003).

Bovine tuberculosis (bTB), caused by *Mycobacterium bovis*, is a cosmopolitan disease that can infect humans, but is primarily of concern as a disease of livestock. bTB infections make up less than 2% of all tuberculosis cases in humans in the UK (de la Rua-Domenech, 2006), but the disease still receives close attention. Worldwide, the pathogen occurs in various wild mammalian reservoirs, typically ungulates in North America and Africa and introduced common brushtail possums in New Zealand. European badgers are of particular concern in Britain, because badger–cattle transmission is estimated to account for nearly half of all cattle infections (Donnelly and Nouvellet, 2013). Efforts to reduce badger–cattle transmission via culling and vaccinating of badgers, and other means, have been explored for decades, but all are costly, and bTB incidence in cattle has steadily increased. Culling exerts contrasting effects at different spatial scales (Vial and Donnelly, 2012), that is, it reduces incidence in cattle in the culled area but increases it in adjacent areas. This apparently results from locally density-dependent effects—infected animals disperse in response to disturbed social structures. Effective measures to reduce badger–cattle transmission remain elusive.

Disease is a critical issue in carnivoran conservation, and domestic dogs and cats exert strong influences, particularly in regions with infrequent vaccination of pets. Rabies, canine distemper, canine adenovirus, and canine parvovirus are major concerns related to African wild dog and Ethiopian wolf conservation (Laurenson *et al.*, 1998), and large numbers of unvaccinated dogs under weak control exacerbate the problem. Canine distemper and sylvatic plague have been major concerns in conserving black-footed ferrets, leading to the near-extinction of the last native population in Wyoming the 1980s, and precluding the recovery of reintroduced populations since then (Section 12.10). *Toxoplasma gondii* is a cosmopolitan protozoan that infects nearly all warm-blooded animals, but sexually reproduces in and sheds infectious oocytes only from felids. It can complicate human pregnancies and is a major source of mortality to

sea otters on the California coast. Domestic cats are suspected of being the primary source of infections (VanWormer *et al.*, 2013). Although free-ranging cats show lower prevalence than do pumas or bobcats, the former are far more numerous in seaside areas, and produce more kittens, throughout the year, than do wild felids. Water runoff provides the needed transport across the sea–land interface (Miller *et al.*, 2008), and marine invertebrates may be the proximal transmission agents, concentrating the oocysts via filter feeding.

11.3.2 Roles in conserving other species

Humans in many cultures have, over the last century, come to understand and accept the positive effects of native carnivorans (Chapter 6) so that in many areas of the world they are no longer perceived as an unmitigated destructive force. Indeed, the pendulum has swung the other way—those in affluent countries imbue carnivorans with positive ecological powers. Many of these effects are documented, but some are known only in certain settings and hypothesized to occur elsewhere. One such example involves dingoes, invasive red foxes, and introduced domestic cats in Australia. Dingoes occupy an interesting gray position on the wild–domestic scale, having been introduced by humans about 4000 years ago. They are largest-bodied eutherian predator in Australia and are suspected of having contributed to the mainland Australian extinctions of the thylacine, Tasmanian devil, and native hen around 3200 years ago (White *et al.*, 2018), which suggests their past role as destroyers of native biodiversity. However, dingoes in modern communities also are credited with limiting the abundance and ecological effects of the red fox and domestic cat. Some authors (e.g. Letnic *et al.*, 2012) see dingoes as uniquely positioned to protect native vertebrates from the strong predatory effects of red foxes and feral domestic cats, whereas others question that potential. The controversy resembles, in several respects, that involving reintroduced wolves and their prey, especially North American elk, in the Yellowstone area (Section 9.2.3):

1. Both systems involve similar taxa of canid, in Australia introduced, and in Yellowstone reintroduced.

2. Both controversies feature a preponderance of scientific publications asserting the importance of wolves/dingoes in protecting ecological integrity, via different mechanisms (preying on native prey vs. competing with invasive carnivorans, respectively). The popular press tends to emphasize the functional importance of these roles in both settings.

3. The literature also includes many criticisms of these types of studies for weaknesses in experimental design and statistical rigor (Section 9.2.3 for Yellowstone wolves, and Allen *et al.*, 2017 and Hayward and Marlow, 2014 for Australian dingoes).

4. These weaknesses are alleged by critics of the putative roles of the dingo and wolf to lead to unfounded confidence in the abilities of either taxon to protect biodiversity or ecological health to the extent that the public is led to believe. Both carnivorans provide ecological benefits, but neither is by itself protective.

The literature is peppered with reports of carnivorans exerting effects that humans value, partly because we need natural communities and ecosystems to function with no human support. Scientists have labeled these functions imprecisely: "apex predators" (various species, Ordiz *et al.*, 2013) are thought to structure communities, but the scale upon which the animal is positioned is unclear. In fact, some scavengers (e.g. American black bear, Engebretsen *et al.*, 2021) dominate nominal "apex predators." "Umbrella species" (lion, Curveira-Santos *et al.*, 2021), "keystone predator" (snow leopard, Lovari *et al.*, 2009), and "flagship carnivore" (tiger and leopard, Li *et al.*, 2019) have even less-precise meanings. Even "mesopredator" is problematic because it is based only on body size, whereas "apex" considers trophic level in some interpretations. The overuse of these terms reduces their semantic precision and utility (Caro, 2010; Miranda, 2018). Further, some authors (e.g. Suraci *et al.*, 2019) consider humans to be a "super predator" in many ecosystems, terrestrial and marine, rendering moot the rankings of carnivoran predators. Some consider apex predator effects, mesopredator release, and trophic cascades to be paradigms or theories applicable across taxa

and contexts, whereas others accept their validity only where specifically documented.

The wide range of processes embraced by these terms describe important natural phenomena, and hypotheses can be developed about whether and how they occur. However, general theories of how these phenomena operate across all taxa, locations, and contexts are nonexistent. The tendency of some scientists to overstate the generality of effects of large (or even mid-sized) carnivorans in communities is common. One important commentary by prominent scientists who study carnivorans on four continents questioned whether carnivoran scientists can be credible advocates for the use of carnivorans in achieving healthy, functioning communities (Allen *et al.*, 2017). The public relies on ecologists to raise alarms regarding environmental threats, but also to describe ecological relationships and expected outcomes of proposed management actions in measured terms, supported by data, with appropriate qualification. Uncritical advocacy for the role of carnivorans in communities may have the unfortunate effect of reducing public confidence in carnivoran science in general.

In addition to their positive values, native carnivorans affect species of conservation concern negatively. These cases commonly involve ecological imbalances of human origin, but in other cases seem to be independent of human influence. An example of the latter is American badger predation on the Agassiz's desert tortoise in southern California. The threatened tortoise suffers high predation losses, some attributable to badgers, which excavate them from their burrows. No contributing human influences have been identified (Emblidge *et al.*, 2015). Imbalances commonly involve human food subsidies, especially provision of prey, which contributes to apparent competition (DeCesare *et al.*, 2010). The example from Kruger National Park involving roan antelope, zebras, wildebeest, and lions (Harrington *et al.*, (1999) described earlier (Section 8.8) is a classic case of indirect human influence over a community, with conservation implications.

The perceived ecological effects of carnivorans introduced outside their native ranges—invasives—are almost universally negative for species of conservation concern. In many cases a species has been introduced to hopefully prey on unwanted vertebrate pests, but few of these hopes were met. Of course, rural dwellers still rely on domestic cats for rodent control near buildings and use dogs in similar ways. Cats were purposefully established over wide areas of rural Australia by the 1890s in an attempt to control erupting invasive European rabbits and mice (Denny and Dickman, 2010). Today, feral domestic cats cause high losses of native mammals across Australia. In the US, cats kill an estimated 1–4 billion birds every year and 6–22 billion mammals every year (Loss *et al.*, 2013). These estimates suggest that domestic cats are the single greatest agent of human-caused wildlife deaths in North America, if not the world (Figure 11.7).

Similarly, the widespread introduction of the small Asian mongoose to subtropical and tropical islands worldwide was intended to reduce rat populations on sugar cane plantations. The mongoose fell short of expectations, and instead caused geographically vast problems on islands rich with endemic vertebrates (Barun *et al.*, 2011). One egregious example involves the Amami rabbit, an island endemic that formerly occurred over most of Amami Island and a small adjacent island in the Ryukyu Island chain of southern Japan. The Amami rabbit is a small-bodied, melanistic, short-eared monophyletic lineage that diverged around 18 Ma (Arregoitia *et al.*, 2015). Since small Asian mongooses were introduced in 1979, the Amami rabbit population has declined, and its range has contracted and fragmented (Ohnishi *et al.*, 2017). It is now critically endangered (Lorenzo *et al.*, 2015).

Also with the expectation of great fortune, the arctic fox was introduced to many of the Aleutian Islands and neighboring Alaskan islands. It briefly supplemented the more lucrative sea otter peltries, which began to be overharvested in the early nineteenth century, but by the time of the Alaska Purchase in 1869, the sea otter fur industry had collapsed. Arctic fox predation drove populations and some subspecies of ground-nesting birds to near extinction before fox control began in the 1940s, and strengthened in the 1980s (Section 9.1). The Aleutian Islands were nearly fox-free, their pre-European state, by 2020 (Ebbert, 2000). The raccoon dog is another example. It also was introduced for the purpose of fur production, from east Asia to Soviet Europe during the first half of the twentieth century. It colonized westward as far as France by 1979, and Italy by 2005. How the raccoon dog affects the native fauna is not well documented, but it is suspected of competing with native carnivorans (Jędrzejewska and Jędrzejewski, 1998; Kauhala and Kowalczyk, 2011).

Figure 11.7 A domestic cat with a hooded warbler it has killed in a city park in North America. Free-ranging domestic cats are estimated to kill billions or tens of billions of birds and mammals in the United States annually, the single greatest agent of human-caused wildlife mortality. Photo: forestpath/CanStock.

Domestic dogs have taken a different invasion trajectory. In contrast to domestic cats and mongooses, they have tended not to persist without human food subsidies, following release for controlling unwanted vertebrates. Instead, they have mostly accompanied human colonization of the world, but in few cases persisted after humans departed. Most commonly, dogs exert influence near human occupations where they are food-subsidized by humans. They adversely affect wildlife via various channels by harassing and killing species of concern, transmitting disease, and competing with native carnivorans (Butler *et al.*, 2004). Banks and Bryant (2007) found that dog-walking displaced birds from natural areas in Australia, and Zapata-Rios and Branch (2018) showed that occupancy rates of native carnivorans in the Ecuadorian Andes responded more negatively to dog occupancy than to loss or fragmentation of habitat. This was so for carnivorans larger than dogs (puma and spectacled bear) as well smaller-bodied species (culpeo fox and striped hog-nosed skunk). These responses tended to be clinal rather than threshold for most species at most sites. However, not all studies reveal strong effects on potential competitors. Dogs in Brazil were recorded in camera traps mostly during daylight hours, as were tayras and coatis. No differences in the activity periods of the latter were observed between where dogs were present vs. absent (Bianchi *et al.*, 2020). In this case, the arboreal abilities of tayras and coatis may have protected them (Box 11.2).

Other invasive species arrive via colonization, but without human-provided transport. The coyote exemplifies this process, following the ecological extinction of the wolf across temperate North America in the late nineteenth century. The generalized diet and behavioral plasticity of the coyote suit it well for filling vacant trophic niches, and it tolerates close human presence. It is implicated in several conservation issues, including those involving the endemic Olympic marmot in Washington, US (Witczuk *et al.*, 2013) and the San Joaquin kit fox in California (Cypher and Spencer, 1998). The expansion of northern raccoons across temperate north America, also attributed to the eradication of wolves, is another example. The invasion of tundra habitats and displacement of arctic foxes by red foxes follows a similar pattern.

11.4 Spiritual and emotional values

Lastly, and in some ways most importantly, carnivorans affect the spiritual lives, behaviors, and psychological well-being of humans. They threaten us, comfort us, entertain us, intrigue us, and are the basis of legends, metaphors, and memes. These effects are difficult to quantify but outweigh some economic values that motivate public opinion and public policy. Both the positive and negative effects of carnivorans become amplified and exaggerated in lore, in popular culture, and in the scientific literature (Figure 11.9). We employ them in symbols, cultural traditions, ancient and modern myths, and advertising. Every pocket of human life on earth for millennia has featured symbols and stories about carnivorans, the only plausible exceptions being remote Pacific Islanders, who lived for generations never witnessing a dog, cat, or mongoose, and only occasionally glimpsing a distant pinniped. For the great majority of humans, carnivorans are key to life experiences with deep emotional significance. If carnivoran taxa were to disappear from Earth's biota, we would be much and permanently changed, with no units of measure to quantify the loss.

Key points

- Carnivorans have posed threats to human safety throughout our evolution, but the numbers of humans killed by carnivorans are smaller than for other vertebrate taxa, including snakes, crocodilians, and hippopotami. Domestic dogs, commonly infected with rabies in some parts of the world, are the chief threat.
- Wild carnivorans kill various domestic animals, including pets and food animals. The latter include ungulates raised under low-intensity husbandry as well as hatchery fish, which are tightly enclosed. In a few cases, large carnivorans are credited with mitigating losses due to depredation.

Box 11.2 The Polynesian dog

The Polynesian dog was one of the few vertebrate species introduced to remote Pacific Islands by humans around 1000 years ago. It was observed in New Zealand by Captain James Cook in 1769, and on Tahiti, Hawaii, and the Marquesas during the same era. The breed was described as fox-sized, short-legged, and varied in color (Figure 11.8). Dog remains are common in archeological deposits on various Pacific islands from Polynesia to the Aleutians (Williams *et al.*, 2018b; Vasyukov *et al.*, 2019). They are thought to have been fed mostly fish and were used mainly for food. The various Polynesian breeds disappeared via introgression with dogs of European origin by the late nineteenth century. Importantly, they seem never to have lived ferally, independent of human food subsidies. This likely is because of the dearth of terrestrial mammalian prey on remote islands, where Polynesian dogs depended on humans for access to marine subsidies.

Figure 11.8 A taxidermic mount of a kuri, or Polynesian dog, collected in 1876 between "Waikava" and Mataura Plains, Catlins, New Zealand, and held in the Museum of New Zealand Te Papa Tongarewa. Dogs were carried by Polynesians to remote Pacific islands but did not become feral or independent of human food subsidies.
Photo: LM000828, Museum of New Zealand Te Papa Tongarewa CC.

- Approaches to limiting or mitigating depredation losses of domestic animals occupy much of the wildlife-conflict literature. These approaches range from the highly specific and mechanistic (e.g. fencing and guarding) to the holistic (e.g. merging conservation and agricultural goals).
- Carnivorans compete with humans for wild foods, particularly marine fish and ungulates. However, parsing the fates of marine organisms in species consumed by humans vs. wild carnivorans is especially difficult because of the complexity of marine food webs.
- Introduced carnivorans have contributed to many extinctions of populations or subspecies during the historical period, especially on islands or where eutherian predators were previously absent.
- Domestic prey can increase carnivoran predation on wild prey via apparent competition.
- Wild carnivorans are the basis of expanding industries of wildlife-based tourism. They also

Figure 11.9 *Infiltrators*, by Walton Ford, 2001. Ford's work in watercolor, gouache, ink, and pencil interprets events in Banbirpur, Uttar Pradesh, India, in 1996. A wolf killed a boy and local people, including the boy's family, blamed the killing on wolf-human creatures that could walk on hind legs and carry victims over their shoulders (Burns, 1996; Ford, 2007). Humans tend to exaggerate and distort—both negatively and positively— the effects of carnivorans on human lives.
Courtesy of the artist and Kasmin Gallery.

Figure 11.9 *continued*

can interfere with these industries. The hunting of wild carnivorans for trophies or food is very restricted now, due to scarcity, social taboos, and legal protections, the best-documented exceptions being bear hunting in Japan and North America.

- Wild carnivorans are used for food in various regions, including West Africa and the Far North. Pinnipeds remain important foods for indigenous people of the Arctic and Subarctic, but many pinnipeds are killed in some areas to reduce competition for wild fish; much of that seal flesh goes uneaten by humans.
- Various carnivoran body parts are used for medicinal, spiritual, or symbolic purposes, leading to conservation issues. Fur was the primary positive economic value of carnivorans to humans for centuries, but the fur industry has contracted sharply in recent decades. Farmed furs have filled the economic niche occupied by wild furs, but even that industry is in decline.
- Carnivorans provide ecological services that narrowly and specifically affect humans. They reduce competition between domestic and wild ungulates, remove human food waste from the landscape, and reduce vehicle collisions with ungulates.
- Carnivorans have important but complex roles in the ecology of diseases important to humans, in some cases completing, but in other cases interrupting, the life cycles of pathogens of humans or species important to humans.
- The growing proximity of carnivorans to humans in cities increases the potential for transmission of several zoonotic diseases.
- Disease is a major concern in the conservation of various carnivoran species, including sea otters, black-footed ferrets, and Ethiopian wolves. Domestic carnivorans are implicated as reservoirs in some of these disease transmission pathways.
- Wild carnivorans kill various species of conservation concern. They also provide ecological benefits that are increasingly valued. Some of the alleged benefits of carnivorans to ecological function are exaggerated or not supported by data. Overstating the benefits of carnivorans to ecological function may reduce the credibility

of carnivoran ecologists among decision makers and the public.

- Carnivorans carry great spiritual and emotional value to humans. We fear them, take comfort in them, and are fascinated by their lives and behaviors. Human cultures are filled with images, symbols, and vocabulary based on carnivorans. We shall be much poorer if we lose carnivoran taxa permanently.

References

Allen, B.L. (2015) "More buck for less bang: reconciling competing wildlife management interests in agricultural food webs," *Food Webs*, **2**, pp. 1–9.

Allen, B.L. *et al.* (2017) "Can we save large carnivores without losing large carnivore science?", *Food Webs*, **12**, pp. 64–75.

Allen, B.L. and West, P. (2013) "The influence of dingoes on sheep distribution in Australia," *Australian Veterinary Journal*, **91**, pp. 261–7.

Allen, B.L. and West, P. (2015) "Dingoes are a major causal factor for the decline and distribution of sheep in Australia," *Australian Veterinary Journal*, **93**, pp. 90–2.

Arregoitia, L.D.V. *et al.* (2015) "Diversity, extinction, and threat status in Lagomorphs," *Ecography*, **38**, pp. 1155–65.

Athreya, V. *et al.* (2016) "A cat among the dogs: leopard *Panthera pardus* diet in a human-dominated landscape in western Maharashtra, India," *Oryx*, **50**, pp. 156–62.

Aubail, A. *et al.* (2011) "Investigation of mercury concentrations in fur of phocid seals using stable isotopes as tracers of trophic levels and geographical regions," *Polar Biology*, **34**, pp. 1411–20.

Baker, P.J. *et al.* (2008) "Terrestrial carnivores and human food production: impact and management," *Mammal Review*, **38**, pp. 123–66.

Banks, P.B. and Bryant, J.V. (2007) "Four-legged friend or foe? Dog walking displaces native birds from natural areas," *Biology Letters*, **3**, pp. 611–3.

Barun, A. *et al.* (2011) "A review of small Indian mongoose management and eradications on islands," in Veitch, C.R., Clout, M.N. and Towns, D.R. (eds.) *Island invasives: eradication and management*. Gland: IUCN, pp. 17–25.

Baynes-Rock, M. (2013) "Local tolerance of hyena attacks in east Hararge Region, Ethiopia," *Anthrozoös*, **26**, pp. 421–33.

BBC News. (2016) "What are the world's deadliest animals?", BBC News [online], June 15, 2016 (Accessed: April 18, 2022).

Bearzi, M. (2017) "Impacts of marine mammal tourism," in Blumstein, D.T. *et al.* (eds.) *Ecotourism's promise and peril: a biological evaluation.* Cham: Springer, pp. 77–96.

Berger, K.M. (2006) "Carnivore-livestock conflicts: effects of subsidized predator control and economic correlates on the sheep industry," *Conservation Biology,* **20,** pp. 751–61.

Bianchi, R.C. *et al.* (2020) "Dog activity in protected areas: behavioral effects on mesocarnivores and the impacts of a top predator," *European Journal of Wildlife Research,* **66,** p. 36.

Black, L.T. (2004) *Russians in Alaska, 1732–1867.* Fairbanks: University of Alaska Press.

Blanco, J.D. *et al.* (2020) "*In silico* mutagenesis of human ACE2 with S protein and translational efficiency explain SARS-CoV-2 infectivity in different species," *PLoS Computational Biology,* **16,** p. e1008450.

Bombieri, G. *et al.* (2019) "Brown bear attacks on humans: a worldwide perspective," *Scientific Reports,* **9,** p. 8573.

Boronyak, L. *et al.* (2022) "Pathways towards coexistence with large carnivores in production systems," *Agriculture and Human Values,* **39,** pp. 47–64.

Boustany, A.M. *et al.* (2021) "Examining the potential conflict between sea otter recovery and Dungeness crab fisheries in California," *Biological Conservation,* **253,** p. 108830.

Braczkowski, A. *et al.* (2020) "Evidence for increasing human–wildlife conflict despite a financial compensation scheme on the edge of a Ugandan national park," *Conservation Science and Practice,* **2,** p. e309.

Braczkowski, A.R. *et al.* (2018) "Leopards provide public health benefits in Mumbai, India," *Frontiers in Ecology and the Environment,* **16,** pp. 176–82.

Brashares, J.S. *et al.* (2004) "Bushmeat hunting, wildlife declines, and fish supply in west Africa," *Science,* **306,** pp. 1180–3.

Buckley, R. (2013) "Tiger tourism: critical issues, general lessons," *Tourism Recreation Research,* **38,** pp. 101–3.

Burns, G.L. and Howard, P. (2003) "When wildlife tourism goes wrong: a case study of stakeholder and management issues regarding dingoes on Fraser Island, Australia," *Tourism Management,* **24,** pp. 699–712.

Burns, J.F. (1996) "In India, attacks by wolves spark old fears and hatreds," New York Times, September 1, p. 1

Butler, J.R.A., du Toit, J.T. and Bingham, J. (2004) "Free-ranging domestic dogs (*Canis familiaris*) as predators and prey in rural Zimbabwe: threats of competition and disease to large wild carnivores," *Biological Conservation,* **115,** pp. 369–78.

Butler, J.R.A. *et al.* (2014) "Dog eat dog, cat eat dog: social-ecological dimensions of dog predation by wild carnivores," in Gompper, M.E. (ed.) *Free-ranging dogs and wildlife conservation.* Oxford: Oxford University Press, pp. 117–43.

Camarós, E. *et al.* (2016) "Large carnivore attacks on hominins during the Pleistocene: a forensic approach with a Neanderthal example," *Archaeological and Anthropological Sciences,* **8,** pp. 635–46.

Carder, G. *et al.* (2016) "The animal welfare implications of civet coffee tourism in Bali," *Animal Welfare,* **25,** pp. 199–205.

Caro, T. (2010) *Conservation by proxy: indicator, umbrella, keystone, flagship, and other surrogate species.* Washington, DC: Island Press.

Carswell, L.P., Speckman, S.G. and Gill, V.A. (2015) "Shellfish fishery conflicts and perceptions of sea otters in California and Alaska," in Larson, S.E., Bodkin, J.L. and VanBlaricom, G.R. (eds.) *Sea otter conservation.* Boston: Academic Press, pp. 333–68.

Charnley, S. (2005) "From nature tourism to ecotourism? The case of the Ngorongoro Conservation Area, Tanzania," *Human Organization,* **64,** pp. 75–88.

Chasco, B.E. *et al.* (2017) "Competing tradeoffs between increasing marine mammal predation and fisheries harvest of Chinook salmon," *Scientific Reports,* **7,** p. 15439.

Cirovic, D., Penevic, A. and Krofel, M. (2016) "Jackals as cleaners: ecosystem services provided by a mesocarnivore in human-dominated landscapes," *Biological Conservation,* **199,** pp. 51–5.

Cleaveland, S., Laurenson, M.K. and Taylor, L.H. (2001) "Diseases of humans and their domestic mammals: pathogen characteristics, host range and the risk of emergence," *Philosophical Transactions of the Royal Society B: Biological Sciences,* **356,** pp. 991–9.

Crudge, B., Nguyen, T. and Trung, C.T. (2020) "The challenges and conservation implications of bear bile farming in Viet Nam," *Oryx,* **54,** pp. 252–9.

Curveira-Santos, G. *et al.* (2021) "Mesocarnivore community structuring in the presence of Africa's apex predator," Proceedings of the Royal Society B: *Biological Sciences,* **288,** p. 20202379.

Cypher, B.L. and Spencer, K.A. (1998) "Competitive interactions between coyotes and San Joaquin kit foxes," *Journal of Mammalogy,* **79,** pp. 204–14.

Dalerum, F., Selby, L.O.K. and Pirk, C.W.W. (2020) "Relationships between livestock damages and large carnivore densities in Sweden," *Frontiers in Ecology and Evolution,* **7,** p. 507.

de la Rua-Domenech, R. (2006) "Human *Mycobacterium bovis* infection in the United Kingdom: incidence, risks, control measures and review of the zoonotic aspects of bovine tuberculosis," *Tuberculosis,* **86,** pp. 77–109.

DeCesare, N.J. *et al.* (2010) "Endangered, apparently: the role of apparent competition in endangered species conservation," *Animal Conservation,* **13,** pp. 353–62.

Denny, E.A. and Dickman, C.R. (2010) *Review of cat ecology and management strategies in Australia.* Canberra: Invasive Animals Cooperative Research Center.

Deplazes, P. *et al.* (2004) "Wilderness in the city: the urbanization of Echinococcus multilocularis," *Trends in Parasitology,* **20,** pp. 77–84.

Doherty, T.S. *et al.* (2016) "Invasive predators and global biodiversity loss," Proceedings of the National Academy of Sciences *of the United States of America,* **113,** pp. 11261–5.

Dolin, E.J. (2010) *Fur, fortune and empire: the epic history of the fur trade in America.* New York: W.W. Norton.

Donnelly, C.A. and Nouvellet, P. (2013) "The contribution of badgers to confirmed tuberculosis in cattle in high-incidence areas in England," *PLoS Currents,* **5.** doi: 10.1371/currents.outbreaks.097a904d3f3619db2fe78d24bc776098

du Plessis, J. *et al.* (2018) "Past and current management of predation on livestock," in Kerley, G.I.H., Wilson, S.L. and Balfour, D. (eds.) *Livestock predation and its management in South Africa: a scientific assessment.* Port Elizabeth: Centre for African Conservation Ecology, pp. 125–77.

Ebbert, S. (2000) "Successful eradication of introduced foxes from large Aleutian Islands," in Salmon, T.P. and Crabb, A.C. (eds.) *Proceedings of the 19th vertebrate pest conference.* Davis: University of California, Davis, pp. 127–31.

Emblidge, P.G. *et al.* (2015) "Severe mortality of a population of threatened Agassiz's desert tortoises: the American badger as a potential predator," *Endangered Species Research,* **28,** pp. 109–16.

Engebretsen, K.N. *et al.* (2021) "Recolonizing carnivores: is cougar predation behaviorally mediated by bears?", *Ecology and Evolution,* **11,** pp. 5331–43.

Feeroz, M.M., Begum, S. and Hasan, M.K. (2011) "Fishing with otters: a traditional conservation practice in Bangladesh," *IUCN Otter Specialist Group Bulletin,* **28**(A), pp. 14–41.

Ford, W. (2007) *Pancha tantra.* Cologne: Taschen.

Garrote, P.J. *et al.* (2017) "Individual attributes and party affect large carnivore attacks on humans," *European Journal of Wildlife Research,* **63,** p. 80.

Gilbert, S.L. *et al.* (2017) "Socioeconomic benefits of large carnivore recolonization through reduced wildlife–vehicle collisions," *Conservation Letters,* **10,** pp. 431–9.

Givhan, R. (2019) "Fur is under attack. It's not going down without a fight," Washington Post [online], December 23, 2019 (Accessed: April 18, 2022).

Gloor, S. *et al.* (2001) "The rise of urban fox populations in Switzerland," *Mammalian Biology,* **66,** pp. 155–64.

Gompper, M.E. (2002) "The ecology of northeast coyotes: current knowledge and priorities for future research," *Wildlife Conservation Society Working Paper,* **17,** pp. 1–49.

Gompper, M.E. and Vanak, A.T. (2008) "Subsidized predators, landscapes of fear and disarticulated carnivore communities," *Animal Conservation,* **11,** pp. 13–4.

Goodrich, J.M. *et al.* (2011) "Conflicts between Amur (Siberian) tigers and humans in the Russian Far East," *Biological Conservation,* **144,** pp. 584–92.

Gordon, I.J. (2009) "What is the future for wild, large herbivores in human-modified agricultural landscapes?", *Wildlife Biology,* **15,** pp. 1–9.

Haemig, P.D. *et al.* (2008). "Red fox and tick-borne encephalitis (TBE) in humans: can predators influence public health?", *Scandinavian Journal of Infectious Diseases,* **40,** pp. 527–32.

Han, B.A. *et al.* (2021) "The ecology of zoonotic parasites in the Carnivora," *Trends in Parasitology,* **37,** pp. 1096–110.

Harrington, R. *et al.* (1999) "Establishing the causes of the roan antelope decline in the Kruger National Park, South Africa," *Biological Conservation,* **90,** pp. 69–78.

Hayward, M.W. and Marlow, N. (2014) "Will dingoes really conserve wildlife and can our methods tell?", *Journal of Applied Ecology,* **51,** pp. 835–8.

Hegglin, D., Ward, P.I. and Deplazes, P. (2003) "Anthelmintic baiting of foxes against urban contamination with Echinococcus multilocularis," *Emerging Infectious Diseases,* **9,** pp. 1266–72.

Home, C., Bhatnagar, Y.V. and Vanak, A.T. (2017) "Canine conundrum: domestic dogs as an invasive species and their impacts on wildlife in India," *Animal Conservation,* **21,** pp. 275–82.

Honey, M. *et al.* (2016) "The comparative economic value of bear viewing and bear hunting in the Great Bear Rainforest," *Journal of Ecotourism,* **15,** pp. 199–240.

Hristienko, H. and McDonald, J.E. (2007) "Going into the 21st century: a perspective on trends and controversies in the management of the American black bear," *Ursus,* **18,** pp. 72–88.

Huijser, M.P. and McGowen, P.T. (2010) "Reducing wildlife–vehicle collisions," in Beckman, J.P. *et al.* (eds.) *Safe passages.* Washington, DC: Island Press, pp. 51–74.

Huygens, O.C. *et al.* (2001) "Asiatic black bear conservation in Nagano Prefecture, central Japan: problems and solutions," *Biosphere Conservation,* **3,** pp. 97–106.

Jacobsen, L. (2005) "Otter (*Lutra lutra*) predation on stocked brown trout (*Salmo trutta*) in two Danish lowland rivers," *Ecology of Freshwater Fish,* **14,** pp. 59–68.

Jędrzejewska, B. and Jędrzejewski, W. (1998) *Predation in vertebrate communities: the Białowieża Primeval Forest as a case study.* New York: Springer Verlag.

Karanth, K.K., Jain, S. and Mariyam, D. (2017) "Emerging trends in wildlife and tiger tourism in India," in

Chen, J.S. and Prebensen, N.K. (eds.) *Nature tourism.* New York: Routledge, pp. 159–71.

Kaschner, K. and Pauly, D. (2005) "Competition between marine mammals and fisheries: food for thought," in Salem, D.J. and Rowan, A.N. (eds.) *The state of the animals III.* Washington, DC: Humane Society Press, pp. 95–117.

Kauhala, K. and Kowalczyk, R. (2011) "Invasion of the raccoon dog *Nyctereutes procyonoides* in Europe: history of colonization, features behind its success, and threats to native fauna," *Current Zoology*, **57**, pp. 584–98.

Keech, M.A. *et al.* (2011) "Effects of predator treatments, individual traits, and environment on moose survival in Alaska," *Journal of Wildlife Management*, **75**, pp. 1361–80.

Kirkwood, R. *et al.* (2003) "Pinniped-focused tourism in the southern hemisphere: a review of the industry," in Gales, N., Hindell, M. and Kirkwood, R. (eds.) *Marine mammals: fisheries, tourism and management issues.* Collingsworth: CSIRO Publishing, pp. 257–76.

Larson, S.D. *et al.* (2013) "Impacts of sea otter (*Enhydra lutris*) predation on commercially important sea cucumbers (*Parastichopus californicus*) in southeast Alaska," *Canadian Journal of Fisheries and Aquatic Sciences*, **70**, pp. 1498–1507.

Laurenson, K. *et al.* (1998) "Disease as a threat to endangered species: Ethiopian wolves, domestic dogs and canine pathogens," *Animal Conservation*, **1**, pp. 273–80.

Lesté-Lasserre, C. (2020) "Pandemic dooms Danish mink—and mink research," *Science*, **370**, p. 754.

Letnic, M., Ritchie, E.G. and Dickman, C.R. (2012) "Top predators as biodiversity regulators: the dingo *Canis lupus dingo* as a case study," *Biological Reviews*, **87**, pp. 390–413.

Li, P.J. (2004) "China's bear farming and long-term solutions," *Journal of Applied Animal Welfare Science*, **7**, pp. 71–81.

Li, Z. *et al.* (2019) "Coexistence of two sympatric flagship carnivores in the human-dominated forest landscapes of Northeast Asia," *Landscape Ecology*, **34**, pp. 291–305.

Linnell, J.D.C. *et al.* (2002) "The fear of wolves: a review of wolf attacks on humans," *NINA Oppdragsmelding*, **731**, pp. 1–65.

Livingstone, E. and Shepherd, C.R. (2016) "Bear farms in Lao PDR expand illegally and fail to conserve wild bears," *Oryx*, **50**, pp. 176–84.

Logan, K.A. and Runge, J.P. (2020) "Effects of hunting on a puma population in Colorado," *Wildlife Monographs*, **209**, pp. 1–35.

Lorenzo, C., Rioja-Paradela, T.M. and Carrillo-Reyes, A. (2015) "State of knowledge and conservation of endangered and critically endangered lagomorphs worldwide," *Therya*, **6**, pp. 11–30.

Loss, S.R., Will, T. and Marra, P.P. (2013) "The impact of free-ranging domestic cats on wildlife of the United States," *Nature Communications*, **4**, p. 1396.

Lovari, S. *et al.* (2009). "Restoring a keystone predator may endanger a prey species in a human-altered ecosystem: the return of the snow leopard to Sagarmatha National Park," *Animal Conservation*, **12**, pp. 559–70.

Ludwig, G.X. *et al.* (2002) "Otter *Lutra lutra* predation on fanned and free-living salmonids in boreal freshwater habitats," *Wildlife Biology*, **8**, pp. 193–9.

Lund, T.A. (1980) *American wildlife law.* Berkeley: University of California Press.

MacMillan, D.C. and Phillip, S. (2008) "Consumptive and non-consumptive values of wild mammals in Britain," *Mammal Review*, **38**, pp. 189–204.

Martone, R.G. *et al.* (2020) "Characterizing tourism benefits associated with top-predator conservation in coastal British Columbia," *Aquatic Conservation: Marine and Freshwater Ecosystems*, **30**, pp. 1208–19.

Miller, J.R.B. *et al.* (2016) "Effectiveness of contemporary techniques for reducing livestock depredations by large carnivores," *Wildlife Society Bulletin*, **40**, pp. 806–15.

Miller, M.A. *et al.* (2008) "Type X *Toxoplasma gondii* in a wild mussel and terrestrial carnivores from coastal California: new linkages between terrestrial mammals, runoff and toxoplasmosis of sea otters," *International Journal for Parasitology*, **38**, pp. 1319–28.

Miller, S.D., Schoen, J.W. and Schwartz, C.C. (2017) "Trends in brown bear reduction efforts in Alaska, 1980–2017," *Ursus*, **28**, 135–49.

Miranda, E.B.P. (2018) "Reintroducing apex predators: the perils of muddling guilds and taxocenoses," *Royal Society Open Science*, **5**, p. 180567.

Morisette, L., Christensen, V. and Pauly, D. (2012) "Marine mammal impacts in exploited ecosystems: would large scale culling benefit fisheries?", *PLoS ONE*, **7**, p. e43966.

Mukeka, J.M. *et al.* (2019) "Human–wildlife conflicts and their correlates in Narok County, Kenya," *Global Ecology and Conservation*, **18**, p. e00620.

National Research Council. (1997) *Wolves, bears, and their prey in Alaska: biological social challenges in wildlife management.* Washington, DC: National Academy Press.

Ng'weno, C.C. *et al.* (2017) "Lions influence the decline and habitat shift of hartebeest in a semiarid savanna," *Journal of Mammalogy*, **98**, pp. 1078–87.

Odden, J., Nilsen, E.B. and Linnell, J.D.C. (2013) "Density of wild prey modulates lynx kill rates on free-ranging domestic sheep," *PLoS ONE*, **8**, p. e79261.

Ogada, M.O. *et al.* (2003) "Limiting depredation by African carnivores: the role of livestock husbandry," *Conservation Biology*, **17**, pp. 1521–30.

Ohnishi, N. *et al.* (2017) "The influence of invasive mongoose on the genetic structure of the endangered

Amami rabbit populations," *Ecological Research*, **32**, pp. 735–41.

Ordiz, A., Bischof, R. and Swenson, J.E. (2013) "Saving large carnivores, but losing the apex predator?", *Biological Conservation*, **168**, pp. 128–33.

Ostfeld, R.S. and Holt, R.D. (2004) "Are predators good for your health? Evaluating evidence for top-down regulation of zoonotic disease reservoirs," *Frontiers in Ecology and the Environment*, **2**, pp. 13–20.

Ostfeld, R.S. *et al.* (2018) "Tick-borne disease risk in a forest food web," *Ecology*, **99**, pp. 1562–73.

Otranto, D. and Deplazes, P. (2019) "Zoonotic nematodes of wild carnivores," *International Journal for Parasitology: Parasites and Wildlife*, **9**, pp. 370–83.

Oude Munnink, B.B. *et al.* (2021) "Transmission of SARS-CoV-2 on mink farms between humans and mink and back to humans," *Science*, **371**, pp. 172–7.

Packer, S. *et al.* (2019) "Species-specific spatiotemporal patterns of leopard, lion and tiger attacks on humans," *Journal of Applied Ecology*, **56**, pp. 585–93.

Penteriani, V. *et al.* (2017) "Humans as prey: coping with large carnivore attacks using a predator–prey interaction perspective," *Human–Wildlife Interactions*, **11**, pp. 192–207.

Pirie, T.J., Thomas, R.L. and Fellowes, M.D.E. (2017) "Increasing game prices may alter farmers' behaviours towards leopards (*Panthera pardus*) and other carnivores in South Africa," *PeerJ*, **5**, p. e3369.

Pobiner, B.L. (2015) "New actualistic data on the ecology and energetics of hominin scavenging opportunities," *Journal of Human Evolution*, **80**, pp. 1–16.

Rastogi, A. *et al.* (2015) "Wildlife-tourism, local communities and tiger conservation: a village-level study in Corbett Tiger Reserve, India," *Foreign Policy and Economics*, **61**, pp. 11–19.

Raynor, J.L., Grainger, C.A. and Parker, D.P. (2021) "Wolves make roadways safer, generating large economic returns to predator conservation," Proceedings of the National Academy of Sciences *of the United States of America*, **118**, p. e2023251118.

Ripple, W.J. *et al.* (2019) "Large carnivores under assault in Alaska," *PLoS Biology*, **17**, p. e3000090.

Robards, M.D. and Reeves, R.R. (2011) "The global extent and character of marine mammal consumption by humans: 1970–2009," *Biological Conservation*, **144**, pp. 2770–86.

Rust, N.A. *et al.* (2016) "Why has human–carnivore conflict not been resolved in Namibia?", *Society and Natural Resources*, **29**, pp. 1079–94.

Shehzad, W. *et al.* (2014) "Forest without prey: livestock sustain a leopard *Panthera pardus* population in Pakistan," *Oryx*, **49**, pp. 248–53.

Sinclair, H.M. (1953) "The diet of Canadian Indians and Eskimos," Proceedings of the Nutrition Society, **12**, pp. 69–82.

Sonne, C. *et al.* (2019) "Human exposure to PFOS and mercury through meat from Baltic harbour seals (*Phoca vitulina*)," *Environmental Research*, **175**, pp. 376–83.

Støen, O.-G. *et al.* (2018) "Brown bear (*Ursus arctos*) attacks resulting in human casualties in Scandinavia 1977–2016; management implications and recommendations," *PLoS ONE*, **13**, p. e0196876.

Suraci, J.P. *et al.* (2019) "Fear of humans as apex predators has landscape-scale impacts from mountain lions to mice," *Ecology Letters*, **22**, pp. 1578–86.

Swanepoel, L.H. *et al.* (2017) "A systematic review of rodent pest research in Afro-Malagasy small-holder farming systems: Are we asking the right questions?", *PLoS ONE*, **12**, p. e0174554.

Thiel, R.P., Merrill, S. and Mech, L.D. (1998) "Tolerance by denning wolves, *Canis lupus*, to human disturbance," *Canadian Field-Naturalist*, **122**, pp. 340–2.

Thinley, P. *et al.* (2018) "The ecological benefit of tigers (*Panthera tigris*) to farmers in reducing crop and livestock losses in the eastern Himalayas: Implications for conservation of large apex predators," *Biological Conservation*, **219**, pp. 119–25.

Tortato, F.R. *et al.* (2017) "The numbers of the beast: Valuation of jaguar (*Panthera onca*) tourism and cattle depredation in the Brazilian Pantanal," *Global Ecology and Conservation*, **11**, pp. 106–14.

Van Duyne, C. *et al.* (2009) "Wolf predation among reintroduced Przewalski horses in Hustai National Park, Mongolia," *Journal of Wildlife Management*, **73**, pp. 836–43.

van Eeden, L.M. *et al.* (2018) "Carnivore conservation needs evidence-based livestock protection," *PLoS Biology*, **16**, p. e2005577.

VanWormer, E. *et al.* (2013) "*Toxoplasma gondii*, source to sea: higher contribution of domestic felids to terrestrial parasite loading despite lower infection prevalence," *EcoHealth*, **10**, pp. 277–89.

Vasyukov, D.D. *et al.* (2019) "Ancient canids of the Aleutian Islands (new archaeological discoveries from the Islands of Four Mountains)," *Quaternary Research*, **91**, pp. 1028–44.

Vial, F. and Donnelly, C.A. (2012) "Localized reactive badger culling increases risk of bovine tuberculosis in nearby cattle herds," *Biology Letters*, **8**, pp. 50–3.

Vilata, J., Oliva, D. and Sepúlveda, M. (2009) "The predation of farmed salmon by South American sea lions (*Otaria flavescens*) in southern Chile," *ICES Journal of Marine Science*, **67**, pp. 475–82.

White, L.C. *et al.* (2018) "High-quality fossil dates support a synchronous, Late Holocene extinction of devils and thylacines in mainland Australia," *Biology Letters,* **14**, p. 20170642.

White, P.C.L. *et al.* (2000) "Fox predation as a cause of lamb mortality on hill farms," *Veterinary Record,* **147**, pp. 33–7.

Williams, C.L. *et al.* (2018b) "What we have lost: domestic dogs of the ancient South Pacific," *Annual Research and Review in Biology,* **25**, pp. 1–11.

Williams, S.T. *et al.* (2018a) "Predation by small mammalian carnivores in rural agro-ecosystems: An undervalued ecosystem service?", *Ecosystem Services,* **30**, pp. 362–71.

Witczuk, J., Pagacz, S. and Mills, L.S. (2013) "Disproportionate predation on endemic marmots by invasive coyotes," *Journal of Mammalogy,* **94**, pp. 702–13.

Yirga, G. *et al.* (2015) "Spotted hyena (*Crocuta crocuta*) concentrate around urban waste dumps across Tigray, northern Ethiopia," *Wildlife Research,* **42**, pp. 563–9.

Zapata-Rios, G. and Branch, L.C. (2018) "Mammalian carnivore occupancy is inversely related to presence of domestic dogs in the high Andes of Ecuador," *PLoS ONE,* **13**, p. e0192346.

CHAPTER 12

How humans affect carnivorans

The most influential community interactor with
carnivorans—the species that has had the strongest
effect on the distributions and abundances of carnivoran species for centuries or millennia—is *Homo
sapiens*. In Chapter 11 I showed that carnivorans
have affected human ecology throughout our evolution and continue to do so today. The reciprocal effects have been incomparably stronger for
millennia, with many carnivoran species expanding or contracting drastically in distribution and
abundance, while humans steadily increased in
numbers and footprint. The topics covered in this
chapter subsume the documented positive and negative effects exerted by humans on carnivoran
behaviors, physiology, distributions, abundances,
and species persistence. They include conservation
efforts, and the results of those efforts, weak or
strong. Clearly, the breadth of this chapter warrants multiple book-length treatments, and many
excellent ones exist. Some treat the Carnivora in
broad phylogenetic context (e.g. Fascione *et al.*,
2004; Purvis *et al.*, 2005). Others focus on the Carnivora (Gittleman 1996) or taxonomic subsets: felids
(Macdonald and Loveridge, 2010), ursids (Penteriani and Melletti, 2021), canids (Sillero-Zubiri *et al.*,
2004), and musteloids (Macdonald *et al.*, (2017). Various treatments of high-profile species are valued
resources for conservationists (e.g. Hayward and
Somers, 2009). Clearly, I cannot distill this extensive literature, which tends toward the taxon- and
site-specific.

Instead, my goal is to create historical, taxonomic,
and geographic context for how humans affect carnivorans and describe the major mechanisms and
relative strength of their effects. I compare those
effects with those exerted on other mammalian

orders, leading to a survey of how humans attempt
to protect the products of evolution—conservation.
This is one of the most intensively studied carnivoran topics—of the about 15,000 Web of Science-
indexed publications with the topic "carnivore"
in 2021, 25% dealt specifically with conservation,
and many others have conservation implications.
Clearly, the scientific community considers conservation and carnivore biology to have strong conceptual overlap. I survey the forms of the conservation
enterprise, ranging from the traditional and holistic
to the technology-based and invasive, concluding
with one of the most recently recognized major
threats to carnivoran biodiversity—global climate
change.

12.1 The state of the carnivoran fauna

How widely distributed and abundant are carnivorans, relative to before or in the absence of human
influence? The first complication in this discussion
is the difficulty of knowing when and where human
influence began. Pre-human hominins are alleged to
have caused carnivoran extinctions 200,000–300,000
years ago. The inferential approaches available to
test such hypotheses are limited, and so the onset
of human effects is difficult to date (Faith *et al.*,
2020). Through prehistory and history, human influence has depended on the technology available, the
economics of human living styles, and recently on
cultural constraints in the form of decrees, statutes,
and treaties. In the twenty-first century, carnivorans
make up a tiny fraction of Earth's biomass, of its
vertebrate biomass (Figure 12.1), and its biomass of
wild mammals (Bar-On *et al.*, 2018). Carnivorans in
the Americas were greatly diminished during the

Carnivoran Ecology. Steven W. Buskirk, Oxford University Press. © Steven W. Buskirk (2023). DOI: 10.1093/oso/9780192863249.003.0012

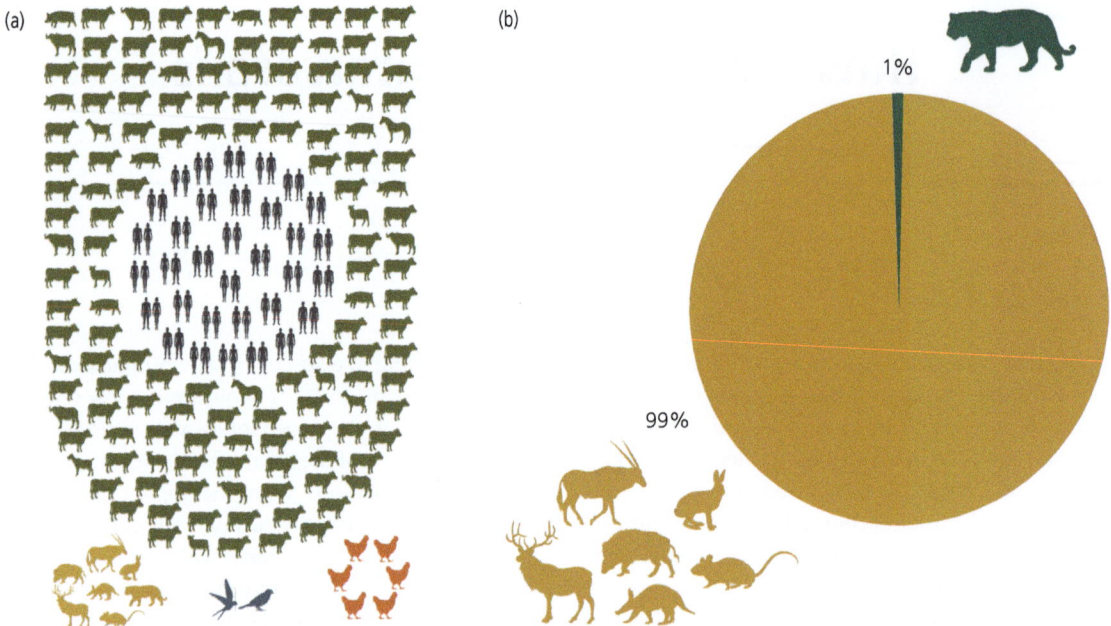

Figure 12.1 The biomass distribution of endothermic vertebrates on Earth. A: Humans (black) and their mammalian livestock (green), mostly pigs and cattle, make up more than twenty times the biomass of wild mammals (orange). Domestic birds (red) make up three times the biomass of wild birds. Reptiles and amphibians make up a tiny fraction and are not depicted (Bar-On et al., 2018). B: Wild carnivorans (green) make up about 1% of the biomass of their prey on land (orange) (estimated from Carbone and Gittleman, 2002).

Late Pleistocene; Bar-On and colleagues (2018) estimated that the biomass density of large (>44 kg) mammals on land 50,000 years ago was sevenfold greater than today. Among the most endangered large carnivorans, median loss of geographic range has been about 60%, with values for some species of over 90%. Earth has relatively few carnivoran individuals, especially large-bodied ones, and declines in abundance and geographic range have accelerated in recent decades (Estes *et al.*, 2011). Although few carnivoran species have become extinct during the historical period (Box 12.1), loss of named subspecies and geographic ranges has been severe. However, not all trends follow this pattern; some carnivorans, including a few large-bodied ones, increased in abundance, or expanded in distribution during 1980–2020. These two opposing, but by no means countervailing, patterns frame the discussion of carnivoran conservation—how they are being removed and lost from lands and waters and how they are returning to formerly occupied ranges, via human efforts and on their own.

12.2 Mechanisms of effects

The idea that humans could exert strong effects on entire species—even leading to extinction—extends back only a century or two. The extinctions or near-extinctions of the dodo, great auk, and American bison became persuasive examples of humans as agents during the historical period, but extinctions caused by prehistoric humans were slower to be recognized as plausible (Koch and Barnosky, 2006). Two lines of evidence became most suggestive. First, the large body sizes of extinct taxa suggested selective human exploitation; small-bodied birds and mammals did not show such high rates of disappearance from the fossil record. Second, the temporal coincidence between the disappearances of taxa and the abrupt arrivals of humans—Neolithic and more modern—was precise (Brook and Bowman, 2002). Regions where humans arrived suddenly included the Americas, Australia, and many oceanic islands. By contrast, across Eurasia the extinction pulse was weaker, consistent with gradual human evolution and colonization. Africa, where humans coevolved with wild vertebrates, showed no Late Pleistocene extinction pulse in its zooarchaeological record (Pushkina and Raia, 2008). The roles of humans in Pleistocene–Holocene large mammal extinctions have been debated vigorously for decades, and current thinking grants humans a central role. Multiple factors and interactions are now considered important, including rapidly changing climates, human alteration of habitat, novel diseases, competition between humans and carnivorans for food, and overhunting per se. Braje and Erlandson (2013) depicted humans as having accelerated extinction rates for plants and animals across a 50,000-year period of human expansion and technological development, extending to the present. Even in some contemporary cases, although humans are recognized as negatively affecting carnivoran populations, the mechanism of action is unclear. Pinnipeds in the North Pacific Ocean present such an example. The harbor seal, northern fur seal, and Steller sea lion all underwent population collapses during 1970–2000, presumably from human-mediated causes. However, whether bottom-up (competition from fisheries harvests) or top-down (killer whale predation) forces were primary remains controversial (Wade *et al.*, 2007; Estes *et al.*, 2009).

The other consideration in attributing carnivoran decline and loss to specific factors is the relative vulnerability of taxa. Large-bodied carnivorans are widely recognized as vulnerable, because of low population densities and low intrinsic rates of increase. However, Jachowski and colleagues (2021) inferred that the smallest carnivorans—weasels—have declined in apparent abundance by around 90% in North America in the years 1960–2020. The causes of the declines are not clear but include several putative factors discussed in this chapter. They are mirrored by suggestive trends in Europe and illustrate that large body size may merely focus our attention on visible and charismatic species, rather than selectively predisposing them to effects. Many covariates also affect vulnerability; Gonzalez-Voyer and colleagues (2016) found that carnivorans with relatively large brains were more predisposed to rarity or extinction. This was not attributed to direct effects, but to constraints on life histories imposed by large brains. In effect, the selective pressures that favored relatively large brains over evolutionary history are not adaptive in the current human-dominated environment.

12.3 Direct mortality: poisoning, hunting, and trapping

Hunting and trapping have had highly variable effects on carnivoran populations. At one extreme, hunting for sea otters eradicated them from most of the northwest coast of North America during the period 1745–1900 (Figure 12.3). Russian fur hunters were the most proficient otter hunters, having learned from Aleuts and other indigenous peoples. Once the tremendous value of otter pelts in the Far East was recognized, various countries sent expeditions to the northwest coast of North America, where their agents used hunting, trading, and coercion to collect pelts. The result was the reduction of the species' numbers from an estimated 100,000–300,000 to fewer than 2000 scattered across several populations, and the contraction of the geographic range by over 90% (Kenyon, 1969; Beichman *et al.*, 2022). In this case, a global market for a luxury item created incentives to harvest the sea otter to near extinction.

By contrast, the coyote in western North America has undergone little or no year-to-year change in abundance in the face of extensive hunting and trapping over the last several decades. Economic incentives are too weak to warrant greater effort, the use of toxicants is regulated, and the species is ecologically adaptable. For decades it was thought that one mechanism underlying this resilience to human harvest is compensatory reproduction, especially if dominant members of social groups are selectively removed. However, intensive modern studies have failed to confirm these beliefs. One study in South Carolina, US showed only weak compensatory effects on reproduction (Kilgo *et al.*, 2017); fecundity was actually lower during the trapping phase than it was before. Compensatory effects have been shown for the black-backed jackal, an African coyote-analogue and the primary cause of small livestock predation in southern Africa. Minnie and colleagues (2016) found that removal of

Box 12.1 The Barbary lion

The now-extinct Barbary lion is recognized as a lineage distinct from southern and eastern African forms and from Asian populations, but lacking an accepted subspecific name (Dubach *et al.*, 2013). This was the lion known to ancient Egyptians, Greeks, and Romans, and the first lion described by European scientists. Ranging across the sub-Mediterranean coast to the Atlas Mountains, this lion was frequently captured and displayed in European zoos and other exhibits, making it familiar to the western world. It became rare by the eighteenth century and extinct by 1980 (Black *et al.*, 2013, (Figure 12.2). Attempts to recover the lineage via advanced genetic engineering now are being considered.

Figure 12.2 *La Dernier Image*, by Walton Ford, 2018. Ford's watercolor interprets the last known photograph of a Barbary lion, taken by Casablanca-based Marcelin Flandrin from a biplane in the Atlas Mountains in 1925. He published the photo on a postcard that sold widely and became iconic across Europe. Although the Barbary lion subspecies was rare by then, a few individuals may have persisted until the 1970s (Black *et al.*, 2013).
Courtesy of the artist and Kasmin Gallery.

Box 12.1 *Continued*

Figure 12.2 *continued*

jackals on farms lowered the median age from 5–6 years to 2–3 years, and increased ingress, pregnancy rates, and litter sizes for young females, relative to populations not subjected to control. Density-dependent recruitment reduced the effectiveness of control efforts.

Between the extremes framed by sea otters and coyotes, most carnivorans hunted or trapped in modern times fall into an intermediate realm, able to withstand hunting and trapping under data-based management and effective enforcement, but potentially vulnerable otherwise. The smallest

Figure 12.3 Adult and juvenile sea otter pelts, Unalaska Island, Alaska, 1892. Sea otters were nearly eradicated as a species by hunting for furs, which continued their population decline until the early twentieth century.
Photo: Wikimedia Commons CC.

carnivorans (i.e. weasels) have too little value and are too difficult to capture to be at risk. The largest carnivorans (i.e. pinnipeds, large cats) have high value and are inherently vulnerable to overharvest. Industrial sealing accounted for high mortality and near depletion of populations of several pinniped species in northern and southern oceans from the eighteenth to the early twentieth centuries. However, these species recovered to varying degrees under protection from treaties and other restrictive frameworks (Bonner, 1968; Paterson, 1977; Busch, 1985). Cultural factors are important as well. Ikanda and Packer (2008) contrasted the effects of retaliatory vs. ritual killings of lions by Maasai pastoralists in Tanzania. The former are carried out in response to livestock depredations by nomadic lions outside of protected areas, whereas the latter are clandestine rites of passage for some young Maasai males. The authors concluded that ritual killings, although difficult to monitor and prevent, are likely trivial compared with the potential effects of retaliatory killings as the pastoral human footprint grows.

The degree of selectivity—for species, for age class, for sex, and for other traits—varies across lethal means of taking in ways that are suggested by Montgomery and colleagues (2020, Fig. 1). Vehicle collisions and poisoning are less selective, whereas hunting for trophies can be highly selective. Population-level effects depend on a host of factors: reproductive rate, market value of the carcass, habitat occupied, and whether the species can be attracted to baits. Laws and treaties dealing with trafficking and enforcement of those rules play important roles in some places and for some species, but not others. Osuri and colleagues (2020) reviewed the relative effects of hunting vs. forest modification on tropical birds and mammals worldwide. They found that hunting had stronger effects on carnivorous than herbivorous mammals, and that large body size predisposed to stronger negative effects of hunting generally, but especially among carnivorous mammals. This contrasts with a view commonly held that wildlife is less affected by hunting than by habitat loss in the north temperate zone (Heffelfinger et al., 2013); the reason for this difference is unclear.

Poisoning has long been problematic for carnivoran conservation, and it is still used today across most of the world in various forms. These range from legally used target-specific toxicants in ways that reduce secondary and non-target effects to illicitly used banned broad-spectrum poisons. It is difficult to isolate how poisoning of herbivores secondarily affects carnivorans, because carnivorans can be intoxicated incidentally, or affected via reduced prey availability. Studying these effects is challenging because poisoned animals tend to die undetected. The literature presents multiple examples of toxicant-related effects with uncertain mechanisms. Multi-decadal poisoning of plateau pikas on the Qinghai–Tibetan Plateau of China caused a nearly hundredfold reduction in pika abundance and a tenfold reduction in carnivoran tracks on transects (Badingqiuying et al., 2016); however, the mechanism of the carnivoran decline was unclear. In California, Gabriel and colleagues (2012) reported on necropsies of fisher carcasses collected opportunistically. Of fifty-eight fishers examined, 79% showed exposure to anticoagulant rodenticides, including four animals inferred to have died from toxicosis. This was unexpected considering that the fishers sampled came from regions mostly comprising national forests and national parks, where toxicant use was regulated. Whether the animals were exposed via ingesting prey or poisoned baits was not clear.

Most continents have long histories of intentionally poisoning mammalian predators. Early toxicants were plant- and animal-based, with low target-specificity; more modern ones have been synthetic, with greater specificity. In southern Africa, European colonists introduced an "extermination mentality" against predators and scavengers (Ogada, 2014), employing poisoned baits as early as 1660. By the late nineteenth century, efforts to eradicate carnivorans of various species were government organized and funded, with incentives for private participation. In other parts of Africa where European-style ranching was introduced, the use of poisons came with it. Strychnine, an alkaloid extract of various tropical plants, was the most common agent, and was used in military conflicts and for political purposes in various African countries. The wide availability of strychnine contributed to regional losses of African wild dogs, lions, spotted hyenas, and cheetahs. Black-backed jackals were particular targets where small livestock (i.e. sheep and goats) were raised (Ogada, 2014). These practices have continued to the present and even may have intensified in recent decades as human populations have grown. South America shows a similar pattern—widespread use of poisons to control carnivorans before 1980, followed by the abandonment of many ranches and recolonization by pumas and their native prey. More recently, ranching has recovered, and pumas now depredate livestock more heavily. A spate of poisoning cases was reported from Argentina and other countries during the 2010s, with important implications for pumas and scavenging species (Pauli *et al.*, 2018).

The US had a similar history of toxicant use against predators until a 1972 executive ban, which restricted uses on public lands. Around the same time, environmental laws were passed that subjected poisons to strict labeling requirements; they could be used only by trained and certified personnel, against particular species, and in specific ways. The result has been population recoveries of some carnivoran species on public lands in the western US since 1972. Wolverines, swift foxes, and other species have all increased their geographic ranges markedly, in part because of reduced and regulated use of toxicants. Public sentiments about predator control continue to evolve; in 2002 the State of New Mexico enacted a new statute prohibiting all trapping, snaring, and poisoning of wildlife on public lands—47% of its area. Predator poisoning is an ongoing issue in western Europe. Monitoring efforts in France, Italy, and Greece show widespread use of illegal toxicants against wild carnivorans (Berny *et al.*, 2015; Chiari *et al.*, 2017; Ntemiri *et al.*, 2018). Clearly the licit and illicit use of poisons against or around carnivorans continues to be a factor in conservation (Lennox *et al.*, 2018).

12.4 Other agents of direct mortality

Section 10.1 showed that vehicle collisions could be major causes of carnivoran deaths, although sampling bias may affect this generalization. Roadway mortality could be high because prey are concentrated on roadside vegetation. Surface runoff of water from paved roads, using salt on roads, or treating vegetation on roadsides can increase abundances of certain prey and lead to increased carrion on the roadway itself. Silva and colleagues (2019) showed this effect for Egyptian mongooses and stone martens in Portugal. In effect, roadways can act as ecological traps, attracting carnivorans to locally high prey or carrion densities, then causing disproportionate mortality. However, patterns across studies are not consistent. Major highways also pose barriers to movement that result in population fragmentation (Section 10.3.3). The combined effects of roads on carnivoran conservation—vehicle collisions, genetic barriers, avoidance, and attraction—are complex but mostly negative for threatened carnivorans. Road construction is expected to increase explosively in developing countries in the twenty-first century, further reducing wild habitats available to some large carnivorans (Carter *et al.*, 2020).

12.5 Physical exclusion

Predator-proof fencing is generally too costly to build and maintain to be effective, except around small concentrations of livestock or other valued resources. Bomas made of thorny shrub stems have been traditional predator barriers used by indigenous peoples across Africa (Lichtenfeld *et al.*, 2015). Used for overnight protection of sheep, goats, and cattle, they may be fortified with artificial materials (Mkonyi *et al.*, 2017) and are cost-effective. Predator-resistant fencing also is common around poultry production; it likewise involves small land areas and little habitat loss. The same is true for fish hatcheries, where fencing or netting to exclude minks, otters, and avian predators is common; sea cages that enclose farmed salmonids are an extension of that practice. The most extensive, costly, and controversial predator-exclusion fence is that used to control dingo predation on livestock in southeastern Australia (Figure 12.4). The fence, which was about 5600 km long in 2020, was built beginning in the 1880s to eventually enclose 820,000 km^2 of land. Poisoning and shooting maintain low dingo densities in the fenced area, which experiences lower sheep depredation losses than areas north of the fence. The cost-effectiveness and ecological effects of the fence are debated vigorously in the scientific and popular presses (e.g. Levy, 2009; Letnic and Koch, 2010). At a much smaller scale, barriers are used for conservation purposes where invasive predators endanger native birds and mammals. The practice is most common in New Zealand, where dozens of fenced enclosures protect areas of less than 10 km^2 each. These fencing enclosures generate controversy as well, primarily because of their high costs (Scofield *et al.*, 2011). Carnivoran-exclusion fencing for conservation likewise sees limited application in Hawaii (Young *et al.*, 2013).

Figure 12.4 The dingo fence (red, in inset) in southeastern Australia extends for about 5600 km and was constructed to exclude dingoes from livestock production areas. South of the fence, dingoes are subjected to lethal control and are uncommon or absent. The fence has reduced depredation on livestock, but also results in increased abundances of macropodids, emus, and invasive European rabbits, which have unwanted ecological effects.
Photo: fotofritz/Can Stock Photo.

12.6 Effects on behavior

Behavior is the main inducible defense used by mammals to reduce risks posed by humans (Montgomery *et al.*, 2020) and humans have pervasive and well-documented effects on many carnivoran behaviors, most of which have the potential to affect trophic ecology, demography, or fitness. Most, but certainly not all, wild carnivorans avoid human voices, human-generated sounds, human odors, and other indicators of human presence. Behavioral responses to human presence are similar to prey reactions to predators, in that they fall into categories of reactive and anticipatory. Smith and colleagues (2017) showed that playing recordings of human voices caused pumas in the southwestern US to react by fleeing a feeding site more quickly, taking longer to return, and spending less total time feeding than playing animal vocalizations—tree frog calls—with no fitness implications. Ditmer and colleagues (2021) showed that artificial light level was the strongest predictor of where pumas preyed on mule deer at the urban–wildland interface. European badgers in Britain reacted negatively to recorded human sounds, although they did not react similarly to recorded sounds of large carnivorans long extinct in Britain (Clinchy *et al.*, 2016). Badgers feared human sounds more than those of domestic dogs or bears, and much more than those

of wolves, showing that the extinction of wolves has altered recognition of that threat in British badgers.

Roads and vehicles provide abundant cues that allow carnivorans to avoid or be attracted to them, and negative carnivoran responses roads are well documented (e.g. Lamb *et al.*, 2018). Two European studies have shown greater avoidance of roads by mustelids than by the red fox, but whether this represents mustelid avoidance of a larger competitor or taxonomic differences in responses to roads per se is not clear (Ruiz-Capillas *et al.*, 2013; Planillo *et al.*, 2018). Some large carnivorans travel parallel to roads or on road surfaces under darkness or with light traffic (Forman and Alexander 1998). For example, wolves in western Canada travelled two to three times faster on linear structures than through native forest, but the strength of preference varied with the type of structure. Wolves preferred long, straight stretches with few physical or visual barriers (e.g. railways, conventional seismic lines, and pipelines) more than narrow, sinuous structures with physical or visual barriers. They also preferred roads with light traffic (Dickie *et al.*, 2017). In some cases, a carnivoran responds oppositely to multiple metrics of proximity to humans. The Molina's hog-nosed skunk in central Argentina occurred less frequently farther from villages, but also less frequently near roads (Zanón Martinez *et al.*, 2021).

Not surprisingly, carnivorans preying on human-sourced foods move differently than those preying on wild ungulates. Brown bears in northeastern Turkey showed two movement types—bears that fed at refuse dumps were nearly sedentary, whereas those that never fed at dumps moved a mean of 166 km before winter denning in search of food (Cozzi *et al.*, 2016). Wolves in the Apennine Mountains of Italy fed mostly on human refuse and livestock carcasses in the 1970s, but various measures eased the recovery of roe deer, wild boar, and red deer over the following forty years. This did not have the expected effect of shifting wolf activities away from human-dominated areas and human-sourced foods and has not reduced wolf-livestock conflicts (Ciucci *et al.*, 2020).

Smith and colleagues (2015, 2017) showed that pumas killed prey at higher rates in urbanized areas than elsewhere, but that feeding times in urban areas were reduced and more interrupted than when away from human activity. In effect, pumas in urban settings used prey carcasses less efficiently near humans, requiring them to kill more prey than otherwise. The net effect on ingestion rates was not clear, suggesting a trade-off between prey availability and efficiency of carcass consumption. This possibility is consistent with the findings by Blecha and colleagues (2018) that pumas in a Colorado study area found high prey availability in urban settings but avoided those places unless food-stressed. Coon and colleagues (2019) also found mixed effects of urbanization on the health of pumas; those in peri-urban environments showed higher body condition scores than those within cities or in wilderness areas beyond the peri-urban belt. A similar pattern has been shown for coyotes living around Edmonton, Canada. Roughly half of the study animals appeared to be healthy, whereas the others were visibly infested with *Sarcoptes* mites (Murray *et al.*, 2015). Mange-infested animals lived in more developed areas, maintained larger home ranges, and assimilated less protein than those that appeared to be healthy. It is not clear whether ill animals seek foods from humans, or animals that seek foods from humans become more ill. So, urbanization affects carnivorans variably on different axes.

Anticipatory responses of carnivorans recognize patterns of human behavior and reduce contact in more general ways. Nocturnality is the best studied of these reactions; across mammalian species there is a trend toward increased nocturnality in response to a wide range of human activities. Carnivorans respond largely as do other mammals, with large-bodied species showing stronger nocturnality responses than small-bodied ones (Gaynor *et al.*, 2018). Exceptions to this pattern have been noted, however. Frey and colleagues (2020) studied the activity responses of carnivorans to human disturbance and other factors in western Canada and found that American martens showed stronger nocturnality in less disturbed than more disturbed landscapes. Coyotes were crepuscular in the low-disturbance landscape but cathemeral in high-disturbance ones, and ermines were more diurnal in disturbed landscapes. Carnivorans in Croatia showed more complex patterns of nocturnality in relation to intraguild interactions. Wolves, Eurasian lynx, and red foxes were all nocturnal or crepuscular, thereby avoiding peak human activity times, but avoided each other to the extent predicted by likelihood of interspecific conflict—human presence mediated the nocturnality effect of wolves and lynx on foxes (Haswell *et al.*, 2020). Brown bears in Norway moderated their foraging times according to risk of being killed by hunters. Hunting mortality was highest during early morning hours, but bears could not afford to remain inactive all day because of their autumnal hyperphagy, which coincided with the hunting season. Instead, they shifted their foraging, primarily for berries, to afternoon, except for some foraging on poor-quality patches in morning (Hertel *et al.*, 2016). This represents a clear trade-off between risk of starvation vs. risk of being killed by hunters. Ordiz and colleagues (2012), also working in Norway, explored activity shifts across sex–age classes, and showed that females with cubs of the year, which were protected from hunting, modified their movement patterns much less than solitary bears. This raises interesting questions about how information regarding risk from humans is conveyed across population segments with variable vulnerability. Reduced foraging to avoid humans has clear implications for the prey of the affected carnivoran. Suraci and colleagues (2019) showed that playing recorded sounds of humans reduced diurnal activity of bobcats and total activity of striped skunks, but had opposing effects on deer mice, increasing their use of space and foraging activity. Deer mice were typical prey of small carnivorans in this Californian study area.

Abundant evidence shows that humans mediate community interactions involving carnivorans. In Chapter 7 I presented several examples of mechanisms for these effects (Sections 7.7.7, 7.8); however, many more studies show correlations between occurrence of two carnivoran species with no known mechanism. In recent years, remote camera studies repeatedly have documented such effects; for example, Zanón Martinez and colleagues (2021) showed that puma occurrence was associated with reduced overall occurrence of Molina's hog-nosed skunk and increased proximity of skunk occurrences to villages, suggestive of skunks seeking proximity to villages to avoid contact with pumas. This resembles the findings of Yovovich and colleagues (2021) regarding human activity protecting deer from puma predation.

12.7 Habitat- and prey-mediated effects

"Habitat" is one of the most widely used, yet diffuse, concepts in wildlife biology, and how human-modified habitats affect carnivoran behavior is difficult to distill. Humans modify habitats in highly diverse ways, and the level of modification ranges from slight (e.g. foot trails passing through protected wilderness areas) to profound (e.g. heavy urbanization). Urbanization is highly multi-factorial, including road density, noise levels, artificial light, structural changes, food subsidies, and other factors, so that isolating individual habitat effects is challenging. Clearly, some carnivorans tolerate human-modified habitats more than others; red foxes have occupied city centers over much of the north temperate zone in recent decades, whereas wolverines require large tracts of forest or tundra with high wilderness character. Blaum and colleagues (2007) showed that African wild cats, striped polecats, cape foxes, and suricates were negatively affected by increases in shrub cover in the Kalahari. Shrub encroachment is a general response to overgrazing by livestock in the region. By contrast, yellow mongooses, bat-eared foxes, and small-spotted genets showed peak abundances

at intermediate shrub densities, and black-backed jackals showed no response. Carnivoran responses to shrub encroachment and most other human-caused habitat changes are taxon specific.

Various effects mediated by prey are known. Serrouya and colleagues (2017) described a system in British Columbia, Canada, in which apparent competition was experimentally reduced via hunting. Here, the caribou is a species of conservation concern; moose are historically recent arrivals associated with shrub vegetation, and now represent the primary, but less-preferred, prey of wolves. Managers increased hunting harvest of moose, resulting in a 70% decline in numbers, and wolf numbers decreased as well. Annual adult survival of caribou increased from 0.78 to 0.88. In this case, experimental reduction of its primary prey led to reduced carnivoran abundance and improved prospects for an endangered, more preferred prey.

Human transport of plant species into new environments creates opportunities for plant invasion, which also affects carnivorans. The invasive grass *Arundo donax* forms dense stands in many areas worldwide and was studied by Hardesty-Moore and colleagues (2020) in relation to coyote and bobcat occupancy in California. They found that small mammals were positively, but carnivorans negatively, associated with *Arundo* stands, which they attributed to prey sheltering in the dense monotypic stands, reducing the value of those stands as hunting sites for carnivorans.

12.8 Disease-related effects

Earlier I described how pathogen life cycles mediate or are influenced by some aspect of carnivoran ecology (Section 6.3; Section 11.3.1). However, a more basic question recognizes that carnivorans have always lived with infectious diseases. How have humans made the mortality and morbidity of wildlife diseases worse or better? Again, domestics—dog and cat populations particularly—are key factors. By being human-subsidized and protected from many stresses of living wild, they represent large reservoirs for enzootic diseases and are able to transmit pathogens to wild carnivorans and between wild carnivorans and humans. A second avenue by which humans affect carnivoran diseases is by transporting pathogens along with wildlife. Pavlin and colleagues (2009) showed that of the hundreds of thousands of mammals imported to the US during 2000–2005 the two taxonomic families with the greatest potential for carrying zoonotic disease organisms were carnivoran. Historically, this explains canine distemper and sylvatic plague introduced to the Americas, and Aleutian mink disease introduced to Europe (Mañas et al., 2001). The same process occurs at smaller spatial scales. Translocating nuisance carnivorans is recognized as a major agent of disease spread in Eurasia (Chipman et al., 2008), but other kinds of translocation—for stocking game populations, for sale as pets, and unintentionally—serve the same function, spreading local disease outbreaks.

A third major mechanism by which humans affect disease prevalence and virulence is via nutrient enrichment—increasing concentrations of elements that normally limit community functions. The most prevalent of these are additions of N and P to soils and waters in agricultural landscapes. Effects of nutrient enrichment are best documented for aquatic and wetland systems and can take two forms: N and P can increase to the point of toxicity (Hernández et al., 2016), or nutrients can facilitate disease transmission (Strandin et al., 2018). Transmission can be affected via increases in host densities, changes in host specificities, and other pathways (Johnson et al., 2010). Most of the documented effects of toxicity on animals have involved aquatic or wetland systems but occur more broadly. Effects of nutrient–disease links are known from various vertebrates, but no carnivorans to date. They seem most likely to appear in species that concentrate at feeding sites for multiple carnivoran and prey species, and in agricultural landscapes. Rohr and colleagues (2019) estimated that over 50% of all zoonotic diseases that emerged in humans since 1940 were associated with agricultural drivers—an effect that is expected to grow along with the human population in coming decades.

A last mechanism by which humans can affect disease ecology of carnivorans is via diseases of their prey. Several human-influenced disease outbreaks in herbivores are known or suspected to affect carnivoran predators. One is the catastrophic die-off of vicuñas (68% reduction) and guanacos (77% reduction) in San Guillermo National Park, Argentina, over two years in the 2010s. Vicuñas were the nearly exclusive prey of pumas and foods of culpeo foxes in that region, and the die-off affected carnivorans accordingly (Monk et al., 2022). The vicuña decline, due to sarcoptic mange, is suspected to have been initiated by the introduction of llamas near the park boundary (Ferreyra et al., 2022).

12.9 Recolonizations, reintroductions, and restorations

Recolonizations represent a hopeful trend monitored and facilitated by conservationists where possible. These are caused by humans in the sense that they typically reflect the relaxation of some previous negative human influence. Some of the most dramatic examples have been recoveries of pinnipeds following the end of industrial sealing (and other factors, Hindell et al., 1994) and include the return of several species of fur seals to breeding sites in the southern oceans (Section 10.3.4), gray seals to northeastern North America (Wood et al., 2020), and northern elephant seals to areas from which they had been absent for decades (Lowry et al., 2014). On land, some of the most dramatic recolonizations by large carnivorans have occurred in Europe, where brown bear, wolf, and Eurasian lynx populations have expanded in geographic range over several decades (Boitani and Linnell, 2015). The giant otter has reappeared in Argentina for the first time in decades (Figure 12.5), which represents either a range expansion of over 1000 km by waterway or the presence of unrecognized relict populations (Leuchtenberger and di Martino, 2021). Other examples are being reported from various continents (Lamb et al., 2020), including wild dogs in northern Kenya (Woodroffe, 2011), pumas in South America (Pauli et al., 2018), and European pine martens in Scotland (Sheehy et al., 2018).

Importantly, the vocabulary used in this subject area is nuanced. "Recolonization" refers to reoccupation of a former range via dispersal, home-range establishment, and reproduction. "Reintroduction" describes human transport and release of animals, and "augmentation" denotes releases into areas where the species still occurs, but at low abundance.

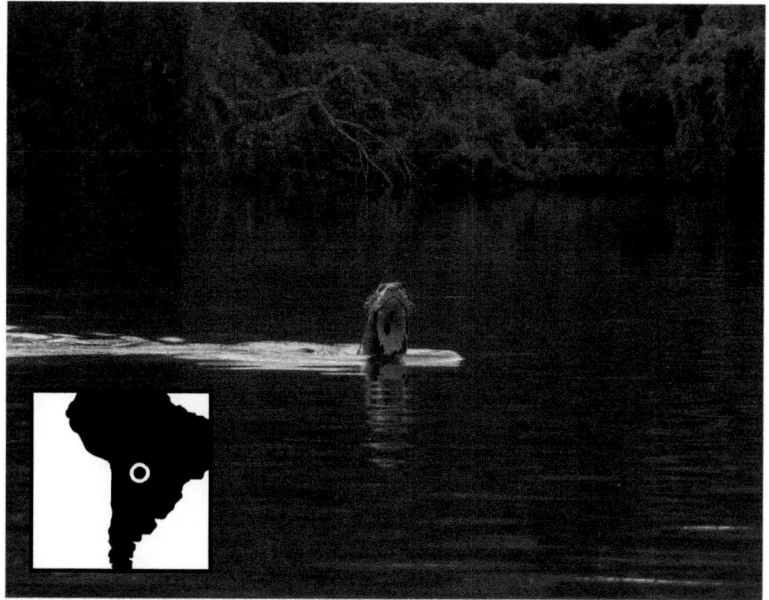

Figure 12.5 A giant otter photographed in El Impenetrable National Park, Argentina in 2021. This was the first such documentation of the presence of the species in Argentina since the 1980s. Photo: Gerardo Cerón, Fundación Rewilding Argentina.

"Genetic rescue" is a special case of augmentation in which a translocation attempts specifically to correct the effects of inbreeding and inbreeding depression. "Restoration" and "rewilding" tend to describe broader efforts of facilitating the success of reintroductions. These efforts can include reducing livestock numbers in a release area, increasing wild herbivore numbers to support carnivorans, educating the public, and facilitating nature-based tourism industries. Disease control is conducted in a few cases.

Some early reintroduction efforts were remarkably successful. The sable is the most widely reintroduced carnivoran, having been trapped to scarcity over much of boreal Eurasia by 1600; it remained rare for three centuries. In the years 1940–1965 more than 19,000 animals were translocated into areas of the Soviet Union with few or no sables, repopulating 1,400,000 km^2 (Bakeyev and Sinitsyn, 1994). Fortuitously, the effort was benefitted by the Great Patriotic War of 1940–1945, which drew many young men away from trapping and into military service early in the reintroduction program. The first documented reintroduction of a large carnivoran for conservation purposes was of the brown bear to Białowieża Forest, Poland and Belarus, beginning in 1937 (Samojlik *et al.*, 2018);

translocations were supplemented by dispersal and recolonization. Several other restoration measures were key to this success. First, the occasional prey of bears recovered over decades of effort, much of it under hunting management frameworks. Ungulate populations had been severely overharvested during the nineteenth century, and twentieth-century efforts to increase the sustainability of hunting also benefitted large predators. Second, deforestation resulting from nineteenth-century practices was reversed by afforestation, following declines in pastoralism. The human population migrated from small rural parcels to towns and cities, and agriculture became more centralized, reducing subsistence behaviors and competition with wild carnivorans. Third, cultural norms became more accepting of the presence of large carnivorans. For the brown bear and other recovering carnivoran species in Europe, translocation is only a part of broader, longer-term efforts that are combined with changing human attitudes and patterns of land occupancy.

A more recent example is provided by the restoration of the jaguar to the Iberá Wetlands. Jaguars had been reduced to 5% of their pre-colonial geographic range in Argentina, and the prey community was likewise depopulated. Prior to any reintroductions, land was purchased by a conservation organization, donated to the federal government, and a large area given protected-area status. Some potential prey species (e.g. capybara) were allowed to recover but others (e.g. Pampas deer, giant anteater, and collared peccary) required reintroduction, in some cases using captive breeding of animals confiscated after being taken from the wild for pets. Wildlife law enforcement was strengthened, and local communities were persuaded that wildlife-based tourism could become an economic asset. A breeding program for jaguars provided stock for releases, but animals required behavioral conditioning to increase survival in the wild. The first releases of jaguars took place in 2020–2021 (Donadio et al., 2022). In such cases as brown bears in eastern Europe and jaguars in Argentina, "reintroduction" does not convey the scale and complexity of the effort required to restore large carnivoran populations.

One of the most remarkable features of carnivoran restoration is the extent to which it has been embraced or contemplated recently by populous, wealthy countries. The environmental Kuznets curve (reviewed by Kojola et al., 2018) predicts that as countries become developed and wealthier, they suffer worse environmental damage, but as they progress still further along the wealth axis, environmental damage declines. They do this by outsourcing resource extraction and other damaging industries and by developing tolerance and positive attitudes toward environmental concerns, including wild carnivorans. Kojola and colleagues (2018) found that this pattern did not apply to large carnivorans in Europe; wolves and brown bears were scarce even in the wealthiest countries. However, it seems to apply to some countries if the focal carnivoran is smaller than wolves and bears and does not conflict strongly with human interests. For example, red foxes raised in a Seoul, South Korea zoo were released into the wild in Gyeongsang Province in 2012, and further releases followed. At the time they were believed to be the only free-ranging red foxes in a country with a human population of 52,000,000, the species having been driven to apparent extinction in the wild (Lee et al., 2013). South Korea also has attempted to restore populations of Asiatic black bears using animals confiscated from bear bile farms (Borzée

et al., 2019). New Jersey has the highest human population density of any of the United States, and yet in 2015 it supported a mean black bear density of 0.12/km² (Gompper et al., 2015). The geographic range of black bears within the state tripled during 1995–2020, without reintroductions or augmentations; restrictions on hunting and education were the primary tools. These examples show that restorations and recoveries of carnivorans—even large-bodied ones—need not be restricted to vast areas of wilderness.

Additional restorations to wealthy countries with high human densities are being considered where they would have been implausible two decades ago. The Eurasian lynx is being considered for restoration to Scotland, where roe deer populations have recovered dramatically in recent years (Ovenden, 2019). Support for reintroduction of the clouded leopard into Taiwan is strong among urban dwellers, but less so among indigenous communities in rural areas. The greater number of urbanites pushes the balance of public sentiment toward support for reintroduction, but the rural dwellers would be affected directly by any likely conflicts—a general socio-economic phenomenon (Greenspan et al., 2020). The tiger, now extinct over most of southeast Asia, is being evaluated for reintroduction to eastern Cambodia (Gray et al., 2017).

12.10 Reintroduction and restoration outcomes

The science of reintroduction has developed rapidly since 1995, consistent with its growing use as a conservation tool. Across vertebrates, carnivorans have been overrepresented in reintroduction attempts, partly reflecting humans impacts on their distributions (Seddon et al., 2005). Still, carnivorans account for a small proportion of all species reintroduced, and generalizations across all vertebrates may not apply (Seddon et al., 2007). The literature on even a single species can be inconclusive because many reintroduction efforts, especially early ones, lacked experimental treatments. Each program was a statistical unit of replication, without co-variates that could be considered post hoc. Jule and colleagues (2008) examined one potentially important factor—captive breeding—in forty-five studies involving seventeen carnivoran species across five families. They found that wild-born animals were more likely to survive reintroduction than captive-born ones, the latter of which were more susceptible to starvation, predation, competition, and other factors.

Evaluating the success of carnivoran restorations also is difficult because criteria for success vary across species and conservation cultures. Robert and colleagues (2015) reviewed how reintroduction success is associated with various criteria defined by the International Union for Conservation of Nature and found that intrinsic growth rate predicted short-term persistence better than founder population size; long-term extinctions were more likely to be due to stochastic events than to growth rate. Conservation culture plays an important role as well. In the US, success tends to be defined in terms of longer-term population persistence than is the case in South Africa, where carnivoran populations that are small and require intensive management are more accepted as successful (Licht *et al.*, 2010). Lions, leopards, cheetahs, African wild dogs, and two species of hyena were reintroduced to private and provincial reserves in Eastern Cape Province, South Africa in the early 2000s, and most of these reintroductions were considered successful over a two-decade study period (Banasiak *et al.*, 2021). South African introductions also require more restrictions on visitors and are more likely to involve multiple landowners and be self-funded. US reintroductions tend to employ less management intervention, be funded by federal programs, and be evaluated on a longer-term basis.

In some cases, managers have translocated carnivorans for conservation purposes only to discover undetected relict populations near the release site. This occurred in the case of the fisher, one of the most widely reintroduced carnivorans in North America. Fishers were translocated to central Alberta from eastern Canada; however, Stewart and colleagues (2017) showed that the population at the release site several years later had descended from undetected adjacent populations, not the putative founders from farther east.

The black-footed ferret illustrates how even a complex and extensive reintroduction program can fall short of its conservation goals. The last wild population of 100–200 animals was discovered in 1981 and studied briefly before it crashed as a result of an epizootic of canine distemper in the mid-1980s. At around the same time, sylvatic plague was identified in prairie dog colonies nearby. This disease, caused by the bacterium *Yersinia pestis*, causes high mortality in prairie dogs, and kills ferrets bitten by infected fleas (Mize *et al.*, 2017). The subsequent captive breeding program was based on eighteen original animals, seven of which participated in breeding. Descendants had been reintroduced to thirty sites in the US, Canada, and Mexico by 2022. Despite this major effort, only several hundred ferrets were estimated to live in the wild at that time, and at only a few sites. The inability of black-footed ferrets to cope with two Old World diseases, along with the dearth of large areas of uninfected prairie dog habitat, account for this reintroduction program not meeting its goals.

Whether through recolonization or reintroduction, carnivoran recoveries raise hopes for restoring lost ecological functions (Wolf and Ripple, 2018), and some of those hopes are realized. However, in some cases a carnivoran population regarded as recovered may be less abundant than its pre-bottleneck predecessor, so that expectations for ecological effects are not met. Kuijper and colleagues (2016) predicted this would limit the potential for recovered carnivoran populations to exert density-mediated trophic cascades, except in some low-productivity habitats. Moll and colleagues (2016) described such an outcome of reintroduced lions and spotted hyenas on prey behaviors in South Africa. The reintroduced lions increased aggregation behaviors in ungulates, especially smaller-bodied species, but only slightly. Lions elicited stronger aggregation responses than did hyenas. Also in South Africa, reintroduction of lions was associated with broadening of the trophic niche of black-backed jackals, reflecting their increased access to large ungulate carrion (Codron *et al.*, 2018). Some subtle effects can be consequential, however. In northwestern Spain, land abandonment has led to shifts in the carnivoran community, so that the red fox has replaced the European pine marten as the main frugivore (López-Bao *et al.*, 2015), with potential implications for seed dispersal.

12.11 Trophic subsidies

While humans have always competed with carnivorans for some foods, they also have subsidized the diets of some species for millennia. Semi-domestic dogs were commensal with humans before they became companions and working animals, and the wildcat was attracted to the rodents that fed on early human grain crops. With time, cats became directly subsidized by humans who recognized the value of their predatory services (Krajcarz *et al.*, 2020). Today, many carnivorans rely on foods made available—directly or indirectly—by humans, in ways that are not always intuitive. Carnivoran consumption of human waste is nearly ubiquitous; where they occupy urban and peri-urban environments, wild carnivorans find various discarded foods, foods left outside for pets, or foods left specifically for them (Figure 12.6).

These subsidies affect multiple aspects of carnivoran ecology profoundly (Gompper and Vanak, 2008). Human food sources tend to be more spatially and temporally predictable than wild ones and tend to benefit opportunistic species over those with specific needs for wild foods (Oro *et al.*, 2013). In Israel, food waste at disposal sites was found to reduce home range sizes and increase annual survivorship of golden jackals—in some villages drastically (Bino

et al., 2010). Carnivorans also consume foods left as subsidies for other wildlife; "vulture restaurants" created to subsidize threatened avian scavengers have the effect of increasing hyena and jackal abundances in South Africa (Yarnell *et al.*, 2014). Some intentional feeding is ancillary to hunting; the diets of black bears in northern Wisconsin were over 40% hunting baits, on a year-long basis. Bear densities in the region were high, which Kirby and colleagues (2017) attributed in part to baiting. In effect, humans nourish black bears via hunting baits and then kill some of the animals produced via this trophic pathway. Anthropologists have long recognized the importance of human feces in the diets of domestic dogs (Coppinger and Coppinger, 2001), which remains high in developing countries and regions with poor sanitation. It has been described for a rural area of Zimbabwe, where feces were an important and consistently available food for free-ranging dogs (Butler *et al.*, 2018). This represents a public health risk in some areas because dogs hold and deposit various helminth parasites that use humans as definitive hosts (Traub *et al.*, 2002). Wild carnivorans have not been documented to use human feces in this way.

Human hunting of ungulates would appear to be an important source of carnivoran carrion in the

Figure 12.6 A South American coati feeds on human refuse at Iguacu Falls, Brazil. Many carnivoran species in human-modified environments receive strong human food subsidies. These subsidies lead to high local densities, increased interspecific competition, and the potential for increased disease spread.
Photo: alatom/iStock.

form of viscera left at kill sites. Clearly, hunter-left ungulate viscera affect the behaviors and short-term nutrition of scavengers (e.g. Lafferty *et al.*, 2016). How longer-term nutrition is affected is less clear. After all, hunter-killed animals would die from other causes otherwise; the question is how the location and timing of viscera deposition affects carnivoran nutrition. Lanszki and colleagues (2018) studied the effect of the removal of hunter-left viscera on golden jackals in Hungary. Fallow deer, red deer, and wild boar were legally hunted, and the study estimated availability of hunter-left viscera before and during experimental removal of viscera. Golden jackals did not show a statistically significant decrease in consumption of viscera following removal, suggesting that other foods and other sources of viscera (e.g. that from wounded animals that died undetected) compensated for the removed viscera.

These examples show that trophic subsidies increase the fitness of some carnivoran populations, although mixed effects have been reported (Newsome *et al.*, 2014). Kirby and colleagues (2019) found that black bears in Colorado that consumed more human foods underwent more interrupted winter denning and showed greater chromosomal evidence of cellular aging. Whether this is due to the human-sourced diet or other stresses associated with proximity to humans is not clear. Ursids seem particularly prone to trade-off effects associated with human food subsidies. Brown bears in Banff, Canada had isotopic signatures consistent with high consumption of ungulate carcasses killed by trains, but train collisions were the largest single cause of bear deaths (Hopkins *et al.*, 2014). Similarly, Lafferty and colleagues (2016) documented moose carrion left by hunters in southeast Alaska and inferred that carnivorans of at least six species benefited from this subsidy. However, black bears and brown bears were occasionally killed when associating with these remainders, so that a net fitness enhancement to bears was not obvious. Lastly, Kirby and colleagues (2016) examined diets of black bears across much of Colorado and found high dependence on human food sources (>30% of assimilated diet) in eastern Colorado. They likewise found strong correlations between use of human foods, housing density, and likelihood of bear–human conflicts. The

frequent reporting of mixed fitness effects of food subsidies on temperate ursids (e.g. Sergiel *et al.*, 2020; Penteriani *et al.*, 2021) suggests that they have especially complex relationships with human foods subsidies. This could result from the omnivorous behaviors of bears, or from the hyperphagy that precedes winter denning.

At the community or ecosystem level, recent stable isotopic studies show how humans can dominate the trophic condition of carnivorans. Manlick and Pauli (2020) examined the enrichment of ^{13}C and ^{15}N in seven species of carnivorans across a gradient of human disturbance in the Great Lakes region of the US. The region is dominated by C_3 primary production, so that consumption of foods derived from crop plants can be detected. They found strong isotopic evidence of human influence on carnivoran diets, even in the wolf and other species generally regarded as wilderness associates. They estimated that over 25% of the carnivoran diet in the most disturbed landscapes was derived from human sources and inferred that greater consumption of human foods increased trophic niche overlap among species. This increases the potential for various competitive and disease-mediated effects. Not all carnivorans were predisposed to use human-sourced foods; one obligate predator, the bobcat, showed little use (<5%) of foods derived from crop sources.

12.12 Other subsidies

Humans also subsidize carnivorans through non-trophic channels. For example, many human-modified habitats favor some species over others. Various small carnivorans, mostly mustelids, herpestids, procyonids, and mephitids, occupy buildings, barns, and other structures across most continents (e.g. Poglayen-Neuwall and Toweill, 1988; Lariviere, 2002). These habitat modifications have expanded the geographic ranges of various species, particularly in regions that lack complex physical structure near the ground. Water can be a limiting resource in arid landscapes and humans alter the spatial availability of water via irrigation, wildlife watering devices, and water left outdoors for pets or wildlife. In the Mojave Desert in California, US, coyotes, American badgers, bobcats, and common gray foxes concentrated their activities near human-provided water sources (Rich *et al.*, 2019). Red foxes are not as well adapted as small desert foxes to arid landscapes and can displace smaller foxes in water-subsidized areas (e.g. Kumar *et al.*, 2019a). Competition between carnivoran species near human water sources, as in so many other cases, is mediated by body size (Hall *et al.*, 2021).

12.13 Mediation of competitive interactions

Section 7.7.7 provides several examples of how humans mediate interactions between carnivoran species. One general pattern is for human proximity or developments to affect a larger-bodied carnivoran more strongly, causing a competitively subordinate species to associate more closely with humans. This has been shown for tigers and leopards in India (Kumar *et al.*, 2019b), leopards and wolves in Iran (Mohammadi *et al.*, 2021), and the bat-eared fox and larger carnivorans in South Africa (Welch *et al.*, 2017). A contrary pattern also appears, in which a larger-bodied but more human-tolerant species is negatively correlated with presence of similar-sized or smaller-bodied carnivorans. The coyote is such a species in Arizona and California (Baker and LeBerg, 2018; Smith *et al.*, 2018), but shows greater avoidance of humans than does the red fox in urban parks in Cleveland, Ohio, US. (Moll *et al.*, 2018). Body size is an imperfect predictor of ability to live closely with humans.

12.14 Conservation genetics

The goal of conservation is to preserve or restore the products of evolution and natural ecological processes. The products of evolution include adaptive phenotypes and the genomes that code for them. Conservation pursues goals commensurate with the length of the evolutionary process and the uniqueness of the adaptive phenotype at issue. Populations are commonly considered less critical conservation issues than are adapted ecotypes, the latter in some cases represented by nominal subspecies. Subspecies tend to be lower conservation priorities than biological species, and so on.

However, in the twenty-first century, this simple characterization is inadequate to depict the goals of conservation. With global climate change and other drastic alteration of the environment by humans, it is no longer safe to assume that phenotypes that were adaptive until recently can be successful in their current geographic ranges into the future. The challenges facing conservation genetics have become much more complex.

Wild populations of sufficient size to avoid inbreeding, inbreeding depression, and drift are susceptible to other genetic processes that affect population persistence in the wild. One of these is introgression, either with another wild taxon or between wild and domestic lineages. Section 10.3.4 gives examples of potential outbreeding depression—reduced fitness in the progeny of matings between distant relative. Examples of naturally occurring carnivoran hybrids that are known or suspected of suffering from reduced fitness include crosses of the European pine marten and the sable, and the Canada lynx and bobcat (Schwartz *et al.*, 2004). Matings facilitated by human intervention—for example via translocation—can dilute locally adapted genomes. Relatedly, wild carnivorans can hybridize with domestic descendants of the wild lineage. Examples of the domestic–wild crosses occur between domestic cats and wildcats, and between domestic dogs and various African *Canis* spp. (Section 10.3.4). Australian dingoes are considered a conservation priority by some and their introgression with domestic dogs is therefore problematic (Randi, 2008).

Genetic and non-genetic processes in small populations can be difficult to parse in application. Demographic stochasticity—unpredictable variation in population size arising from environmental variation—can lead to sudden population declines over small geographic areas, initiating population bottlenecks. Allee effects can lead to negative density dependence—lower recruitment or increased mortality with decreasing density. For example, animals in low-density populations can fail to breed because they cannot find mates during a restricted mating season, or they may mate but require social facilitation to feed or defend young. Lacking enough conspecific helpers, reproduction may fail. By contrast, genetic processes are probabilistic or selective in nature. Inbreeding—matings between close relatives—is more likely in small populations and leads to the expression of deleterious recessive alleles, one basis of inbreeding depression. Genetic drift also occurs in small populations, tending to permanently remove uncommon alleles from the gene pool. In combination, inbreeding and drift lower genetic variability, reducing the ability of affected animals to respond to environmental variability via adaptation. These three processes were central to the "extinction vortex" paradigm that was a central tenet of small population biology in the 1980s—the tendency for small populations to shrink, in ways that are difficult to reverse via intervention (Soulé, 1986).

A fourth process is human-imposed selection for traits that are not adaptive in the wild. This is encountered in captive propagation, which has been employed with various vertebrates (Kleiman et al., 2010) but few critically endangered carnivorans (Wildt et al., 2003 for the giant panda). Captive propagation led by zoos was hailed in the 1980s as a lifeline for endangered taxa, but its promise faded as practical limitations became apparent. Of the few programs to propagate carnivorans in captivity for release into the wild (see Rodriguez et al., 1995 for Iberian lynx example), the black-footed ferret is by far the largest and best-documented, with over 10,000 ferrets born in captivity during 1987–2022, thousands of them released to the wild. Over this span of time, no large free-ranging population has been able to be maintained, despite strong interagency collaboration, so that captive animals have represented the primary gene pool throughout the program. During this time, ferrets have undergone a number of phenotypic changes, including decreased body size (Wisely et al., 2002). There is a tendency in various mammalian species for captivity to lead to greater docility and tolerance of human proximity, some of which are not adaptive in the wild (Frankham, 2008). Some of these trait changes are reversible via reintroduction (Wisely et al., 2005).

The typical goals of genetic management of captive and wild populations are to minimize all four of these effects. Managers aim to increase effective population size (N_e) and to minimize inbreeding and inbreeding depression (Ballou et al., 2010). Increasing N_e includes managing genetic

contributions to progeny based on known relatedness of candidate parents—studbook management, which has been used for various carnivorans, including the African wild dog, red panda, snow leopard, and California sea lion (Princée, 2016). Contact with humans may be limited, and exposure to vigorous prey helps to mitigate the loss of predatory behaviors. The main goal, however, is to limit the generational duration of captive breeding—to return the population to the wild as quickly as possible.

Population augmentation is used to accomplish both demographic and genetic objectives. The former include correcting skewed sex ratios and reducing the likelihood of stochastic population declines. Attempted genetic rescue via augmentation is common, although successful examples involving carnivorans are limited (Whiteley et al., 2014); it can take the form of translocation or facilitated dispersal. A well-documented case of translocation involves the puma in Florida (locally called the "Florida panther"). This subspecies had been isolated from other puma populations for decades and declined to an estimated twenty-two individuals by the 1990s. The population showed typical symptoms of inbreeding depression: cryptorchidism, heart defects, and impaired reproduction (Wisely et al., 2015). Augmentation using eight females from Texas in 1995 resulted in a positive population response and reduced physiological abnormalities by the 2010s. However, using the practice involves uncertainties. First, short-term increases in population growth immediately following augmentation, reported in multiple studies, could reflect heterosis, which is inherently transient over generational time. Evaluations of success after only one or two generations are likely to exaggerate any longer-term positive effects. Second, translocations from distant or different environments can lead to genetic swamping or outbreeding depression. These related concepts refer to diluting locally adapted genomes with those from ecotypically dissimilar populations. Third, augmentation of a highly inbred population by a single animal or group of closely related ones may cause new arrivals to dominate matings, resulting in unintended reduction of N_e and accelerated inbreeding depression (Bell et al., 2019).

Facilitated dispersal is accomplished in various ways, including building structures that reduce barriers to dispersal. Increasingly in developed countries, highway barriers are spanned by wildlife underpasses and overpasses (Corlatti *et al.*, 2009). These tend to be costly but can be funded via highway use taxes rather than wildlife-specific sources. They are constructed to benefit various vertebrate quadrupeds—not just single species—which strengthens the case for their use. Comparisons of design features that benefit carnivorans are not available, but the strong dispersal abilities of many carnivorans make them ready users of these structures. Asari and colleagues (2020) described carnivoran use of underpasses in Hokkaido, Japan and found that the species that used underpasses the most were mid-sized carnivorans.

Genetic engineering for conservation is a strategy of last resort, being costly and employing emergent technologies. Some of the methods have been applied in the case of the black-footed ferret. This carnivoran presents a perfect storm of conservation challenges; it is the only ferret species native to the New World, and an obligate predator of a narrow range of prey—prairie dogs of Genus *Cynomys*. These ground squirrels have undergone severe population loss and range reduction as a result of poisoning, land conversion to crop production, and recreational shooting. Little land in protected status occurs in habitats occupied by the ferret, and it and its prairie dog prey are tolerated poorly by most agricultural producers. Indeed, many US jurisdictions specifically prohibit introduction of prairie dogs, and some criminalize failure to control their numbers on private lands. Ferrets are highly vulnerable to canine distemper, and both ferrets and prairie dogs succumb to sylvatic plague. The application of pesticides to prairie dog colonies to control fleas, a cost-prohibitive approach, has shown promise, and vaccines for both prairie dogs and ferrets are being tested. The black-footed ferret has lost much of its genetic variability so that it now compares with the cheetah, southern sea otter, and Amur tiger (Beichman *et al.*, 2019).

For these reasons, planned genetic engineering of the ferret now addresses two goals—restoration of the genomic variability that was present in the last wild population in the 1980s and correcting the incompetency of the ferret's immune system against canine distemper and plague. The first process is essentially population augmentation via generations that lived in the distant past, preserved cryogenically. Campbell and colleagues (1996) described such an approach, which involves somatic cell nuclear transfer. It has been used on several carnivoran species with limited success, in terms of numbers of neonates that survive for long: African wildcat, sand cat, and coyote (Wisely *et al.*, 2015, Table 1). In the case of the ferret, frozen somatic tissue of a female that died in the 1980s, with no descendants living in 2021, was used to extract a nucleus, which was inserted into the denucleated egg of a domestic ferret. The zygote was inserted into the uterus of the surrogate mother, which gave birth to four neonates, two of the mother's species and one living black-footed ferret (Figure 12.7). This success adds a new individual to the very small breeding population of black-footed ferrets, and one without a long bottleneck in its ancestry. This approach improves the prospects for genetic restoration of the ferret, and surely will be applicable to other endangered carnivorans in coming decades.

Newly plausible next steps in improving the genetic prospects of the black-footed ferret involve genome editing. This approach has been in development for decades but became newly practical when Knott and Doudna (2018) introduced the clustered regularly interspaced short palindromic repeats (CRISPR-Cas9) system. The mechanism underlying CRISPR enables targeting functional proteins to specific genome locations to modify the coding sequence or expression of genes (Segelbacher *et al.*, 2021). This system can produce CRISPR-associated molecules to modify most DNA sequences, using short RNA molecules. In the case of the ferret, knowing the sequences responsible for its immune incompetency against canine distemper and plague would allow editing those sequences to match the sequences of the steppe polecat, the Old-World sibling species of the black-footed ferret, which can mount effective immune responses to both distemper and plague. Other genome-manipulating methods are newly available, and the breadth of their applicability to conservation issues is just beginning to be appreciated (Phelps *et al.*, 2019).

Figure 12.7 The first cloned Western Hemisphere carnivoran, a black-footed ferret named Elizabeth Ann. The female was born to a domestic ferret and was healthy on her first birthday in 2021, when this photo was taken.
US Fish and Wildlife Service photo by Kim Fraser.

While such approaches are being considered, the public's acceptance of them must also be weighed. In an era when stigma exists regarding genetically modified foods, similarly modifying wild carnivorans for conservation purposes is unlikely to receive wide public support without major education efforts. Kohl and colleagues (2019) showed that US adults perceived more risk than potential benefit in using gene editing for conservation purposes. Respondents tended to fear that such methods could be used inappropriately, but those concerns were affected by the respondents' confidence in scientific methods. Sandler and colleagues (2021) analyzed the ethical issues involved in whether to clone the black-footed ferret; they found several key factors that affect the public's acceptance of the approach. Clearly, using cloning, gene editing, and other biotechnologies for conservation purposes requires considering and explaining to the public why the methods are needed, possible unintended consequences of using them, and the alternatives to using them.

12.15 Global climate change

Arguably the most pressing and pervasive environmental issue of the twenty-first century, climate change poses existential threats to major portions of the earth's biota, including carnivorans. The mechanisms by which a warming and otherwise changing climate can affect carnivoran populations are numerous and complex, with many interactions possible. While a number of modeling efforts have predicted how various mammalian taxa are likely to respond, only a few such mechanisms specific to carnivorans are based on strong data. They provide a glimpse of an uncertain future for many carnivoran populations and taxa, which are affected via various pathways. A warming climate interacting with human food sources has been shown to affect black bear winter denning in Colorado, US. Warmer temperatures reduced the duration of winter denning and increased bear use of human-sourced foods while active (Johnson *et al.*, 2017). The African wild dog, seemingly adapted to subtropical-tropical climates, shows reduced reproductive success at high ambient temperatures, inferred to result from reduced foraging times. Apparently, the species foregoes foraging to feed pups when temperatures are too high (Woodroffe *et al.*, (2017). Wild dogs also suffered increased human-inflicted mortality at high ambient temperatures for various hypothesized reasons (Rabaiotti *et al.*, 2021). In many cases, scientists predict future changes in species abundances based on past year-to-year variation in weather. For example, the distributions of three species of foxes in Tunisia were predicted best by elevation, temperature, and precipitation, and these

correlations were taken to suggest future climate-caused changes (Karssene *et al.*, 2017). The encroachment of red foxes into tundra habitats and displacement of arctic foxes across the Holarctic could be driven by habitat changes or differences in the species' differing cold tolerances (Figure 12.8). The mechanisms underlying many of these correlations are not known and could be physiological or mediated by habitat.

Other examples of predicted climate-mediated habitat change include snow leopards preying on blue sheep in the Himalayas of Nepal (Aryal *et al.*, 2016), where habitat is expected to diverge from the current elevational distribution of the snow leopard's primary prey. A changing climate also is expected to benefit some species or exert net effects that are difficult to characterize as positive or negative. For example, several small and mid-sized carnivorans have increased in size over the past several decades, by 24% for European pine martens in central Europe, and by 6% for stone martens in the same region (Wereszczuk *et al.*, 2021). American martens in mainland Alaska increased in body size by an estimated 1.5% from 1950 to 2000 (Yom-Tov *et al.*, 2008). Other carnivorans are showing comparable increases in body size, attributed to a warming climate. Carnivoran species with low cold tolerance are specifically benefitted by a warming climate. The northern raccoon, escaped from fur farms across most of the Palearctic from Europe to

Figure 12.8 A red fox holds the remains of an arctic fox that it has killed and eaten near Churchill Falls, Manitoba, Canada. A warming climate has facilitated the invasion of tundra habitats by red foxes, increasing competition between the two species across North America and Eurasia. Arctic foxes now are conservation concerns in several countries.
Photo: Don Gutoski, by permission.

Japan, has become highly invasive. The species is not especially cold-adapted, and a warming climate is expected to expand its geographic range northward dramatically by 2050 (Louppe *et al.*, 2019). Some native carnivoran species will benefit as well. Baltensperger and colleagues (2017) showed that the distribution of American martens in southcentral Alaska had expanded in 1980–2010, attributed to an increase in snowfall, which improved opportunities for martens to forage in the subnivean space and rest in protected microenvironments. Climate effects are exerted via disease ecology as well. Geographic range shifts are predicted to increase disease transmission between wildlife species by bringing populations into sympatry that were previously allopatric (Carlson *et al.*, 2022).

Among the most intuitive of current and forecasted effects are those based on a physical constant—the melting temperature of water. The presence, abundance, and physical properties of snow and ice affect many carnivoran species from temperate to polar environments. Changes in snow characteristics on land are affecting carnivorans via several mechanisms. First, snow provides a protective thermal environment important to northern small rodents and shrews—important prey for weasels and martens (Section 4.4.1), and reducing the thickness of the snow layer increases exposure to cold. Second, midwinter thawing events create snow crusts, which are barriers to species that enter and exit the subnivean layer and provide support for species that walk on its surface (Section 7.7.6; Pauli *et al.*, 2013). The effect of the latter is to increase the mobility of species that are not adapted to walking in deep, soft snow. Suffice and colleagues (2020) showed that increases in trapper harvests of the fisher in the southern part of its range in Quebec, Canada have coincided with a shift in forest type, with lower harvests of martens, and with increased spring rains. Rains become ice crusts on the snow surface, and restrict marten access to the subnivean space, where they mostly forage. Snow is specifically required for reproductive denning by wolverines in North America. In the southern part of the geographic range they give birth exclusively in dens covered by snow that persists late into spring. They also remain above or near altitudinal or latitudinal tree line through most of the

rest of the year, possibly because of limited heat tolerance. Forecasted warming will translate into dramatic decreases in snowpacks in the contiguous United States and southern Canada (McKelvey *et al.*, 2011), with implications for the persistence of wolverine populations.

Some of the highest-profile climate-related wildlife issues—involving the polar bear and its principal prey species—center on disappearing sea ice. Scientists recognized potential climate-linked impacts to polar bears and ice-loving seals decades ago and recent research and accelerated climate-related environmental change have strengthened the case for this link. The primary mechanism is via effects on ice, which is important for hauling out and breeding for ringed seals, bearded seals, and walrus. Sea ice cover also excludes other seal species that compete with ice-loving seals, another abiotic factor mediating carnivoran coexistence (Moore and Huntington, 2008). The overall effect has been to force polar bears to travel farther—in some cases by swimming across the open ocean—to forage. Ice-free summers are becoming longer, during which polar bears are shifting their diets to land-dwelling species. Predation on nests of barnacle geese, common eiders, and glaucous gulls has resulted in regional nesting failures for those species (Prop *et al.*, 2015). Few prospects are possible for mitigation of effects related to snow or ice characteristics. Short of the reversal of climate change, no measures can ensure persistence of snow- and ice-dependent species, including arctic and subarctic pinnipeds, polar bears, some weasels, and boreal forest martens (Ragen *et al.*, 2008).

Great uncertainty accompanies predicting how humans will affect carnivorans over the remaining Anthropocene Epoch; however, a few trends seem likely. First, if we consider extinction to be an evolutionary process, then humans and the Carnivora will continue to coevolve, as they have for millennia. In the near term, carnivorans will account for disproportionate public attention to endangered vertebrate species because of their inherent vulnerability to population extinction and high public interest. Second, rates of environmental change are increasing so that effects on carnivoran ranges and abundances will emerge on decadal—rather than much longer—time scales. Contributing to this trend will be non-linear environmental effects that appear as thresholds in population responses to various stressors, in some cases interacting synergistically. Gradual effects will occur, but so will threshold ones. Third, effects will be geographically variable; reductions and losses of carnivorans will accelerate where human populations grow or where the human footprint on native vegetation intensifies. By contrast, regions where human populations decline or where resource extraction and agriculture are sourced from elsewhere will continue to see some carnivoran recoveries. These will involve restorations and recolonizations, particularly of species that do not threaten human interests. Fourth, disease interactions between humans and carnivorans will diversify and intensify, as geographic ranges shift in response to climate. Fifth, genetic methods will offer the potential for creating individuals with the genomes of animals long-deceased and editing those genomes to facilitate reintroduction or recovery in an altered wild environment. However, the costs, complexity, and public acceptance of those methods are barriers to success, and the availability of suitable release sites is not assured. The future of Order Carnivora will only be resolved when the human population is brought into balance with sustainable production systems, which may coincide with the end of the Anthropocene, and the beginning of a new epoch.

Key points

- Carnivorans make up a tiny fraction of Earth's biomass, and their distributions and abundances have been affected by human actions for hundreds of thousands of years.
- Understanding that humans could cause extinctions of entire species is historically recent.
- Hunting, trapping, and other direct mortality agents have played important roles in some declines; other species show remarkable resilience to persistence human persecution. Poisoning has been a particularly pernicious cause of carnivoran declines, and continues in some regions today.
- Vehicle collisions on roads remain an important cause of carnivoran mortality, and multi-lane roads with fenced margins and high traffic can serve as nearly complete barriers to various quadrupeds.
- Physical exclusion of carnivorans is employed to varying degrees, the most dramatic example of which involves the 5600-km-long dingo fence of southeastern Australia, which is designed to protect sheep and cattle. Its use results in diverse ecological effects and sustained public and scientific debate.
- The wide range of reported behavioral effects on carnivorans are fully context- and species-specific.
- Trophic effects on wild carnivorans are pervasive and multifaceted. Species differ strongly in their dependence on human foods and possible negative consequences of dependence.
- Humans affect the incidence and severity of many carnivoran diseases. Domestic dogs and cats are frequent participants in disease transmission, and humans transporting carnivorans serve as dispersal agents.
- Effects on carnivorans from human-exacerbated diseases of prey species are increasingly recognized.
- Some carnivorans are returning to former geographic ranges via reintroduction or dispersal-colonization. Reintroduction science has expanded rapidly, and many measures are employed to enhance reintroduction success.
- Even wealthy, populous countries are conducting or considering restoration of carnivoran populations to former geographic ranges.
- The perceived success of reintroduction programs depends on biological factors as well as cultural standards.
- The ability of reintroduced carnivoran populations to provide desired ecological services depends on population density. The most consistently restored ecological effect of reintroducing carnivorans is suppression of smaller carnivoran competitors.
- Humans subsidize diets of many carnivorans, including some presumed wilderness dwellers. Subsidies have the expected effects of increasing abundances but can lead to niche overlap and competition.
- Rare carnivorans face many genetic challenges, including introgression with commoner species or with domestics. Inbreeding and drift in small

populations contribute to extinction vortices. Outbreeding likewise can impair the vigor of small populations.

- Extraordinary measures to restore carnivoran genomes include cloning of somatic cells of long-deceased individuals and gene editing to facilitate survival in an altered world.
- Climate change presents a profound and complex challenge to carnivoran conservation; effects are exerted via physiological, habitat, behavioral, and prey-mediated pathways.
- Effects mediated through altered regimes of sea ice and snow cover on land are affecting a range of snow- and ice-adapted species. Changes in sea ice cover loom as major threats to polar bears and the ice-loving seals that they eat. Alterations in the snow environment stand to affect mammals that live beneath the snow and on its surface. Short of reversing changes in climate, no known measures can mitigate these impacts.

References

Aryal, A. *et al.* (2016) "Predicting the distributions of predator (snow leopard) and prey (blue sheep) under climate change in the Himalaya," *Ecology and Evolution*, 6, pp. 4065–75.

Asari, Y. *et al.* (2020) "Overpasses intended for human use can be crossed by middle- and large-size mammals," *Landscape and Ecological Engineering*, 16, pp. 63–8.

Badingqiuying, Smith, A.T., Senko, J. and Siladan, M.U. (2016) "Plateau pika *Ochotona curzoniae* poisoning campaign reduces carnivore abundance in southern Qinghai, China," *Mammal Study*, 41, pp. 1–8.

Baker, A.D. and LeBerg, P.L. (2018) "Impacts of human recreation on carnivores in protected areas," *PLoS ONE*, 13, p. e0195436.

Bakeyev, N.N. and Sinitsyn, A.A. (1994) "Status and conservation of sables in the Commonwealth of Independent States," in Buskirk, S.W. *et al.* (eds.) *Martens, sables, and fishers: biology and conservation*. Ithaca: Cornell University Press, pp. 246–54.

Ballou, J.D. *et al.* (2010) "Demographic and genetic management of captive populations," in Kleiman, D.G., Thompson, K.V. and Baer, C.K. (eds.) *Wild mammals in captivity: principles and techniques for zoo management*. 2nd edn. Chicago: University of Chicago Press, pp. 219–52.

Baltensperger, A.P., Morton, J.M. and Huettmann, F. (2017) "Expansion of American marten (*Martes americana*) distribution in response to climate and landscape change

on the Kenai Peninsula, Alaska," *Journal of Mammalogy*, 98, pp. 703–14.

Banasiak, N.M., Hayward, M.W. and Kerley, G.I.H. (2021) "Ten years on: have large carnivore reintroductions to the Eastern Cape Province, South Africa, worked?", *African Journal of Wildlife Research*, 51, pp. 111–26.

Bar-On, Y.M., Phillips, R. and Milo, R. (2018) "The biomass distribution on earth," *Proceedings of the National Academy of Sciences of the United States of America*, 115, pp. 6506–11.

Beichman, A.C. *et al.* (2022) "Genomic analyses reveal range-wide devastation of sea otter populations," *Molecular Ecology* [online], preprint.

Beichman, A.C. *et al.* (2019) "Aquatic adaptation and depleted diversity: a deep dive into the genomes of the sea otter and giant otter," *Molecular Biology and Evolution*, 36, pp. 2631–55.

Bell, D.A. (2019) "The exciting potential and remaining uncertainties of genetic rescue," *Trends in Ecology and Evolution*, 34, pp. 1070–9.

Berny, P. *et al.* (2015) "Vigilance poison: illegal poisoning and lead intoxication are the main factors affecting avian scavenger survival in the Pyrenees (France)," *Ecotoxicology and Environmental Safety*, 118, pp. 71–82.

Bino, G. *et al.* (2010) "Abrupt spatial and numerical responses of overabundant foxes to a reduction in anthropogenic resources," *Journal of Applied Ecology*, 47, pp. 1262–71.

Black, S.A. *et al.* (2013) "Examining the extinction of the Barbary lion and its implications for felid conservation," *PLoS ONE*, 8, p. e60174.

Blaum, N. *et al.* (2007) "Shrub encroachment affects mammalian carnivore abundance and species richness in semiarid rangelands," *Acta Oecologica*, 31, pp. 86–92.

Blecha, K.A., Boone, R.B. and Alldredge, M.W. (2018) "Hunger mediates apex predator's risk avoidance response in wildland–urban interface," *Journal of Animal Ecology*, 87, pp. 609–22.

Boitani, L. and Linnell, J.D.C. (2015) "Bringing large mammals back: large carnivores in Europe," in Pereira H.M., and Navarro, L.M. (eds.) *Rewilding European landscapes*. New York: Springer Open, pp. 67–84.

Bonner, W.N. (1968) "The fur seal of South Georgia," *British Antarctic Survey Scientific Report*, 56, pp. 1–81.

Borzée, A. *et al.* (2019) "First dispersal event of a reintroduced Asiatic black bear (*Ursus thibetanus*) in Korea," *Russian Journal of Theriology*, 18, pp. 51–5.

Braje, T.J. and Erlandson, J.M. (2013) "Human acceleration of animal and plant extinctions: a Late Pleistocene, Holocene, and Anthropocene continuum," *Anthropocene*, 4, pp. 14–23.

Brook, B.W. and Bowman, D.M.J.S. (2002) "Explaining the Pleistocene megafaunal extinctions: models, chronologies, and assumptions," *Proceedings of the National Academy of Sciences of the United States of America*, 99, pp. 14624–7.

Busch, B.C. (1985) *The war against the seals*. Kingston: McGill-Queen's University Press.

Butler, J.R.A., Brown, W.Y. and du Toit, J.T. (2018) "Anthropogenic food subsidy to a commensal carnivore: the value and supply of human faeces in the diet of free-ranging dogs," *Animals*, 8, p. 67.

Campbell, K.H. *et al.* (1996) "Sheep cloned by nuclear transfer from a cultured cell line," *Nature*, 380, pp. 64–6.

Carbone, C. and Gittleman, J.L. (2002) "A common rule for the scaling of carnivore density," *Science*, 295, pp. 2273–5.

Carlson, C.J. *et al.* (2022) "Climate change increases cross-species viral transmission risk," *Nature*, 607, pp. 555–62.

Carter, N. *et al.* (2020) "Road development in Asia: assessing the range-wide risks to tigers," *Science Advances*, 6, p. eaaz9619.

Chiari, M. *et al.* (2017) "Pesticide incidence in poisoned baits: a 10-year report," *Science of the Total Environment*, 601, pp. 285–92.

Chipman, R. *et al.* (2008) "Downside risk of wildlife translocation," in Dodet, B. *et al.* (eds.) *Towards the elimination of rabies in Eurasia: Joint OIE/WHO/EU International Conference, Paris, May 2007: Proceedings. Developments in Biologicals series*, Vol. 131. Basel: Karger, pp. 223–32.

Ciucci, P. *et al.* (2020) "Anthropogenic food subsidies hinder the ecological role of wolves: insights for conservation of apex predators in human-modified landscapes," *Global Ecology and Conservation*, 21, p. e00841.

Clinchy, M. *et al.* (2016) "Fear of the human 'super predator' far exceeds the fear of large carnivores in a model mesocarnivore," *Behavioral Ecology*, 27, pp. 1826–32.

Codron, D. *et al.* (2018) "Meso-carnivore niche expansion in response to an apex predator's reintroduction— a stable isotope approach," *African Journal of Wildlife Research*, 48, p. 013004.

Coon, C.A.C. *et al.* (2019) "Effects of land-use change and prey abundance on the body condition of an obligate carnivore at the wildland-urban interface," *Landscape and Urban Planning*, 192, p. 103648.

Coppinger, R. and Coppinger, L. (2001) *Dogs: a startling new understanding of canine origin, behavior, and evolution*. New York: Scribner.

Corlatti, L., Hackländer, K. and Frey-Roos, F. (2009) "Ability of wildlife overpasses to provide connectivity and prevent genetic isolation," *Conservation Biology*, 23, pp. 548–56.

Cozzi, G. *et al.* (2016) "Anthropogenic food resources foster the coexistence of distinct life history strategies: year-round sedentary and migratory brown bears," *Journal of Zoology*, 300, pp. 142–50.

Dickie, M. *et al.* (2017) "Faster and farther: wolf movement on linear features and implications for hunting behaviour," *Journal of Applied Ecology*, 54, pp. 253–63.

Ditmer, M.A. *et al.* (2021) "Artificial nightlight alters the predator–prey dynamics of an apex carnivore," *Ecography*, 44, pp. 149–61.

Donadio, E., Di Martino, S. and Heinonen, S. (2022) "Rewilding Argentina: lessons for the 2030 biodiversity targets," *Nature*, 603, pp. 225–7.

Estes, J.A. *et al.* (2009) "Causes and consequences of marine mammal population declines in southwest Alaska: a food-web perspective," *Philosophical Transactions of the Royal Society B: Biological Sciences*, 364, pp. 1647–58.

Estes, J.A *et al.* (2011) "Trophic downgrading of planet Earth," *Science*, 333, pp. 301–6.

Faith, J.T. *et al.* (2020) "The uncertain case for human-driven extinctions prior to Homo sapiens," *Quaternary Research*, 96, pp. 88–104.

Fascione, N., Delach, A. and Smith, M.E. (2004) *People and predators*. Washington, DC: Island Press.

Ferreyra, H.d.V. *et al.* (2022) "Sarcoptic mange outbreak decimates South American wild camelid populations in San Guillermo National Park, Argentina," *PLoS ONE*, 17, p. e0256616.

Forman, R.T.T. and Alexander, L.E. (1998) "Roads and their major ecological effects," *Annual Review of Ecology and Systematics*, 29, pp. 207–31.

Frankham, R. (2008) "Genetic adaptation to captivity in species conservation programs," *Molecular Ecology*, 17, pp. 325–33.

Frey, S. *et al.* (2020) "Move to nocturnality not a universal trend in carnivore species on disturbed landscapes," *Oikos*, 129, pp. 1128–40.

Gabriel, M.W *et al.* (2012) "Anticoagulant rodenticides on our public and community lands: spatial distribution of exposure and poisoning of a rare forest carnivore," *PLoS ONE*, 7, p. e40163.

Gaynor, K.M. *et al.* (2018) "The influence of human disturbance on wildlife nocturnality," *Science*, 360, pp. 1232–5.

Gittleman, J.L. (ed.) (1996) *Carnivore behavior, ecology, and evolution*. Vol. 2. Ithaca: Cornell University Press.

Gompper, M.E., Belant, J.L. and Kays, R. (2015) "Carnivore coexistence: America's recovery," *Science*, 347, pp. 382–3.

Gompper, M E. and Vanak, A.T. (2008) "Subsidized predators, landscapes of fear and disarticulated carnivore communities," *Animal Conservation*, 11, pp. 13–14.

Gonzalez-Voyer, A. *et al.* (2016) "Larger brain size indirectly increases vulnerability to extinction in mammals," *Evolution*, 70, pp. 1364–75.

Gray, T.N.E. *et al.* (2017) "A framework for assessing readiness for tiger *Panthera tigris* reintroduction: a case study from eastern Cambodia," *Biodiversity Conservation*, 26, pp. 2383–99.

Greenspan, E. *et al.* (2020) "Evaluating support for clouded leopard reintroduction in Taiwan: insights from surveys of indigenous and urban communities," *Human Ecology*, 48, pp. 733–47.

Hall, L.K. *et al.* (2021) "The influence of predators, competitors, and habitat on the use of water sources by a small desert carnivore," *Ecosphere*, 12, p. e03509.

Hardesty-Moore, M., Orr, D. and McCauley, D.J. (2020) "Invasive plant *Arundo donax* alters habitat use by carnivores," *Biological Invasions*, 22, pp. 1983–95.

Haswell, P.M. *et al.* (2020) "Fear of the dark? A mesopredator mitigates large carnivore risk through nocturnality, but humans moderate the interaction," *Behavioral Ecology and Sociobiology*, 74, p. 62.

Hayward, M.W. and Somers, M.J. (eds.) (2009). *Reintroduction of top-order predators*. London: Wiley-Blackwell.

Heffelfinger, J.R., Geist, V. and Wishart, W. (2013) "The role of hunting in North American wildlife conservation," *International Journal of Environmental Studies*, 70, pp. 399–413.

Hernández, D.L. *et al.* (2016) "Nitrogen pollution is linked to U.S. listed species declines," *BioScience*, 66, pp. 213–22.

Hertel, A.G. *et al.* (2016) "Temporal effects of hunting on foraging behavior of an apex predator: do bears forego foraging when risk is high?", *Oecologia*, 182, pp. 1019–29.

Hindell, M.A., Slip, D.J. and Burton, H.R. (1994) "Possible causes of the decline of southern elephant seal populations in the southern Pacific and southern Indian Oceans," in LeBoeuf, B.J. and Laws, R.M. (eds.) *Elephant seals*. Berkeley: University of California Press, pp. 66–84.

Hopkins, J.B. III. *et al.* (2014) "Stable isotopes reveal rail-associated behavior in a threatened carnivore," *Isotopes in Environmental and Health Studies*, 50, pp. 322–31.

Ikanda, K. and Packer, C. (2008) "Ritual vs. retaliatory killing of African lions in the Ngorongoro Conservation Area, Tanzania," *Endangered Species Research*, 6, pp. 67–74.

Jachowski, D. *et al.* (2021) "Tracking the decline of weasels in North America," *PLoS ONE*, 16, p. e0254387.

Johnson, H.E. *et al.* (2017) "Human development and climate affect hibernation in a large carnivore with implications for human–carnivore conflicts," *Journal of Applied Ecology*, 55, pp. 663–72.

Johnson, P.T.J. *et al.* (2010) "Linking environmental nutrient enrichment and disease emergence in humans and wildlife," *Ecological Applications*, 20, pp. 16–29.

Jule, K.R., Leaver, L.A. and Lea, S.E.G. (2008) "The effects of captive experience on reintroduction survival in carnivores: a review and analysis," *Biological Conservation*, 141, pp. 355–63.

Karssene, Y. *et al.* (2017) "Global warming drives changes in carnivore communities in the North Sahara Desert," *Climate Research*, 72, pp. 153–62.

Kenyon, K.W. (1969) "The sea otter in the eastern Pacific Ocean," *North American Fauna*, 68, pp. 1–352.

Kilgo, J.C. *et al.* (2017) "Reproductive characteristics of a coyote population before and during exploitation," *Journal of Wildlife Management*, 81, pp. 1386–93.

Kirby, R., Alldredge, M.W. and Pauli, J.N. (2016) "The diet of black bears tracks the human footprint across a rapidly developing landscape," *Biological Conservation*, 200, pp. 51–9.

Kirby, R. *et al.* (2019) "The cascading effects of human food on hibernation and cellular aging in free-ranging black bears," *Scientific Reports*, 9, p. 2197.

Kirby, R., MacFarland, D.M. and Pauli, J.N. (2017) "Consumption of intentional food subsidies by a hunted carnivore," *Journal of Wildlife Management*, 81, pp. 1161–9.

Kleiman, D.G., Thompson, K.V. and Baer, C.K. (eds.) (2010) *Wild mammals in captivity: principles and techniques for zoo management*. 2nd edn. Chicago: University of Chicago Press.

Knott, G.J. and Doudna, J.A. (2018) "CRISPR-Cas guides the future of genetic engineering," *Science*, 361, pp. 866–9.

Koch, P.L. and Barnosky, A.D. (2006) "Late Quaternary extinctions: state of the debate," *Annual Review of Ecology, Evolution, and Systematics*, 37, pp. 215–50.

Kohl, P.A. *et al.* (2019) "Public views about editing genes in wildlife for conservation," *Conservation Biology*, 33, pp. 1286–95.

Kojola, I. *et al.* (2018) "Can only poorer European countries afford large carnivores?", *PLoS ONE*, 13, p. e0194711.

Krajcarz, M. *et al.* (2020) "Ancestors of domestic cats in Neolithic Central Europe: isotopic evidence of a synanthropic diet," *Proceedings of the National Academy of Sciences of the United States of America*, 117, pp. 17710–9.

Kuijper, D.P.J. *et al.* (2016) "Paws without claws? Ecological effects of large carnivores in anthropogenic landscapes," *Proceedings of the Royal Society B: Biological Sciences*, 283, p. 20161625.

Kumar, S., Magar, T. and Dhamala, M.K. (2019a) "Habitat selection and threats of red fox (*Vulpes vulpes*) in Rara National Park, Nepal," *Review of Environment and Earth Sciences*, 6, pp. 1–13.

Kumar, U. *et al.* (2019b) "Do conservation strategies that increase tiger populations have consequences for other wild carnivores like leopards?", *Scientific Reports*, 9, p. 14673.

Lafferty, D.J.R. *et al.* (2016) "Moose (*Alces alces*) hunters subsidize the scavenger community in Alaska," *Polar Biology*, 39, pp. 639–47.

Lamb, C.T. *et al.* (2020) "The ecology of human–carnivore coexistence," *Proceedings of the National Academy of Sciences of the United States of America*, 117, pp. 17876–83.

Lamb. C.T. *et al.* (2018) "Effects of habitat quality and access management on the density of a recovering grizzly bear population," *Journal of Applied Ecology*, 55, pp. 1406–17.

Lanszki, J., Hayward, M.W. and Nagyapáti, N. (2018) "Feeding responses of the golden jackal after reduction of anthropogenic food subsidies," *PLoS ONE*, 13, p. e0208727.

Lariviere, S. (2002) "*Ictonyx striatus*," *Mammalian Species*, 698, pp. 1–5.

Lee, H.-J. *et al.* (2013) "Release strategy for the red fox (*Vulpes vulpes*) restoration project in Korea based on population viability analysis," *Korean Journal of Environment and Ecology*, 27, pp. 417–28.

Lennox, R.J. *et al.* (2018) "Evaluating the efficacy of predator removal in a conflict-prone world," *Biological Conservation*, 224, pp. 277–89.

Letnic, M. and Koch, F. (2010) "Are dingoes a trophic regulator in arid Australia? A comparison of mammal communities on either side of the dingo fence," *Australian Ecology*, 35, pp. 167–75.

Leuchtenberger, C. *et al.* (2021) "Hope for an apex predator: giant otters rediscovered in Argentina," *Oryx*, 55, pp. 810–11.

Levy, S. (2009) "The dingo dilemma," *Bioscience*, 59, pp. 465–9.

Licht, D.S. *et al.* (2010) "Using small populations of wolves for ecosystem restoration and stewardship," *BioScience*, 60, pp. 147–53.

Lichtenfeld, L.L., Trout, C. and Kisimir, E.L. (2015) "Evidence-based conservation: predator-proof bomas protect livestock and lions," *Biodiversity Conservation*, 24, pp. 483–91.

López-Bao, J.V., González-Varo, J.P. and Guitián, J. (2015) "Mutualistic relationships under landscape change: carnivorous mammals and plants after 30 years of land abandonment," *Basic and Applied Ecology*, 16, pp. 152–61.

Louppe, V. *et al.* (2019) "Current and future climatic regions favourable for a globally introduced wild carnivore, the raccoon Procyon lotor," *Scientific Reports*, 9, p. 9174.

Lowry, M.S. *et al.* (2014) "Abundance, distribution, and population growth of the northern elephant seal (*Mirounga angustirostris*) in the United States from 1991 to 2010," *Aquatic Mammals*, 40, pp. 20–31.

Macdonald, D.W. and Loveridge, A.J. (2010) *Biology and conservation of wild felids.* Oxford: Oxford University Press.

Macdonald, D.W., Newman, C. and Harrington, L.A. (2017) *Biology and conservation of musteloids.* Oxford: Oxford University Press.

Mañas, S. *et al.* (2001) "Aleutian mink disease parvovirus in wild carnivores in Spain," *Journal of Wildlife Diseases*, 37, pp. 138–44.

Manlick, P.J. and Pauli, J.N. (2020) "Human disturbance increases trophic niche overlap in terrestrial carnivore communities," *Proceedings of the National Academy of Sciences of the United States of America*, 117, pp. 26842–8.

McKelvey, K.S. *et al.* (2011) "Climate change predicted to shift wolverine distributions, connectivity, and dispersal corridors," *Ecological Applications*, 21, pp. 2882–97.

Minnie, L., Gaylard, A. and Kerley, G.I.H. (2016) "Compensatory life-history responses of a mesopredator may undermine carnivore management efforts," *Journal of Applied Ecology*, 53, pp. 379–87.

Mize, E.L., Grassel, S.M. and Britten, H.B. (2017) "Fleas of black-footed ferrets (*Mustela nigripes*) and their potential role in the movement of plague," *Journal of Wildlife Diseases*, 53, pp. 521–53.

Mkonyi, F.J. *et al.* (2017) "Fortified bomas and vigilant herding are perceived to reduce livestock depredation by large carnivores in the Tarangire–Simanjiro ecosystem, Tanzania," *Human Ecology*, 45, pp. 513–23.

Mohammadi, A. *et al.* (2021) "Contrasting responses of large carnivores to land use management across an Asian montane landscape in Iran," *Biodiversity and Conservation*, 30, pp. 4023–37.

Moll, R.J. *et al.* (2018) "Humans and urban development mediate the sympatry of competing carnivores," *Urban Ecosystems*, 21, pp. 765–78.

Moll, R.J. *et al.* (2016) "Spatial patterns of African ungulate aggregation reveal complex but limited risk effects from reintroduced carnivores," *Ecology*, 97, pp. 1123–34.

Monk, J.D. *et al.* (2022) "Cascading effects of a disease outbreak in a remote protected area," *Ecology Letters*, 25, pp. 1152–63.

Montgomery, R.A., Macdonald, D.W. and Hayward, M.W. (2020) "The inducible defences of large mammals to human lethality," *Functional Ecology*, 34, pp. 2426–41.

Moore, S.E. and Huntington, H.P. (2008) "Arctic marine mammals and climate change: impacts and resilience," *Ecological Applications*, 18(2) supplement, pp. S157–65.

Murray, M. *et al.* (2015) "Poor health is associated with use of anthropogenic resources in an urban carnivore," *Proceedings of the Royal Society B: Biological Sciences*, 282, p. 20150009.

Newsome, T.M. *et al.* (2014) "The ecological effects of providing resource subsidies to predators," *Global Ecology and Biogeography*, 24, pp. 1–11.

Ntemiri, K. *et al.* (2018) "Animal mortality and illegal poison bait use in Greece," *Environmental Monitoring and Assessment*, 190, p. 488.

Ogada, D.L. (2014) "The power of poison: pesticide poisoning of Africa's wildlife," *Annals of the New York Academy of Sciences*, 1322, pp. 1–20.

Ordiz, A. *et al.* (2012) "Do bears know they are being hunted?", *Biological Conservation*, 152, 21–8.

Oro, D. *et al.* (2013) "Ecological and evolutionary implications of food subsidies from humans," *Ecology Letters*, 16, pp. 1501–14.

Osuri, A.M. *et al.* (2020) "Hunting and forest modification have distinct defaunation impacts on tropical mammals and birds," *Frontiers in Forests and Global Change*, 2, p. 87.

Ovenden, T. *et al.* (2019) "Improving reintroduction success in large carnivores through individual-based modelling: how to reintroduce Eurasian lynx (*Lynx lynx*) to Scotland," *Biological Conservation*, 234, pp. 140–53.

Paterson, D.G. (1977) "The North Pacific seal hunt, 1886–1910: rights and regulations," *Explorations in Economic History*, 14, pp. 97–119.

Pauli, J.N., Donadio, E. and Labertucci, S.A. (2018) "The corrupted carnivore: how humans are rearranging the return of the carnivore-scavenger relationship," *Ecology*, 99, pp. 2122–4.

Pauli, J.N. *et al.* (2013) "The subnivium: a deteriorating seasonal refugium," *Frontiers in Ecology and the Environment*, 11, pp. 260–7.

Pavlin, B.I., Schloegel, L.M. and Daszak, P. (2009) "Risk of importing zoonotic diseases through wildlife trade, United States," *Emerging Infectious Diseases*, 15, pp. 1721–6.

Penteriani, V. *et al.* (2021) "Does artificial feeding affect large carnivore behaviours? The case study of brown bears in a hunted and tourist exploited subpopulation," *Biological Conservation*, 254, p. 108949.

Penteriani, V. and Melletti, M. (eds.) (2021) *Bears of the world: ecology, conservation and management*. Cambridge: Cambridge University Press.

Phelps, M.P., Seeb, L.W. and Seeb, J.E. (2019) "Transforming ecology and conservation biology through genome editing," *Conservation Biology*, 34, pp. 54–65.

Planillo, A. *et al.* (2018) "Carnivore abundance near motorways related to prey and roadkills," *Journal of Wildlife Management*, 82, pp. 319–27.

Poglayen-Neuwall, I. and Toweill, D.E. (1988) "*Bassariscus astutus*," *Mammalian Species*, 327, pp. 1–8.

Princée, F.P.G. (2016) *Exploring studbooks for wildlife management and conservation*. Cham: Springer.

Prop, J. *et al.* (2015) "Climate change and the increasing impact of polar bears on bird populations," *Frontiers in Ecology and Evolution*, 3, p. 33.

Purvis, A., Gittleman, J.L. and Brooks, T. (2005) *Phylogeny and conservation*. Cambridge: Cambridge University Press.

Pushkina, D. and Raia, P. (2008) "Human influence on distribution and extinctions of the late Pleistocene Eurasian megafauna," *Journal of Human Evolution*, 54, pp. 769–82.

Rabaiotti, D. *et al.* (2021) "High temperatures and human pressures interact to influence mortality in an African carnivore," *Ecology and Evolution*, 11, pp. 8495–506.

Ragen, T.J., Huntington, H.P. and Hovelsrud, G.K. (2008) "Conservation of arctic marine mammals faced with climate change," *Ecological Applications*, 18(2 Suppl), pp. S166–74.

Randi, E. (2008) "Detecting hybridization between wild species and their domesticated relatives," *Molecular Ecology*, 17, pp. 285–93

Rich, L.N. *et al.* (2019) "Artificial water catchments influence wildlife distribution in the Mojave Desert," *Journal of Wildlife Management*, 83, pp. 855–65.

Robert, A. *et al.* (2015) "Defining reintroduction success using IUCN criteria for threatened species: a demographic assessment," *Animal Conservation*, 18, pp. 397–406.

Rodriguez, A., Barrios, L. and Delibes, M. (1995) "Experimental release of an Iberian lynx (*Lynx pardinus*)," *Biodiversity and Conservation*, 4, pp. 382–94.

Rohr, J.R. *et al.* (2019) "Emerging human infectious diseases and the links to global food production," *Nature Sustainability*, 2, pp. 445–56.

Ruiz-Capillas, P., Mata, C. and Malo, J.E. (2013) "Community response of mammalian predators and their prey to motorways: implications for predator–prey dynamics," *Ecosystems*, 16, pp. 617–26.

Samojlik, T. *et al.* (2018) "Lessons from Białowieża Forest on the history of protection and the world's first reintroduction of a large carnivore," *Conservation Biology*, 32, pp. 808–16.

Sandler, R.L., Moses, L. and Wisely, S.M. (2021) "An ethical analysis of cloning for genetic rescue: case study of the black-footed ferret," *Biological Conservation*, 257, p. 109118.

Schwartz, M.K. *et al.* (2004) "Hybridization between Canada lynx and bobcats: genetic results and management implications," *Conservation Genetics*, 5, pp. 349–55.

Scofield. R.P., Cullen, R. and Wang, M. (2011) "Are predator-proof fences the answer to New Zealand's terrestrial faunal biodiversity crisis?", *New Zealand Journal of Ecology*, 35, pp. 312–7.

Seddon, P.J., Armstrong, D.P. and Maloney, R.F. (2007) "Developing the science of reintroduction biology," *Conservation Biology*, 21, pp. 303–12.

Seddon, P.J., Soorae, P.S. and Launay, F. (2005) "Taxonomic bias in reintroduction projects," *Animal Conservation*, 8, pp. 51–8.

Segelbacher, G. *et al.* (2021) "New developments in the field of genomic technologies and their relevance to conservation management," *Conservation Genetics*, 23, pp. 217–42.

Sergiel, A. *et al.* (2020) "Losing seasonal patterns in a hibernating omnivore? Diet quality proxies and faecal cortisol metabolites in brown bears in areas with and without artificial feeding," *PLoS ONE*, 15, p. e0242341.

Serrouya, R. *et al.* (2017) "Experimental moose reduction lowers wolf density and stops decline of endangered caribou," *PeerJ*, 5, p. e3736.

Sheehy, E., Sutherland, C., O'Reilly, C., and Lambin, X. 2018. The enemy of my enemy is my friend: native pine marten recovery reverses the decline of the red squirrel by suppressing grey squirrel populations. *Proceedings of the Royal Society B* 285:20172603.

Sillero-Zubiri, C., Hoffmann, M. and Macdonald, D.W. (eds.) (2004) *Canids: foxes, wolves, jackals and dogs. Status survey and conservation action plan*. Cambridge: IUCN/SSC Canid Specialist Group.

Silva, C. *et al.* (2019) "Factors influencing predator roadkills: the availability of prey in road verges," *Journal of Environmental Management*, 247, pp. 644–50.

Smith, J. A. *et al.* (2017) "Fear of the human 'super predator' reduces feeding time in large carnivores," *Proceedings of the Royal Society B: Biological Sciences*, 284, p. 2017 0433.

Smith, J.A. *et al.* (2018) "Human activity reduces niche partitioning among three widespread mesocarnivores," *Oikos*, 127, pp. 890–901.

Smith, J. A., Wang, Y. and Wilmers, C. C. (2015) "Top carnivores increase their kill rates on prey as a response to human-induced fear," *Proceedings of the Royal Society B: Biological Sciences*, 282, p. 20142711.

Soulé, M.E. (1986) *Conservation biology: the science of scarcity and diversity*. Sunderland: Sinauer Associates.

Stewart, F.E.C. *et al.* (2017) "Distinguishing reintroduction from recolonization with genetic testing," *Biological Conservation*, 214, pp. 242–9.

Strandin, T., Babayan, S.A. and Forbes, K.M. (2018) "Reviewing the effects of food provisioning on wildlife immunity," *Philosophical Transactions of the Royal Society B: Biological Sciences*, 373, p. 20170088.

Suffice, P. *et al.* (2020) "Habitat, climate, and fisher and marten distributions," *Journal of Wildlife Management*, 84, pp. 277–92.

Suraci, J.P. *et al.* (2019) "Fear of humans as apex predators has landscape-scale impacts from mountain lions to mice," *Ecology Letters*, 22, pp. 1578–86.

Traub, R. J. *et al.* (2002) "The role of dogs in transmission of gastrointestinal parasites in a remote tea-growing community in northeastern India," *American Journal of Tropical Medicine and Hygiene*, 67, pp. 539–45.

Wade, P.R. *et al.* (2007) "Killer whales and marine mammal trends in the North Pacific—a re-examination of evidence for sequential megafaunal collapse and the prey switching hypothesis," *Marine Mammal Science*, 23, pp. 766–802.

Welch, R.J. *et al.* (2017) "Hunter or hunted? Perceptions of risk and reward in a small mesopredator," *Journal of Mammalogy*, 98, pp. 1531–7.

Wereszczuk, A. *et al.* (2021) "Different increase rate in body mass of two marten species due to climate warming potentially reinforces interspecific competition," *Scientific Reports*, 11, p. 24164.

Whiteley, A.R. *et al.* (2014) "Genetic rescue to the rescue," *Trends in Ecology and Evolution*, 30, 42–9.

Wildt, D.E. *et al.* (2003). "Toward a more effective reproductive science for conservation," in: Holt, W.V. *et al.* (eds.) *Reproductive science and integrated conservation*. Cambridge: Cambridge University Press, pp. 2–20.

Wisely, S.M., Ososky, J.J. and Buskirk, S.W. (2002) "Morphological changes to black-footed ferrets (*Mustela nigripes*) resulting from captivity," *Canadian Journal of Zoology*, 80, pp. 1562–8.

Wisely, S.M. *et al.* (2015) "A road map for 21st-century genetic restoration: gene pool enrichment of the black-footed ferret," *Journal of Heredity*, 106, pp. 581–92.

Wisely, S.M. *et al.* (2005) "Environment influences morphology and development for *in situ* and *ex situ* populations of the black-footed ferret (*Mustela nigripes*)," *Animal Conservation*, 8, pp. 321–8.

Wolf, C. and Ripple, W.J. (2018) "Rewilding the world's large carnivores," *Royal Society Open Science*, 5, p. 172235.

Wood, S.A. *et al.* (2020) "Rates of increase in gray seal (*Halichoerus grypus atlantica*) pupping at recolonized sites in the United States, 1988–2019," *Journal of Mammalogy*, 101, pp. 121–8.

Woodroffe, R. 2011. Demography of a recovering African wild dog (Lycaon pictus) population. *Journal of Mammalogy* 92:305–15.

Woodroffe, R., Groom, R. and McNutt, J.W. (2017) "Hot dogs: high ambient temperatures impact reproductive success in a tropical carnivore," *Journal of Animal Ecology*, 86, pp. 1329–38.

Yarnell, R.W. *et al.* (2014) "Evidence that vulture restaurants increase the local abundance of mammalian carnivores in South Africa," *African Journal of Ecology*, 53, pp. 287–94.

Yom-Tov, Y., Yom-Tov, S. and Jarrell, G. (2008) "Recent increase in body size of the American marten *Martes americana* in Alaska," *Biological Journal of the Linnean Society*, 93, pp. 701–7.

Young, L.C. *et al.* (2013) "Multi-species predator eradication within a predator-proof fence at Ka'ena Point, Hawai'i," *Biological Invasions*, 15, pp. 2627–38.

Yovovich, V., Thomsen, M. and Wilmers, C.C. (2021) "Pumas' fear of humans precipitates changes in plant architecture," *Ecosphere*, 12, p. e03309.

Zanón Martínez, J.I. *et al.* (2021) "Assessing carnivore spatial co-occurrence and temporal overlap in the face of human interference in a semiarid forest," *Ecological Applications*, 32, p. e02482.

List of extant carnivoran species

Extant and recently extinct species of Carnivora (modified from Wilson and Reeder, 2005 and the IUCN Red List of Threatened Species). Extinctions during the historical period are from IUCN Red List of Threatened Species (online, various dates). Domesticated lineages are not given species status. The putative species in the *Mustela erminea* complex (Section 10.3.3) are not recognized here; nomenclature of the American clade of weasels and American mink follows Patterson *et al.*, 2021. Other species recognized here (e.g. *Martes gwatkinsii*) are likely to be synonymized.

Order Carnivora
  Suborder Feliformia
    Family Felidae (38 spp.)
      Genus *Acinonyx*

Acinonyx jubatus	cheetah

      Genus *Caracal*

Caracal caracal	caracal

      Genus *Catopuma*

Catopuma badia	bay cat
Catopuma temminckii	Asian golden cat

      Genus *Felis*

Felis bieti	Chinese mountain cat
Felis chaus	jungle cat
Felis manul	Pallas' cat
Felis margarita	sand cat
Felis nigripes	black-footed cat
Felis sylvestris	wildcat
Felis sylvestris catus	domestic cat

      Genus *Leopardus*

Leopardus braccatus	pantanal cat
Leopardus colocolo	colocolo
Leopardus geoffroyi	Geoffroy's cat
Leopardus guigna	kodkod
Leopardus jacobitus	Andean mountain cat
Leopardus pajeros	pampas cat
Leopardus pardalis	ocelot
Leopardus tigrinus	oncilla
Leopardus wiedii	margay

      Genus *Leptailurus*

Leptailurus serval	serval

      Genus *Lynx*

Lynx canadensis	Canadian lynx
Lynx lynx	Eurasian lynx
Lynx pardinus	Iberian lynx
Lynx rufus	bobcat

      Genus *Pardofelis*

Pardofelis marmorata	marbled cat

      Genus *Prionailurus*

Prionailurus bengalensis	leopard cat
Prionailurus iriomotensis	Iriomote cat
Prionailurus planiceps	flat-headed cat
Prionailurus rubiginosus	rusty-spotted cat
Prionailurus vivverinus	fishing cat

      Genus *Profelis*

Profelis aurata	African golden cat

      Genus *Puma*

Puma concolor	puma, mountain lion, panther
Puma yagouaroundi	jaguarundi

      Genus *Neofelis*

Neofelis nebulosa	clouded leopard

      Genus *Panthera*

Panthera leo	lion
Panthera onca	jaguar
Panthera pardus	leopard
Panthera tigris	tiger

      Genus *Uncia*

Uncia uncia	snow leopard

    Family Viverridae (35 spp.)
      Genus *Arctictis*

Arctictis binturong	binturong

      Genus *Arctogalidia*

Arctogalidia trivirgata	small-toothed palm civet

      Genus *Macrogalidia*

Macrogalidia musschenbroekii	Sulawesi palm civet

      Genus *Paguma*

Paguma larvata	masked palm civet

Genus *Paradoxurus*
  Paradoxurus hermaphroditus — Asian palm civet
  Paradoxurus jerdoni — Jerdon's palm civet
  Paradoxurus zeylonensis — Ceylon palm civet
Genus *Chrotogale*
  Chrotogale owstoni — Owston's palm civet
Genus *Cynogale*
  Cynogale bennettii — otter civet
Genus *Diplogale*
  Diplogale hosei — Hose's palm civet
Genus *Hemigalus*
  Hemigalus derbyanus — banded palm civet
Genus *Prionodon*
  Prionodon linsang — banded linsang
  Prionodon pardicolor — spotted linsang
Genus *Civettictis*
  Civettictis civetta — African civet
Genus *Genetta*
  Genetta abyssinica — Abyssinian genet
  Genetta angolensis — Angolan genet
  Genetta bourloni — Bourlon's genet
  Genetta cristata — crested servaline genet
  Genetta genetta — common genet, small-spotted genet
  Genetta johnstoni — Johnston's genet
  Genetta maculata — rusty-spotted genet
  Genetta pardina — pardine genet
  Genetta piscivora — aquatic genet
  Genetta poensis — king genet
  Genetta servalina — servaline genet
  Genetta thierryi — Haussa genet
  Genetta tigrina — Cape genet
  Genetta victoriae — giant forest genet
Genus *Poiana*
  Poiana leightoni — Leighton's genet
  Poiana richardsonii — African genet
Genus *Viverra*
  Viverra civettina — Malabar large-spotted civet
  Viverra megaspila — large-spotted civet
  Viverra tangalunga — Malayan civet
  Viverra zibetha — large Indian civet
Genus *Viverricula*
  Viverricula indica — small Indian civet

Family Eupleridae (8 spp.)
Genus *Cryptoprocta*
  Cryptoprocta ferox — fossa
Genus *Eupleres*
  Eupleres goudotii — falanouc
Genus *Fossa*
  Fossa fossana — Malagasy civet
Genus *Galidia*
  Galidia elegans — ring-tailed mongoose

Genus *Galidictis*
  Galidictis fasciata — broad-striped Malagasy mongoose
  Galidictis grandidieri — Grandidier's mongoose
Genus *Mungotictis*
  Mungotictis decemlineata — narrow-striped mongoose
Genus *Salanoia*
  Salanoia concolor — brown-tailed mongoose

Family Nandiniidae (1 sp.)
Genus *Nandinia*
  Nandinia binotata — African palm civet

Family Herpestidae (33 spp.)
Genus *Atilax*
  Atilax paludinosus — marsh mongoose
Genus *Bdeogale*
  Bdeogale crassicauda — bushy-tailed mongoose
  Bdeogale jacksoni — Jackson's mongoose
  Bdeogale nigripes — black-footed mongoose
Genus *Crossarchus*
  Crossarchus alexandri — Alexander's kusimanse
  Crossarchus ansorgei — Angolan kusimanse
  Crossarchus obscurus — common kusimanse
  Crossarchus platycephalus — flat-headed kusimanse
Genus *Cynictis*
  Cynictis penicillata — yellow mongoose
Genus *Dologale*
  Dologale dybowskii — Pousargues' mongoose
Genus *Galerella*
  Galerella flavescens — Angolan slender mongoose
  Galerella ochracea — Somalian slender mongoose
  Galerella pulverulenta — Cape gray mongoose
  Galerella sanguinea — slender mongoose
Genus *Helogale*
  Helogale hirtula — Ethiopian dwarf mongoose
  Helogale parvula — common dwarf mongoose
Genus *Herpestes*
  Herpestes brachyurus — short-tailed mongoose
  Herpestes edwardsi — Indian gray mongoose

Herpestes fuscus	Indian brown mongoose
Herpestes ichneumon	Egyptian mongoose
Herpestes javanicus	small Asian mongoose
Herpestes naso	long-nosed mongoose
Herpestes semitorquatus	collared mongoose
Herpestes smithii	ruddy mongoose
Herpestes urva	crab-eating mongoose
Herpestes vitticollis	stripe-necked mongoose

Genus *Ichneumia*
Ichneumia albicauda — white-tailed mongoose

Genus *Liberiictis*
Liberiictis kuhni — Liberian mongoose
Genus *Mungos*
Mungos gambianus — Gambian mongoose
Mungos mungo — banded mongoose
Genus *Paracynictis*
Paracynictis selousi — Selous' mongoose
Genus *Rhynchogale*
Rhynchogale melleri — Meller's mongoose
Genus *Suricata*
Suricata suricatta — meerkat, suricate

Family Hyaenidae (4 spp.)
Genus *Crocuta*
Crocuta crocuta — spotted hyena
Genus *Hyaena*
Hyaena brunnea — brown hyena
Hyaena hyaena — striped hyena
Genus *Proteles*
Proteles cristata — aardwolf

Suborder Caniformia
Family Canidae (36 spp.)
Genus *Atelocynus*
Atelocynus microtis — short-eared dog
Genus *Canis*
Canis adustus — side-striped jackal
*Canis anthus*** — African golden wolf
Canis aureus — golden jackal
Canis latrans — coyote
Canis lupus — wolf
Canis lupus domesticus — domestic dog
Canis lupus dingo — dingo
Canis mesomelas — black-backed jackal
Canis rufus (or *Canis lupus* ssp.) — red wolf
Canis simensis — Ethiopian wolf
Genus *Cerdocyon*
Cerdocyon thous — crab-eating fox

Genus *Chrysocyon*
Chrysocyon brachyurus — maned wolf
Genus *Cuon*
Cuon alpinus — dhole
Genus *Dusicyon*
*Dusicyon australis** — Falkland Islands wolf
Genus *Lycalopex*
Lycalopex culpaeus — culpeo fox
Lycalopex fulvipes — Darwin's fox
Lycalopex griseus — chilla, South American gray fox
Lycalopex gymnocercus — Pampas fox
Lycalopex sechurae — Sechuran fox
Lycalopex vetulus — hoary fox
Genus *Lycaon*
Lycaon pictus — African wild dog
Genus *Nyctereutes*
Nyctereutes procyonoides — raccoon dog
Genus *Otocyon*
Otocyon megalotis — bat-eared fox
Genus *Speothos*
Speothos venaticus — bush dog
Genus *Urocyon*
Urocyon cinereoargenteus — common gray fox
Urocyon littoralis — island fox
Genus *Vulpes*
Vulpes bengalensis — Bengal fox
Vulpes cana — Blanford's fox
Vulpes chama — Cape fox
Vulpes corsac — Corsac fox
Vulpes ferrilata — Tibetan sand fox
Vulpes lagopus — arctic fox
Vulpes macrotis — kit fox
Vulpes pallida — pale fox
Vulpes rueppellii — Rüppell's fox
Vulpes velox — swift fox
Vulpes vulpes — red fox
Vulpes zerda — fennec fox

Family Ursidae (8 spp.)
Genus *Ailuropoda*
Ailuropoda melanoleuca — giant panda
Genus *Helarctos*
Helarctos malayanus — sun bear
Genus *Melursus*
Melursus ursinus — sloth bear
Genus *Tremarctos*
Tremarctos ornatus — spectacled bear
Genus *Ursus*
Ursus americanus — American black bear
Ursus arctos — brown bear
Ursus maritimus — polar bear
Ursus thibetanus — Asiatic black bear

Family Otariidae (16 spp.)
 Genus *Arctocephalus*
  Arctocephalus australis   South American fur seal
  Arctocephalus forsteri   Australasian fur seal
  Arctocephalus galapagoensis   Galapagos fur seal
  Arctocephalus gazella   Antarctic fur seal
  Arctocephalus philippii   Juan Fernández fur seal
  Arctocephalus pusillus   brown fur seal
  Arctocephalus townsendii   Guadalupe fur seal
  Arctocephalus tropicalis   subantarctic fur seal
 Genus *Callorhinus*
  Callorhinus ursinus   northern fur seal
 Genus *Eumetopias*
  Eumetopias jubatus   Steller sea lion
 Genus *Neophoca*
  Neophoca cinerea   Australian sea lion
 Genus *Otaria*
  Otaria flavescens   South American sea lion
 Genus *Phocarctos*
  Phocarctos hookeri   New Zealand sea lion
 Genus *Zalophus*
  Zalophus californianus   California sea lion
  *Zalophus japonicus**   Japanese sea lion
  Zalophus wollebaeki   Galapagos sea lion

Family Odobenidae (1 sp.)
 Genus *Odobenus*
  Odobenus rosmarus   walrus

Family Phocidae (19 spp.)
 Genus *Cystophora*
  Cystophora cristata   hooded seal
 Genus *Erignathus*
  Erignathus barbatus   bearded seal
 Genus *Halichoerus*
  Halichoerus grypus   gray seal
 Genus *Histriophoca*
  Histriophoca fasciata   ribbon seal
 Genus *Hydrurga*
  Hydrurga leptonyx   leopard seal
 Genus *Leptonychotes*
  Leptonychotes weddellii   Weddell seal
 Genus *Lobodon*
  Lobodon carcinophaga   crabeater seal
 Genus *Mirounga*
  Mirounga angustirostris   northern elephant seal
  Mirounga leonina   southern elephant seal
 Genus *Monachus*
  Monachus monachus   Mediterranean monk seal
  Monachus schauinslandi   Hawaiian monk seal
  *Monachus tropicalis**   Caribbean monk seal
 Genus *Ommatophoca*
  Ommatophoca rossii   Ross seal
 Genus *Pagophilus*
  Pagophilus groenlandicus   harp seal
 Genus *Phoca*
  Phoca largha   spotted seal
  Phoca vitulina   harbor seal
 Genus *Pusa*
  Pusa caspica   Caspian seal
  Pusa hispida   ringed seal
  Pusa sibirica   Baikal seal

Family Mustelidae (60 spp.)
 Genus *Aonyx*
  Aonyx capensis   African clawless otter
  Aonyx cinerea   oriental small-clawed otter
 Genus *Enhydra*
  Enhydra lutris   sea otter
 Genus *Hydrictis*
  Hydrictis maculicollis   spotted-neck otter
 Genus *Lontra*
  Lontra canadensis   North American river otter
  Lontra felina   marine otter
  Lontra longicaudis   neotropical otter
  Lontra provocax   southern river otter
 Genus *Lutra*
  Lutra lutra   European otter
  Lutra nippon   Japanese otter
  Lutra sumatrana   hairy-nosed otter
 Genus *Lutrogale*
  Lutrogale perspicillata   smooth-coated otter
 Genus *Pteronura*
  Pteronura brasiliensis   giant otter
 Genus *Arctonyx*
  Arctonyx collaris   hog badger
 Genus *Eira*
  Eira barbara   tayra
 Genus *Galictis*
  Galictis cuja   lesser grison
  Galictis vittata   greater grison
 Genus *Gulo*
  Gulo gulo   wolverine
 Genus *Ictonyx*
  Ictonyx libyca   Saharan striped polecat
  Ictonyx striatus   striped polecat
 Genus *Lyncodon*
  Lyncodon patagonicus   Patagonian weasel
 Genus *Martes*
  Martes americana   American marten

*Martes caurina***	Pacific marten
Martes flavigula	yellow-throated marten
Martes foina	beech marten, stone marten
Martes gwatkinsii	Nilgiri marten
Martes martes	European pine marten
Martes melampus	Japanese marten
Martes zibellina	sable

Genus *Meles*
Meles anakuma	Japanese badger
Meles leucurus	Asian badger
Meles meles	European badger

Genus *Mellivora*
| *Mellivora capensis* | honey badger |

Genus *Melogale*
Melogale everetti	Bornean ferret-badger
Melogale moschata	Chinese ferret-badger
Melogale orientalis	Javan ferret-badger
Melogale personata	Burmese ferret-badger

Genus *Mustela*
Mustela altaica	mountain weasel
Mustela erminea	stoat, ermine, short-tailed weasel
Mustela eversmanii	steppe polecat
Mustela itatsi	Japanese weasel
Mustela kathiah	yellow-bellied weasel
Mustela lutreola	European mink
Mustela lutreolina	Indonesian mountain weasel
Mustela nigripes	black-footed ferret
Mustela nivalis	weasel, least weasel
Mustela nudipes	Malayan weasel
Mustela putorius	European polecat
Mustela putorius furo	domestic ferret
Mustela siberica	Siberian weasel
Mustela strigidorsa	back-striped weasel
Mustela subpalmata	Egyptian weasel

Genus *Neogale*
Neogale africana	Amazon weasel
Neogale felipei	Colombian weasel
Neogale frenata	long-tailed weasel
*Neogale macrodon**	sea mink
Neogale vison	American mink

Genus *Pekania*
| *Pekania pennanti* | fisher |

Genus *Poecilogale*
| *Poecilogale albinucha* | African striped weasel |

Genus *Taxidea*
| *Taxidea taxus* | American badger |

Genus *Vormela*
| *Vormela peregusna* | marbled polecat |

Family Mephitidae (12 spp.)
Genus *Conepatus*
Conepatus chinga	Molina's hog-nosed skunk
Conepatus humboldtii	Humboldt's hog-nosed skunk
Conepatus leuconotus	American hog-nosed skunk
Conepatus semistriatus	striped hog-nosed skunk

Genus *Mephitis*
| *Mephitis macroura* | hooded skunk |
| *Mephitis mephitis* | striped skunk |

Genus *Mydaus*
| *Mydaus javanensis* | Sunda stink badger |
| *Mydaus marchei* | Palawan stink badger |

Genus *Spilogale*
Spilogale angustifrons	southern spotted skunk
Spilogale gracilis	western spotted skunk
Spilogale putorius	eastern spotted skunk
Spilogale pygmeaea	pygmy spotted skunk

Family Procyonidae (15 spp.)
Genus *Bassaricyon*
Bassaricyon alleni	Allen's olingo
Bassaricyon beddardi	Beddard's olingo
Bassaricyon gabii	olingo
Bassaricyon lasius	Harris's olingo
*Bassaricyon neblina***	olinguito
Bassaricyon pauli	Chiriqui olingo

Genus *Bassariscus*
| *Bassariscus astutus* | ringtail |
| *Bassariscus sumichrasti* | cacomistle |

Genus *Nasua*
| *Nasua narica* | white-nosed coati |
| *Nasua nasua* | South American coati |

Genus *Nasuella*
| *Nasuella olivacea* | mountain coati |

Genus *Potos*
| *Potos flavus* | kinkajou |

Genus *Procyon*
Procyon cancrivorus	crab-eating raccoon
Procyon lotor	raccoon
Procyon pygmaeus	Cozumel raccoon

Family Ailuridae (1 sp.)
Genus *Ailurus*
| *Ailurus fulgens* | red panda |

* Became extinct during the historical period
** Recognized since Wilson and Reeder, 2005.

References

Patterson, B.D. *et al.* (2021) "On the nomenclature of the American clade of weasels (Carnivora: Mustelidae)," *Journal of Animal Diversity*, 3, pp. 1–8.

Wilson, D.E. and Reeder, D.M. (eds.) (2005) *Mammal species of the world: a taxonomic and geographic reference*. 3rd edn. Baltimore: Johns Hopkins University Press.

List of non-carnivoran species mentioned

Plants

Avocado	*Persea americana*
Beech	*Fagus* sp.
Blueberry/huckleberry	*Vaccinium* sp.
Chestnut	*Castanea* sp.
Devil's club	*Oplopanax horridus*
Hazel	*Corylus* sp.
Iberian pear	*Pyrus bourgaeana*
Kiwi fruit	*Actinidia* sp.
Liana	*Actinidia* sp.
Liaodong oak	*Quercus wutaishanica*
Limber pine	*Pinus flexilis*
Oak	*Quercus* sp.
Paw paw	*Asimina triloba*
Prickly pear	*Opuntia* sp.
Siberian dwarf pine	*Pinus pumila*
Whitebark pine	*Pinus albicaulis*

Invertebrates

Antarctic krill	*Euphausia superba*
Guinea worm	*Dracunculus medinensis*

Fish

Chum salmon	*Oncorhynchus keta*
Cutthroat trout	*Oncorhynchus clarkii*
Lake trout	*Salvelinus namaycush*
Northern anchovy	*Engraulis mordax*

Reptiles

African rock python	*Python sebae*
Agassiz's desert tortoise	*Gopherus agassizii*
Indian cobra	*Naja naja*
Komodo dragon	*Varanus komodoensis*

Birds

American robin	*Turdus migratorius*
Barnacle goose	*Branta leucopsis*
Black grouse	*Tetrao tetrix*
Boreal owl/Tengmalm's owl	*Aegolius funereus*
Common eider	*Somateria mollissima*
Common raven	*Corvus corax*
Common waxbill	*Estrilda astrild*
Dodo	*Raphus cucullatus*
European blackbird	*Turdus merula*
Glaucous gull	*Larus hyperboreus*
Golden eagle	*Aquila chrysaetos*
Great auk	*Pinguinus impennis*
Great horned owl	*Bubo virginianus*
Hazel grouse	*Bonasa bonasia*
Hooded warbler	*Setophaga citrina*
Macquarie Island parakeet	*Cyanoramphus erythrotis*
Native hen	*Tribonyx mortierii*
Nutcracker	*Nucifraga* sp.
Red-billed hornbill	*Tockus erythrochynchus*
Turkey vulture	*Cathartes aura*
Weka	*Gallirallus australis*
Western gull	*Larus occidentalis*
Wood thrush	*Hylocichla mustelina*

Mammals

African buffalo	*Syncerus caffer*
African crested rat	*Lophiomys imhausi*
Amami rabbit	*Pentalagus furnessi*
American beaver	*Castor canadensis*
American bison	*Bison bison*
Bank vole	*Myodes glareolus*
Blue sheep	*Pseudois nayaur*
Blue wildebeest	*Connochaetes taurinus*
Brown rat	*Rattus norvegicus*
Burchell's zebra/plains zebra	*Equus burchellii*
Bushbuck	*Tragelaphus scriptus*
Capybara	*Hydrochoerus hydrochaeris*
Caribou/reindeer	*Rangifer rangifer*
Chital	*Axis axis*
Collared peccary	*Pecari tajacu*
Columbian ground squirrel	*Spermophilus columbianus*
Common brushtail possum	*Trichosurus vulpecula*
Common eland	*Taurotragus oryx*
Common hippopotamus	*Hippopotamus amphibius*

Common muskrat	*Ondatra zibethicus*	Wood mouse	*Apodemus sylvaticus*
Common opossum	*Didelphis marsupialis*	Middle East blind mole rat	*Spalax ehrenbergi*
Common shrew	*Sorex araneus*	Moose	*Alces alces*
Common vampire bat	*Desmodus rotundus*	Mountain hare	*Lepus timidus*
Eurasian red deer, North American elk	*Cervus elaphus*	Mountain vizcacha	*Lagidium viscicia*
		North American porcupine	*Erethizon dorsata*
European hare	*Lepus europaeus*	Olympic marmot	*Marmota olympus*
European rabbit	*Oryctolagus. cuniculus*	Oribi	*Ourebia ourebi*
Fallow deer	*Dama dama*	Pampas deer	*Ozotoceros bezoarticus*
Gaur	*Bos frontalis*	Pine squirrel	*Tamiasciurus* sp.
Gelada monkey	*Theropithecus gelada*	Przewalski horse	*Equus caballus przewalskii*
Gerbil	*Gerbillus* sp.	Rhesus macaque	*Macaca mulatta*
Giant anteater	*Myrmecophaga tridactyla*	Roan antelope	*Hippotragus equinus*
Giant kangaroo rat	*Dipodomys ingens*	Roe deer	*Capreolus capreolus*
Giraffe	*Giraffa camelopardalis*	Snowshoe hare	*Lepus americanus*
Grant's gazelle	*Nanger granti*	South African ground squirrel	*Xerus inauris*
Greater kudu	*Tragelaphus strepsiceros*		
Guanaco	*Lama guanicoe*	Tasmanian devil	*Sarcophilus harrisii*
Günther's dik-dik	*Madoqua guentheri*	Thomson's gazelle	*Eudorcas thomsonii*
Hartebeest	*Alcelaphus bucelaphus*	Thylacine	*Thylacinus cynocephalus*
Impala	*Aepyceros melampus*	Vicuña	*Vicugna vicugna*
Jackson's hartebeest	*Alcelaphus bucelaphus lelwel*	Warthog	*Phacochoerus africanus*
Javan rusa deer	*Rusa timorensis*	White rhinoceros	*Ceratotherium simum*
Killer whale	*Orcinus orca*	White-tailed deer	*Odocoileus virginianus*
Laboratory rat	*Rattus rattus*	Wild boar	*Sus scrofa*
Llama	*Lama glama*		

Glossary

Abduction Increasing the angle of a joint. Movement of a body part away from the midline.

Adduction Reducing the angle of a joint. Movement of a body part toward the midline.

Allopatry The occurrence of two populations in separate, non-overlapping geographic areas. Not sympatric.

Altricial Born in an undeveloped state, requiring extended parental care.

Amylase An enzyme that catalyzes the hydrolysis of starch to disaccharides or trisaccharides.

Amylose A digestion-resistant starch made of glucose sub-units.

Aposematism Signaling by an animal to a potential predator that it is dangerous or noxious.

Bradycardia Reduced heart rate.

Brown adipose tissue Also called brown fat. Adipose tissue found in almost all mammals, particularly neonates and hibernators, that has higher densities of mitochondria and capillaries than white adipose tissue.

Carnivora (adj. carnivoran) A mammalian order characterized by morphological and genetic traits. A branch of the tree of life.

Carnivore (adj. carnivorous) A plant or animal that consumes the flesh of multi-cellular animals, either by predation or scavenging.

Cathemeral Irregularly active at any time of day or night.

Chyme The semi-fluid that passes from the stomach to the small intestine, made up of partially digested food and digestive juices.

Circadian Recurring on a daily basis. Related to 24-hr cycles.

Commensal Referring to a relationship in which one species benefits but the other neither benefits nor experiences harm.

Coursing Cantering or trotting for long distances at intermediate speeds, as opposed to sprinting or walking.

Crepuscular Related to dawn and dusk. Active at dawn or dusk.

Crypsis Camouflage, either visual or olfactory.

Cryptorchidism Failure of the testes to descend into the scrotum.

Definitive host The host in which a parasite normally undergoes sexual reproduction.

Digitigrade A stance in which only the digits contact the substrate, not the metapodials.

Ecomorphotype (syn. ecomorph) A general body plan, emphasizing locomotor and dental adaptations, suited for a particular niche.

Ecotype A locally adapted form that may not be recognized in its taxonomic status.

Effective population size (N_e) The number of individuals in an idealized population would have the same amount of dispersion of allele frequencies under genetic drift or the same rate of inbreeding as the population in question. Idealized populations have random matings, non-overlapping generations, and other traits that are not realistic for wild populations. Various subtypes of N_e are recognized. N_e is commonly much smaller than population size.

Eltonian niche An ecological niche, emphasizing an organism's contributions to as well as requirements from the community.

Enzootic Referring to a disease that regularly occurs in animal populations but does not affect a high percentage of the population or cause large demographic effects.

Epipubic bones Paired slender bones that extend forward from the pelvis in modern marsupials and early mammals.

Fissiped A non-pinniped carnivoran.

Folivory The habit of eating leaves and herbaceous stems.

Functional response The intake rate of a consumer as a function of food density.

Genetic drift Change in the frequency of an allele in response to random sampling of participants in matings. In small populations, drift tends to reduce the frequency of uncommon alleles.

Genetic swamping Gene flow from large populations into small populations of a different ecotype, so that the adaptive suite of the small population is impaired. Also called outbreeding depression.

Grinnellian niche The habitat and food requirements of a species. The requirement aspects of the ecological niche (cf. Eltonian niche).

Helminth Referring to worms, including round worms, tape worms, and segmented worms.

Heterosis Hybrid vigor. The tendency for crossbred individuals to show improved vigor or function.

Hippocampus A portion of the temporal lobe of the cortex that has important functions in consolidating short-term memory into long-term memory, and in spatial memory.

Holarctic The Palearctic and Nearctic biogeographic zones. North America and Eurasia.

Homodont The condition of having post-canine teeth of the same size and shape.

Homolog Similar phenotypic traits derived from the same ancestral trait.

Inbreeding depression Impaired function or reduced survival of the offspring of closely related parents.

Limbic system Primitive structures on both sides of the thalamus that support emotions, behavior, long-term memory, and olfaction.

Ma Mega-annum, one million years before the present.

Medial Referring to or closer to the mid-sagittal plane of the vertebrate body.

Metapodial The metacarpal and metatarsal bones, collectively.

Metatheria A mammalian lineage that includes all modern marsupials and a number of extinct relatives.

Metazoa Multicellular animals having cells differentiated into tissues and organs.

Mucosa Mucous membrane, such as that lining the digestive tract.

Musteloidea A superfamily of the Carnivora, comprising the Mustelidae, Procyonidae, Ailuridae, and Mephitidae.

Myoglobin The iron-based, oxygen-binding protein that is found in skeletal and cardiac muscle cells.

Neocortex A set of layers of the mammalian cerebral cortex involved in higher order functions, such as cognition and spatial reasoning.

Numerical response A change in consumer density in response to a change in food density.

Osmolality The concentration of osmotically active molecules in a solution per unit weight of solvent.

Outbreeding depression See genetic swamping.

Panmixia Random mating within a population. The absence of genetic structure, even over long distances.

Paraphyletic A named lineage that has a common ancestor but does not include all of the descendants of that ancestor.

Peptidase An enzyme that breaks down peptides into amino acids.

Piloerection The raising of hairs by contraction of smooth muscles, causing fur to be thicker, or parts of the pelage to stand erect.

Podials The carpal and tarsal bones, collectively.

Polyandry A mating system in which a female has more than one male mate, particularly within a single biological year.

Polygyny A mating system in which a male has more than one female mate, particularly within a single biological year.

Precocial Being born in an advanced state of development, requiring only brief parental care.

Protraction Extension of some part of the body.

Pylorus The opening from the stomach into the duodenum.

Rostrum The projection of the mammalian skull anterior to the orbits.

Sagittal Referring to the plane of the mammalian body that divides it into left and right halves.

Sanguivory The dietary habit of feeding on blood.

Sarcoptic Referring to a genus of parasitic louse, *Sarcoptes*.

Scansorial Facultatively tree-climbing. Semi-arboreal.

Scapholunate The fused scaphoid and lunate carpal bones.

Semiochemical A chemical used by animals to communicate with animals of the same or different species.

Speciose Having many species.

Squamata The order of reptiles that includes lizards, snakes, and worm lizards.

Stereopsis The perception of depth by the reception in the brain of signals from two eyes, or by moving the position of a single eye, or by the tracking of a moving object.

Subnivean Below the snow surface.

Syntopy Local sympatry. Occupying the same habitats in the same geographic area.

Taxon (pl. taxa) A collection of kinds of organisms related at a specified level—subspecies, species, genus, family, or higher level.

Telencephalon The portion of the forebrain comprising the cerebral hemispheres.

Turbinate The curved sheet-like bone in the mammalian nasal cavity that supports the nasal olfactory membrane.

Vibrissa The long, stiff hairs on the faces of mammals that connect to organs of the tactile sense. Whiskers.

Vital rate Basic rates of components of population change, for example birth rate and death rate.

Xylanase An enzyme that breaks down the polysaccharide xylan, found in cellulose, into xylose, a monosaccharide.

Zoonosis A disease of non-human animals that can be transmitted to humans.

Index

Tables, figures, and boxes are indicated by an italic *t*, *f*, and *b*, respectively, following the page number.